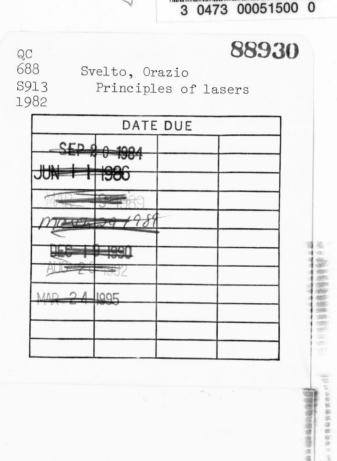

Principles of Lasers
SECOND EDITION

Principles of Lasers
SECOND EDITION

Orazio Svelto
Polytechnic Institute of Milan
Milan, Italy

Translated from Italian and edited by
David C. Hanna
Southampton University
Southampton, England

Plenum Press · New York and London

Library of Congress Cataloging in Publication Data

Svelto, Orazio.
   Principles of lasers.

   Translation of: Principi dei laser.
   Includes bibliographical references and index.
   1. Lasers. I. Hanna, D. C. (David C.), 1941-   . II. Title.
QC688.S913   1982   535.5′8   82-484
   ISBN 0-306-40862-7   AACR2

© 1982 Plenum Press, New York
A Division of Plenum Publishing Corporation
233 Spring Street, New York, N.Y. 10013

Printed in the United States of America

Laser . . . inter eximia naturae dona numeratum
plurimis compositionibus inseritur

The Laser is numbered among the most miraculous gifts
of nature and lends itself to a variety of applications

Plinius, Naturalis Historia, XXII, 49 (1st century A.D.)

An Elaboration of the Quotation from Pliny the Elder:
The Laser during the Graeco-Roman Civilization

During the Graeco-Roman civilization (roughly from the 6th century B.C. to the 2nd century A.D.) the Laser was well known and much celebrated. It was in fact a plant (perhaps belonging to the *Umbelliferae*) which grew wild over a large area around Cyrene (in present-day Libya). Sometimes also called *Laserpitium*, it was considered to be a gift of God due to its almost miraculous properties. It was used to cure a variety of diseases, from pleurisy to various epidemic infections. It was an effective antidote against the poison of snakes, scorpions, and enemy arrows. Its delicate flavor led to its use as an exquisite dressing in the best cuisine. It was so valuable as to be the main source of Cyrenaean prosperity and it was exported to both Greeks and Romans. During the period of Roman domination, it was the only tribute paid by the Cyrenaeans to the Romans, who kept the Laser in their coffers together with their golden ingots. What is perhaps the best testimony to the Laser of those days is to be found on the celebrated Arcesilao cup (now in the Cyrene Museum), where porters can be seen loading the Laser onto a ship under the supervision of King Arcesilao. Both Greeks and Romans tried hard, but without success, to grow the Laser in various parts of *Apulia* and *Ionia* (in the south of Italy). Consequently the Laser became more and more rare and seems to have disappeared around the 2nd century A.D. Ever since then, despite several efforts, no one has been able to find the Laser in the deserts south of Cyrene, and so it remains the lost treasure of the Graeco-Roman civilization.

Preface to the Second Edition

This second edition, appearing about twenty years after the discovery of the laser is a substantially revised version of the first edition. It is, like the first, aimed at both classroom teaching and self-study by technical personnel interested in learning the principles of laser operation. In preparing the second edition the hope has been that both these aims will be better served as a result of the various improvements made.

The main changes have been made with the following aims in mind: (*i*) To update the book. Thus new topics have been added (in particular on various new types of lasers, e.g., rare-gas–halide excimer lasers, color-center lasers, and free-electron lasers), while on the other hand some topics have been given less emphasis (again this applies particularly to some types of lasers, e.g., the ruby laser). Updating is especially important in the area of laser applications, and the chapter on this topic has therefore been completely rewritten. (*ii*) To make some improvements to the logical consistency of the book by rearranging material and adding new material. Thus a few topics have been moved from one section to another and a new chapter entitled *Laser Beam Transformation* has been added. (*iii*) To further reduce the mathematical content, placing greater emphasis on physical descriptions of phenomena. Thus the previous Chapter 9, *Advances in Laser Physics*, has been omitted as it was felt to be too mathematical and specialized for the scope of this book. Finally the problems at the end of each chapter have been substantially revised, and answers have been given to some of them in a separate section at the end of the book.

I wish to acknowledge the following friends and colleagues whose suggestions and encouragement have certainly contributed toward improving the book in a number of ways: Rodolfo Bonifacio, Richard K. Chang, Richard H. Pantell, Fritz P. Schäfer, Ian J. Spalding, and Charles H. Townes. The remarks made by various reviewers of the first edition have also proved most useful and are gratefully acknowledged. I wish also to acknowledge the critical editing of David C. Hanna who has acted as much more than simply a translator. Finally I wish to thank Dante Cigni for kindly providing me with the material concerning the history of the Laser during the Graeco-Roman civilization.

O. Svelto

Preface to the First Edition

This book is the result of more than ten years of research and teaching in the field of quantum electronics. The purpose of the book is to introduce the principles of lasers, starting from elementary notions of quantum mechanics and electromagnetism. Because it is an introductory book, an effort has been made to make it self-contained to minimize the need for reference to other works. For the same reason, the references have been limited (whenever possible) either to review papers or to papers of historical importance.

The organization of the book is based on the fact that a laser can be thought of as consisting of three elements: (*i*) an active material, (*ii*) a pumping system, and (*iii*) a suitable resonator. Accordingly, after an introductory chapter, the next three chapters deal, respectively, with the interaction of radiation with matter, pumping processes, and the theory of passive optical resonators. The concepts introduced in this way are then used in Chapter 5 to develop a theory for the cw and transient behavior of lasers. The theory is developed within the lowest-order approximation, i.e., using the rate-equation approach. This treatment is, in fact, capable of describing most laser characteristics. Obviously, lasers based upon different types of active media have somewhat different characteristics. It is therefore natural that next, Chapter 6 should discuss the characteristic properties of each type of laser. At this point, the reader will have acquired a sufficient understanding of laser operation to go on to a study of the characteristic properties of the output beam (monochromaticity, coherence, directionality, brightness). These properties are dealt with in Chapter 7, which leads to a discussion of the applications for which the laser is potentially suited (Chapter 8). Finally, in Chapter 9, which can be considered as a supplementary chapter, a more advanced treatment is given to the problem of the interaction of radiation and matter. The approach in Chapter 9 is based on the so-called semiclassical approximation. With this more advanced approach, a better physical insight into laser behavior can be obtained. In particular, this treatment is able to account for new physical phenomena (e.g., the production of π, 2π pulses, photon echo, etc.) which could not be described under the rate-equation approximation. A few appendixes, added for completeness, constitute the final part of the book.

Although every effort has been made to present the subject matter in a coherent way, the book must represent some sort of compromise. In some cases, for instance, only the most basic features of a problem are discussed, with very little detail given. Furthermore, as previously mentioned, the treatment is limited to the semiclassical approximation. It is, therefore, not possible to give a full discussion of important phenomena such as spontaneous emission and laser noise. These limitations are, however, often dictated by the need to give the book a wide appeal. In the author's opinion, in fact, the subject matter presented here constitutes the minimum knowledge required for a correct understanding of laser behavior.

O. Svelto

Contents

5 Continuous Wave and Transient Laser Behavior

6 Types of Lasers

1

Introductory Concepts

1.1 SPONTANEOUS AND STIMULATED EMISSION, ABSORPTION

Quantum electronics can be defined as that branch of electronics where phenomena of a quantum nature play a fundamental role. This book will deal with a particular aspect of quantum electronics, namely, the physical principles of lasers and their behavior. Before going into a detailed discussion of the subject, it seems appropriate to devote a little space to an explanation, in a very simple way, of the ideas behind the laser.

A laser exploits three fundamental phenomena which occur when an electromagnetic (e.m.) wave interacts with a material, namely, the processes of spontaneous and stimulated emission and the process of absorption.

1.1.1 Spontaneous Emission (Fig. 1.1a)

Let us consider two energy levels, 1 and 2, of some given material, their energies being E_1 and E_2 ($E_1 < E_2$). As far as the following discussion is concerned, the two levels could be any two out of the infinite set of levels possessed by the material. It is convenient, however, to take level 1 to be the ground level. Let us now assume that an atom (or molecule) of the material is initially in level 2. Since $E_2 > E_1$, the atom will tend to decay to level 1. The corresponding energy difference $(E_2 - E_1)$ must therefore be released by the atom. When this energy is delivered in the form of an e.m. wave, the process will be called spontaneous (or radiative) emission. The frequency ν of the radiated wave is then given by the expression (due to

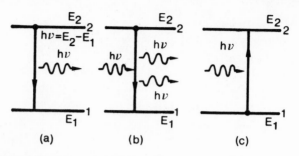

FIG. 1.1. Schematic illustration of the three processes: (a) spontaneous emission; (b) stimulated emission; (c) absorption.

Planck)

$$\nu = (E_2 - E_1)/h \qquad (1.1)$$

where h is Planck's constant. Spontaneous emission is therefore characterized by the emission of a photon of energy $h\nu = E_2 - E_1$, when the atom decays from level 2 to level 1 (Fig. 1.1a). Note that radiative emission is just one of the two possible ways for the atom to decay. The decay can also occur in a nonradiative way. In this case the energy difference $E_2 - E_1$ is delivered in some form other than e.m. radiation (e.g., it may go into kinetic energy of the surrounding molecules).

The probability of spontaneous emission can be characterized in the following way: Let us suppose that, at time t, there are N_2 atoms (per unit volume) in level 2. The rate of decay of these atoms due to spontaneous emission, i.e., $(dN_2/dt)_{sp}$, will obviously be proportional to N_2. We can therefore write

$$\left(\frac{dN_2}{dt} \right)_{sp} = -AN_2 \qquad (1.2)$$

The coefficient A is called the spontaneous emission probability or the Einstein A coefficient (an expression for A was in fact first obtained by Einstein from thermodynamic considerations). The quantity $\tau_{sp} = 1/A$ is called the spontaneous emission lifetime. The numerical value of A (and τ_{sp}) depends on the particular transition involved.

1.1.2 Stimulated Emission (Fig. 1.1b)

Let us again suppose that the atom is found initially in level 2 and that an e.m. wave of frequency ν given by equation (1.1) (i.e., equal to that of the spontaneously emitted wave) is incident on the material. Since this wave has the same frequency as the atomic frequency, there is a finite probability that this wave will force the atom to undergo the transition $2 \rightarrow 1$. In this case the energy difference $E_2 - E_1$ is delivered in the form of an e.m. wave which adds to the incident one. This is the phenomenon of stimulated emission. There is, however, a fundamental distinction between the spontaneous and stimulated emission processes. In the case of spontaneous emission, the atom emits an e.m. wave which has no definite phase relation with that emitted by another atom. Furthermore, the wave can be emitted in any direction. In the case of stimulated emission, since the process is forced by the incident e.m. wave, the emission of any atom adds in phase to that of the incoming wave. This wave also determines the direction of the emitted wave.

In this case, too, we can characterize the process by means of the equation

$$\left(\frac{dN_2}{dt} \right)_{st} = - W_{21}N_2 \tag{1.3}$$

where $(dN_2/dt)_{st}$ is the rate at which transitions $2 \rightarrow 1$ occur as a result of stimulated emission and W_{21} is called the stimulated transition probability. Just as in the case of the A coefficient defined by (1.2), the coefficient W_{21} also has the dimension of $(time)^{-1}$. Unlike A, however, W_{21} not only depends on the particular transition but also on the intensity of the incident e.m. wave. More precisely, for a plane e.m. wave, it will be shown that we can write

$$W_{21} = \sigma_{21}F \tag{1.4}$$

where F is the photon flux of the incident wave and σ_{21} is a quantity having the dimensions of area (it is called the stimulated-emission cross section) and depending only on the characteristics of the given transition.

1.1.3 Absorption (Fig. 1.1c)

Let us now assume that the atom is initially lying in level 1. If this is the ground level, the atom will remain in this level unless some external stimulus is applied to it. We shall assume, then, that an e.m. wave of frequency v given again by (1.1) is incident on the material. In this case there is a finite probability that the atom will be raised to level 2. The energy difference $E_2 - E_1$ required by the atom to undergo the transition is obtained from the energy of the incident e.m. wave. This is the absorption process.

In a similar fashion to (1.3), we can define an absorption rate W_{12} by means of the equation

$$\frac{dN_1}{dt} = -W_{12}N_1 \qquad (1.5)$$

where N_1 is the number of atoms (per unit volume) which, at the given time, are lying in level 1. Furthermore, just as in (1.4), we can write

$$W_{12} = \sigma_{12}F \qquad (1.6)$$

where σ_{12} is some characteristic area (the absorption cross section) which depends only on the particular transition.

In the preceding sections, the fundamental principles of the processes of spontaneous emission, stimulated emission, and absorption have been described. In terms of photons, these processes can be described as follows (see Fig. 1.1): (i) In the spontaneous emission process, the atom decays from level 2 to 1 through the emission of a photon. (ii) In the stimulated process, the incident photon stimulates the $2 \rightarrow 1$ transition and we then have two photons (the stimulating plus the stimulated one). (iii) In the absorption process, the incident photon is simply absorbed to produce the $1 \rightarrow 2$ transition. Finally, it should be noted that $\sigma_{12} = \sigma_{21}$, as Einstein showed at the beginning of the century. This shows that the probabilities of stimulated emission and absorption are equal. From now on, therefore, we will write $\sigma_{12} = \sigma_{21} = \sigma$, and σ will be referred to as the transition cross section[†]. The number of atoms per unit volume in some given level will be called the *population* of that level.

1.2 THE LASER IDEA

Consider two arbitrary energy levels 1 and 2 of a given material and let N_1 and N_2 be their respective populations. If a plane wave with an intensity corresponding to a photon flux F is traveling along the z direction in the material, the elemental change of this flux due to both the stimulated emission and absorption processes in the shaded region of Fig. 1.2, according to equations (1.3)–(1.6), is given by

$$dF = \sigma F(N_2 - N_1)\,dz \qquad (1.7)$$

Equation (1.7) shows that the material behaves as an amplifier (i.e., $dF/dz > 0$) if $N_2 > N_1$, while it behaves as an absorber if $N_2 < N_1$. Now, it

---

[†]The discussion here only applies to nondegenerate levels. For the case of degenerate levels the reader is referred to Section 2.7

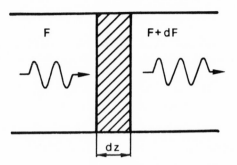

FIG. 1.2. Elemental change dF in the photon flux F for a plane e.m. wave in traveling a distance dz through the material.

is known that, in the case of thermal equilibrium, the energy-level populations are described by Boltzmann statistics. So, if N_1^e and N_2^e are the thermal equilibrium populations of the two levels, we have

$$\frac{N_2^e}{N_1^e} = \exp\left[-\frac{(E_2 - E_1)}{kT} \right] \tag{1.8}$$

where k is Boltzmann's constant and T the absolute temperature of the material. We then see that, for the case of thermal equilibrium, we have $N_2 < N_1$. According to (1.7), the material then acts as an absorber at frequency ν, and this is what happens under ordinary conditions. If, however, a nonequilibrium condition is achieved for which $N_2 > N_1$, then the material will act as an amplifier. In this case we will say that there exists a *population inversion* in the material, by which we mean that the population difference $(N_2 - N_1 > 0)$ is opposite in sign to that which exists under ordinary conditions $(N_2^e - N_1^e < 0)$. A material having a population inversion will be called an *active material*.

If the transition frequency $\nu = (E_2 - E_1)/h$ falls in the microwave region, this type of amplifier is called a *maser* amplifier. The word *maser* is an acronym for "*m*icrowave *a*mplification by *s*timulated *e*mission of *r*adiation." If the transition frequency ν falls in the optical region, the amplifier is called a *laser* amplifier. The word laser is again an acronym, obtained by substituting the letter l (*l*ight) for the letter m (*m*icrowave). The word laser is, however, commonly used not only for frequencies of visible light but for any frequency falling in the far- or near-infrared, in the ultraviolet, and even in the x-ray region. In these cases we will refer to infrared, ultraviolet, or x-ray lasers, respectively.

To make an oscillator from an amplifier, it is necessary to introduce a suitable positive feedback. In the microwave range this is done by placing the active material in a resonant cavity having a resonance at the frequency ν. In the case of a laser, the feedback is often obtained by placing the active

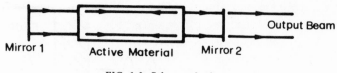

Mirror 1   Active Material   Mirror 2

FIG. 1.3. Scheme of a laser.

material between two highly reflecting mirrors (e.g., plane-parallel mirrors, see Fig. 1.3). In this case, a plane e.m. wave traveling in a direction orthogonal to the mirrors will bounce back and forth between the two mirrors and be amplified on each passage through the active material. If one of the two mirrors is made partially transparent, a useful output beam can be extracted. It is important to realize that for both masers and lasers, a certain threshold condition must be fulfilled. In the laser case, for instance, the oscillation will start when the gain of the active material compensates the losses in the laser (e.g., the losses due to output coupling). According to (1.7), the gain per pass in the active material (i.e., the ratio between the output and input photon flux) is $\exp[\sigma(N_2 - N_1)l]$, where l is the length of the active material. If the only losses present in the cavity are those due to transmission losses, the threshold will be reached when $R_1 R_2 \exp[2\sigma(N_2 - N_1)l] = 1$, where R_1 and R_2 are the power reflectivities of the two mirrors. This equation shows that the threshold is reached when the population inversion reaches a critical value $(N_2 - N_1)_c$ known as the *critical inversion* and given by

$$(N_2 - N_1)_c = -\frac{\ln(R_1 R_2)}{2\sigma l} \tag{1.9}$$

Once the critical inversion is reached, oscillation will build up from the spontaneous emission. The photons which are spontaneously emitted along the cavity axis will, in fact, initiate the amplification process. This is the basis of a laser oscillator, or laser, as it is more simply known.

1.3 PUMPING SCHEMES

We will now consider the problem of how a population inversion can be produced in a given material. At first sight, it might seem that it would be possible to achieve this through the interaction of the material with a sufficiently strong e.m. field at the frequency ν given by (1.1). Since, at thermal equilibrium, level 1 is more populated than level 2, absorption will in fact predominate over stimulated emission. The incoming wave would produce more transitions $1 \to 2$ than transitions $2 \to 1$ and we would hope in this way to end up with a population inversion. We see immediately, however, that such a system would not work (at least in the steady state). When in fact the condition is reached such that the populations are equal ($N_2 = N_1$), then the absorption and stimulated processes will compensate one another and, according to (1.7), the material will then be transparent. This situation is often referred to as two-level *saturation*.

With the use of just two levels 1 and 2, it is impossible therefore to produce a population inversion. It is then natural to question whether this is possible by some suitable use of more than two levels out of the infinite set of levels of a given atomic system. As we shall see, the answer in this case is positive, and we will accordingly talk of a three- or four-level laser, depending upon the number of levels used (Fig. 1.4). In a three-level laser (Fig. 1.4a), the atoms are in some way raised from the ground level 1 to level 3. If the material is such that, after an atom has been raised to level 3, it decays rapidly to level 2, then in this way a population inversion can be obtained between levels 2 and 1. In a four-level laser (Fig. 1.4b), atoms are again raised from the ground level (for convenience we now call this level 0) to level 3. If the atom then decays rapidly to level 2, a population inversion can again be obtained between levels 2 and 1. Once oscillation starts in such a four-level laser, however, the atoms will then be transferred to level 1 (due to stimulated emission). For cw operation of a four-level laser it is necessary, therefore, that the transition $1 \to 0$ should also be very fast.

We have now seen how one can use three or four levels of a given material to produce population inversion. Whether a system will work in a

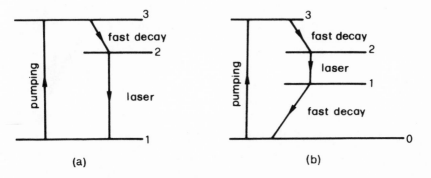

FIG. 1.4. (a) Three-level and (b) four-level laser schemes.

three- or four-level scheme (or whether it will work at all!) depends on whether the various conditions given above are fulfilled. We could of course ask why one should bother with a four-level scheme when a three-level scheme already seems to offer a suitable way of producing a population inversion. The answer is that one can, in general, produce a population inversion much more easily in a four-level than in a three-level laser. To see this, we begin by noting that the energy differences between the various levels of Fig. 1.4 are usually much greater than kT. According to Boltzmann statistics [see, e.g., equation (1.8)] we can then say that essentially all atoms are initially (i.e., at equilibrium) in the ground level. If we now let N_t be the total number of atoms per unit volume of material, these will initially all be in level 1 for the three-level case. Let us now begin raising atoms from level 1 to level 3. They will then decay to level 2, and if this decay is sufficiently fast, level 3 will remain more or less empty. In this case, we first have to raise half of the total population N_t to level 2 in order to equalize the populations of levels 1 and 2. From this point on, any other atom which is raised will then contribute to population inversion. In a four-level laser, however, since level 1 is also initially empty, any atom which has been raised is immediately available for a population inversion. The above discussion shows that, whenever possible, we should look for a material which can operate as a four-level system rather than as a three-level system. The use of more than four levels is, of course, also possible.

The process by which atoms are raised from level 1 to level 3 (in a three-level scheme) or from 0 to 3 (in a four-level scheme) is known as *pumping*. There are several ways in which this process can be realized in practice, e.g., by some sort of lamp of sufficient intensity or by an electrical discharge in the active medium. We refer the reader to Chapter 3 for a more detailed discussion of the various pumping processes. We note here, however, that, if the upper pump level is empty, the rate at which the upper laser level 2 becomes populated by the pumping, $(dN_2/dt)_p$, can in general be written as

$$\left(\frac{dN_2}{dt} \right)_p = W_p N_g \tag{1.10}$$

Here N_g is the population of the ground level [i.e., level 1 or 0 in Fig. 1.4a and b respectively] and W_p is a coefficient which will be called the *pump rate*. To achieve the threshold condition, the pump rate must reach a threshold or critical value that we shall indicate by W_{cp}. Specific expressions for W_{cp} will be obtained in Chapter 5.

1.4 PROPERTIES OF LASER BEAMS

Laser radiation is characterized by an extremely high degree of: (i) monochromaticity, (ii) coherence, (iii) directionality, and (iv) brightness. We shall now consider these properties.

1.4.1 Monochromaticity

Without entering into too much detail, we can say that this property is due to the following two circumstances: (i) Only an e.m. wave of frequency ν given by (1.1) can be amplified. (ii) Since the two-mirror arrangement forms a resonant cavity, oscillation can occur only at the resonant frequencies of this cavity. The latter circumstance leads to the laser linewidth being often much narrower (by as much as six orders of magnitude!) than the usual linewidth of the transition $2 \to 1$ as observed in spontaneous emission.

1.4.2 Coherence

To first order, for any e.m. wave, one can introduce two concepts of coherence, namely spatial and temporal coherence.

To define spatial coherence, let us consider two points P_1 and P_2 which, at time $t = 0$, lie on the same wavefront of some given e.m. wave and let $E_1(t)$ and $E_2(t)$ be the corresponding electric fields at these points. By definition, the difference between the phases of the two fields at time $t = 0$ is zero. Now, if this difference remains zero at any time $t > 0$, we will say that there is a perfect coherence between the two points. If this occurs for any two points of the e.m. wavefront, we will say that the wave has *perfect spatial coherence*. In practice, for any point P_1, the point P_2 must lie within some finite area around P_1 if we want to have a good phase correlation. In this case we will say that the wave has a *partial spatial coherence* and, for any point P, we can introduce a suitably defined coherence area $S_c(P)$.

To define temporal coherence, we now consider the electric field of the e.m. wave at a given point P, at times t and $t + \tau$. If, for a given time delay τ, the phase difference between the two field values remains the same for any time t, we will say that there is temporal coherence over a time τ. If this occurs for any value of τ, the e.m. wave will be said to have perfect time coherence. If this occurs for a time delay τ such that $0 < \tau < \tau_0$, the wave will be said to have partial temporal coherence, with a coherence time equal

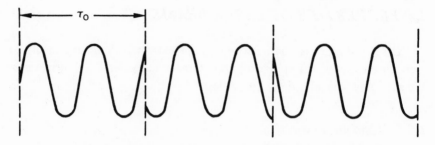

FIG. 1.5. Example of an e.m. wave with a coherence time of approximately τ_0.

to τ_0. An example of an e.m. wave with a coherence time equal to τ_0 is shown in Fig. 1.5. This shows a sinusoidal electric field undergoing phase jumps at time intervals equal to τ_0. We see that the concept of temporal coherence is directly connected with that of monochromaticity. We will in fact show, although this is already obvious from the example shown in Fig. 1.5, that an e.m. wave with a coherence time τ_0 has a bandwidth $\Delta\nu \simeq 1/\tau_0$.

It is worth noting that the two concepts of temporal and spatial coherence are indeed independent of each other. In fact, examples can be given of a wave having perfect spatial coherence but only a limited temporal coherence (or vice versa). If for instance the wave shown in Fig. 1.5 were to represent the electric fields at points P_1 and P_2 mentioned earlier, the spatial coherence between these points would be complete while the wave would have a limited temporal coherence.

We conclude this section by emphasizing that the concepts of spatial and temporal coherence provide only a first-order description of the laser's coherence. Higher-order coherence properties will be discussed in Chapter 7. Such a discussion is essential for a full appreciation of the difference between an ordinary light source and a laser. It will be shown in fact that, by virtue of the differences between the corresponding higher-order coherence properties, a laser beam is fundamentally different from an ordinary light source.

1.4.3 Directionality

This property is a direct consequence of the fact that the active material is placed in a resonant cavity such as the plane parallel one of Fig. 1.3. In fact, only a wave propagating along the cavity direction (or in a direction very near to it) can be sustained in the cavity. To gain a deeper understanding of the directional properties of laser beams (or, in general, of

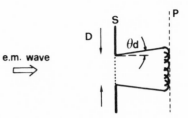

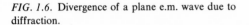

FIG. 1.6. Divergence of a plane e.m. wave due to diffraction.

any e.m. wave), it is convenient to consider, separately, the case of a beam with perfect spatial coherence and the case of partial spatial coherence.

We first consider the case of perfect spatial coherence. Even for this case, a beam of finite aperture has an unavoidable divergence due to diffraction. This can be understood with the help of Fig. 1.6, where a beam of uniform intensity and plane wavefront is assumed to be incident on a screen S containing an aperture D. According to Huygens' principle the wavefront at some plane P behind the screen can be obtained from the superposition of the elementary waves emitted by each point of the aperture. We see that, on account of the finite size D of the aperture, the beam has a finite divergence θ_d. Its value can be obtained from diffraction theory. For an arbitrary amplitude distribution we get

$$\theta_d = \beta\lambda/D \tag{1.11}$$

where λ and D are the wavelength and the diameter of the beam. In equation (1.11) β is a numerical coefficient of the order of unity whose value depends on the shape of the amplitude distribution and on the way in which both the divergence and the beam diameter are defined. A beam whose divergence is given by equation (1.11) is described as being *diffraction limited*.

If the wave has partial spatial coherence, its divergence will be larger than the minimum value set by diffraction. Indeed, for any point P' of the wavefront, the Huygens argument of Fig. 1.6 can only be applied for points lying within the coherence area S_c around point P'. The coherence area thus acts as a limiting aperture for the coherent superposition of the elementary wavelets. The beam divergence will now be given by

$$\theta_c = \beta\lambda/[S_c]^{1/2} \tag{1.12}$$

where again β is a numerical coefficient of the order of unity whose exact value depends on the way in which both the divergence θ_c and the coherence area S_c are defined.

We conclude this general discussion of the directional properties of e.m. waves by pointing out that, given suitable operating conditions, the output beam of a laser can be made diffraction limited.

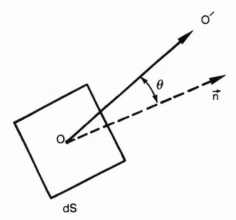

FIG. 1.7. Surface brightness at the point
O for a source of e.m. waves.

1.4.4 Brightness

We define the brightness of a given source of e.m. waves as the power emitted per unit surface area per unit solid angle. To be more precise, let dS be the elemental surface area at point 0 of the source (Fig. 1.7). The power dP emitted by dS into a solid angle $d\Omega$ around the direction OO' can be written as

$$dP = B \cos\theta \, dS \, d\Omega \qquad (1.13)$$

where θ is the angle between OO' and the normal **n** to the surface. The quantity B will generally depend on the polar coordinates θ and ϕ of the direction OO' and on the point O. This quantity B is called the source brightness at the point O in the direction of OO'. In equation (1.13) the factor $\cos\theta$ arises simply from the fact that the physically important quantity is the projection of dS onto a plane orthogonal to the OO' direction. When B is independent of θ and ϕ, the source is said to be isotropic (a Lambert source). A laser of even moderate power (e.g., a few milliwatts) has a brightness which is orders of magnitude greater than that of the brightest conventional sources. This is mainly due to the highly directional properties of the laser beam.

PROBLEMS

1.1. That part of the e.m. spectrum which is of interest in the laser field starts from the submillimeter wave region and goes down in wavelength to the x-ray region. This covers the following regions in succession: (i) far infrared; (ii) near infrared; (iii) visible; (iv) ultraviolet (uv); (v) vacuum ultraviolet (vuv); (vi) soft x-ray; (vii) x-ray. From standard textbooks find the wavelength intervals of the above regions. Memorize or record these intervals since they are frequently used in this book.

1.2. As a particular case of Problem 1.1, memorize or record the wavelengths corresponding to blue, green, and red light.

1.3. If levels 1 and 2 of Fig. 1.1 are separated by an energy $E_2 - E_1$ such that the corresponding transition frequency falls in the middle of the visible range, calculate the ratio of the populations of the two levels in thermal equilibrium at room temperature.

1.4. When in thermal equilibrium (at $T = 300°K$), the ratio of the level populations N_2/N_1 for some particular pair of levels is given by $1/e$. Calculate the frequency ν for this transition. In what region of the e.m. spectrum does this frequency fall?

1.5. A laser cavity consists of two mirrors with reflectivity $R_2 = 1$ and $R_1 = 0.5$. The length of the active material is $l = 7.5$ cm and the transition cross section is $\sigma = 8.8 \times 10^{-19} \text{cm}^2$. Calculate the threshold inversion.

1.6. The beam from a ruby laser ($\lambda = 0.694$ μm) is sent to the moon after passing through a telescope of 1 m diameter. Calculate the beam diameter D on the moon assuming that the beam has perfect spatial coherence (the distance between earth and moon is approximately $384,000$ km).

2

Interaction of Radiation with Matter

2.1 SUMMARY OF BLACKBODY RADIATION THEORY [1]

Let us consider a cavity which is filled with a homogeneous and isotropic dielectric medium. If the walls of the cavity are kept at a constant temperature T, they will continuously emit and receive power in the form of electromagnetic (e.m.) radiation. When the rates of absorption and emission become equal, an equilibrium condition is established at the walls of the cavity as well as at each point of the dielectric. This situation can be described by introducing the energy density ρ, which represents the electromagnetic energy contained in unit volume of the cavity. Since we are dealing with electromagnetic radiation, the energy density can be expressed as a function of the electric field $E(t)$ and magnetic field $H(t)$ according to the well-known formula

$$\rho = \tfrac{1}{2}\epsilon E^2(t) + \tfrac{1}{2}\,\mu H^2(t) \qquad (2.1)$$

where ϵ and μ are, respectively, the dielectric constant and the magnetic permeability of the medium inside the cavity.

We will represent the spectral energy distribution of this radiation by the function ρ_ν, which is a function of the frequency ν. This is defined as follows: $\rho_\nu\,d\nu$ represents the energy density of radiation in the frequency range from ν to $\nu + d\nu$. The relationship between ρ and ρ_ν is obviously

$$\rho = \int_0^\infty \rho_\nu\,d\nu \qquad (2.1a)$$

It can be shown that the spectral energy distribution ρ_ν is a universal function, independent of either the nature of the walls or of the cavity shape, and dependent only on the frequency ν and temperature T of the cavity. This property of ρ_ν can be proved through a simple thermodynamic argument. Let us suppose we have two cavities of arbitrary shape, whose walls are at the same temperature T. To ensure that the temperature remains constant, we may imagine that the walls of the two cavities are in thermal contact with two thermostats at temperature T. Let us suppose that, at a given frequency ν, the energy density ρ_ν' in the first cavity is greater than the corresponding value ρ_ν'' in the second cavity. We now optically connect the two cavities by making a hole in each of them and imaging, through a suitable optical system, each hole on the other. We also insert an ideal filter in the optical system, which lets through only a small frequency range around the frequency ν. If $\rho_\nu' > \rho_\nu''$, there will be a net flow of electromagnetic energy from cavity one to cavity two. This flow of energy, however, violates the second law of thermodynamics, since the two cavities are at the same temperature. Therefore one must have $\rho_\nu' = \rho_\nu''$ for all frequencies.

The problem of calculating this universal function $\rho(\nu, T)$ was a very challenging one for the physicists of the time. Its complete solution is, however, due to Planck, who in order to find a correct solution of the problem had to introduce the so-called hypothesis of light quanta. The blackbody theory is therefore one of the fundamental bases of modern physics.

Since the function ρ_ν is independent of the form of the cavity and of the nature of the dielectric medium, we will consider a rectangular cavity uniformly filled with dielectric and having perfectly conducting walls (see Fig. 2.1). To calculate ρ_ν, we begin by calculating the standing electromag-

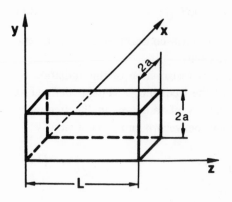

FIG. 2.1. Rectangular cavity with perfectly conducting walls kept at temperature T.

netic field distributions that can exist in this cavity. According to Maxwell's equations, the electric field $E(x, y, z, t)$ must satisfy the wave equation

$$\nabla^2 E - \frac{1}{c^2} \frac{\partial^2 E}{\partial t^2} = 0 \tag{2.2}$$

where ∇^2 is the Laplacian operator and c is the velocity of light in the medium considered. In addition, the field must satisfy the following boundary condition at each wall

$$E \times n = 0 \tag{2.3}$$

where n is the normal to the particular wall under consideration. This condition expresses the fact that the tangential component of the electric field must vanish on the walls of the cavity.

It can easily be shown that the problem is soluble by separation of the variables. Thus, if we put

$$E = u(x, y, z)A(t) \tag{2.4}$$

and substitute (2.4) in (2.2), we have

$$\nabla^2 u = -k^2 u \tag{2.5a}$$

$$\frac{d^2 A}{dt^2} = -(ck)^2 A \tag{2.5b}$$

where k is a constant. Equation (2.5b) has the general solution

$$A = A_0 \sin(\omega t + \phi) \tag{2.6}$$

where A_0 and ϕ are arbitrary constants and where

$$\omega = ck \tag{2.7}$$

With $A(t)$ given by (2.6), we see that the solution (2.4) corresponds to a standing wave configuration of the e.m. field within the cavity. In fact the amplitude of oscillation at a given point of the cavity is constant in time. A solution of this type is called an e.m. *mode* of the cavity.

We are now left with the task of solving equation (2.5a), known as the Helmholtz equation, subject to the boundary condition given by (2.3). It can readily be verified that the expressions

$$u_x = e_x \cos k_x x \sin k_y y \sin k_z z$$
$$u_y = e_y \sin k_x x \cos k_y y \sin k_z z \tag{2.8}$$
$$u_z = e_z \sin k_x x \sin k_y y \cos k_z z$$

satisfy equation (2.5a) for any value of e_x, e_y, e_z, provided that

$$k_x^2 + k_y^2 + k_z^2 = k^2 \tag{2.9}$$

Furthermore, the solution (2.8) already satisfies the boundary condition

(2.3) on the three planes $x = 0$, $y = 0$, $z = 0$. If we now impose the condition that equation (2.3) should also be satisfied on the other walls of the cavity, we have

$$k_x = \frac{l\pi}{2a}$$

$$k_y = \frac{m\pi}{2a} \tag{2.10}$$

$$k_z = \frac{n\pi}{L}$$

where l, m, and n are arbitrary positive integers. Their physical significance can be seen immediately: they represent the number of nodes that the standing wave mode has along the directions x, y and z, respectively. For fixed values of l, m, and n, it follows that k_x, k_y, and k_z will also be fixed, and according to (2.7) and (2.9) the frequency ω of the mode will also be fixed and given by

$$\omega^2_{l,m,n} = c^2 \left[\left(\frac{l\pi}{2a} \right)^2 + \left(\frac{m\pi}{2a} \right)^2 + \left(\frac{n\pi}{L} \right)^2 \right] \tag{2.11}$$

where we have explicitly indicated that the frequency of the mode will depend on the indices l, m, and n. The mode is still not completely determined, however, since e_x, e_y, and e_z are still arbitrary. However, Maxwell's equations provide another condition that must be satisfied by the electric field, i.e., $\nabla \cdot \mathbf{u} = 0$, from which, with the help of (2.8), we get

$$\mathbf{e} \cdot \mathbf{k} = 0 \tag{2.12}$$

In (2.12) we have introduced the two vectors $\mathbf{e}$ and $\mathbf{k}$, whose components along the x, y, and z axes are, respectively, e_x, e_y, e_z and k_x, k_y, k_z. Equation (2.12) therefore shows that, out of the three quantities e_x, e_y, and e_z, only two are independent. In fact, once we fix l, m, n (i.e., once $\mathbf{k}$ is fixed), the vector $\mathbf{e}$ is bound to lie in a plane perpendicular to $\mathbf{k}$. In this plane only two degrees of freedom are left for the choice of the vector $\mathbf{e}$, and only two modes are thus possible. Any other vector $\mathbf{e}$ lying in this plane can in fact be expressed as a linear combination of the previous two vectors.

Let us now calculate the number of modes N_ν resonating in the cavity, with frequency between 0 and ν. This will be the same as the number of modes whose wave vector $\mathbf{k}$ has a magnitude k between 0 and $2\pi\nu/c$. From (2.10) we see that, in a system of coordinates k_x, k_y, k_z, the possible values for $\mathbf{k}$ are given by vectors connecting the origin with the nodal points of the three-dimensional lattice shown in Fig. 2.2. There is obviously a one-to-one correspondence between these points and the possible values of $\mathbf{k}$. Since, however, k_x, k_y, and k_z are positive quantities, we must count only those

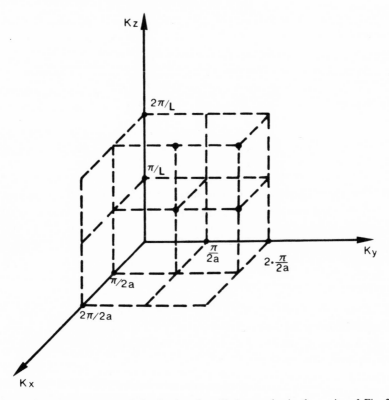

FIG. 2.2. Pictorial illustration of the density of oscillating modes in the cavity of Fig. 2.1. Each point of the lattice corresponds to two cavity modes.

points lying in the positive octant. The number of such points having k between 0 and $2\pi\nu/c$ is $1/8$ of the ratio between the volume of the sphere of radius $2\pi\nu/c$ centered at the origin and the volume of the unit cell of dimensions $\pi/2a$, $\pi/2a$, π/L. Since, as we have previously said, there are two modes possible for each value of k, we have

$$N_\nu = 2\frac{\frac{1}{8}\cdot\frac{4}{3}\pi\left(\frac{2\pi\nu}{c}\right)^3}{\frac{\pi}{2a}\frac{\pi}{2a}\frac{\pi}{L}} = \frac{8\pi\nu^3}{3c^3}V \tag{2.13}$$

where V is the total volume of the cavity. If we now define $p(\nu)$ as the number of modes per unit volume and per unit frequency range, we have

$$p(\nu) = \frac{1}{V}\frac{dN}{d\nu} = \frac{8\pi\nu^2}{c^3} \tag{2.14}$$

Let us now calculate the average energy associated with each of the resonant modes of the cavity. To this end, let us assume that the temperature of the cavity walls is T. According to Boltzmann's statistics, the probability dp that the energy of a given cavity mode lies between E and $E + dE$ is expressed by $dp = C \exp[-(E/kT)]dE$, where C is a constant. The average energy of the mode $\langle E \rangle$ is therefore given by

$$\langle E \rangle = \frac{\int_0^\infty E \exp[-(E/kT)] dE}{\int_0^\infty \exp[-(E/kT)] dE} = kT \tag{2.15}$$

The energy density ρ_ν is then obtained by multiplying the number of modes per unit volume and per unit frequency range $p(\nu)$ by the average energy $\langle E \rangle$ of each mode:

$$\rho_\nu = \left(\frac{8\pi\nu^2}{c^3} \right) kT \tag{2.16}$$

This is the well-known Rayleigh–Jeans radiation formula. It is, however, in complete disagreement with experimental results. Indeed, it is immediately obvious that equation (2.16) must be wrong since it would imply an infinite total energy density ρ [equation 2.1a]. Equation (2.16) does, however, represent the inevitable conclusion of the previous classical arguments.

The problem remained unsolved until, at the beginning of this century, Planck introduced the hypothesis of light quanta. The fundamental hypothesis of Planck was that the energy in a given cavity mode could not have any arbitrary value between 0 and ∞, as was implicitly assumed in equation (2.15), but that the allowed values of this energy should be integral multiples of a fundamental quantity, proportional to the frequency of the mode. In other words, Planck assumed that the energy of the mode could be written as $E = nh\nu$, where n is a positive integer and h a constant (which was later called Planck's constant). Without entering into too many details on this fundamental hypothesis, we merely wish to note here that this essentially implies that energy exchange between the inside of the cavity and its walls involves a discrete amount of energy $h\nu$. This minimum quantity which can be exchanged is called a light quantum or photon. According to this hypothesis, the average energy of the mode is

$$\langle E \rangle = \frac{\sum_{n=0}^{\infty} nh\nu \exp[-(nh\nu/kT)]}{\sum_{n=0}^{\infty} \exp[-(nh\nu/kT)]} = \frac{h\nu}{\exp(h\nu/kT) - 1} \tag{2.17}$$

This formula is quite different from the classical expression (2.15). Obviously for $h\nu \to 0$, (2.17) reduces to (2.15). From (2.14) and (2.17) we get the Planck formula,

$$\rho_\nu = \frac{8\pi\nu^2}{c^3} \frac{h\nu}{\exp(h\nu/kT) - 1} \tag{2.18}$$

which is now in perfect agreement with experimental results, provided that we choose for h the value $h \simeq 6.62 \times 10^{-34}$ J s. The above expression can also be written in terms of the function ρ_ω defined in such a way that $\rho_\omega d\omega$ represents the energy density of radiation with angular frequency between ω and $\omega + d\omega$. By setting $\rho_\omega d\omega = \rho_\nu d\nu$, we get from (2.18)

$$\rho_\omega = \frac{\rho_\nu}{2\pi} = \frac{4\nu^2}{c^3} \frac{\hbar\omega}{\left[\exp(\hbar\omega/kT) - 1\right]} \tag{2.18a}$$

where, following the widely accepted convention, we have set $\hbar = h/2\pi$. Figure 2.3 shows the behavior of ρ_ν as a function of frequency for two values of temperature T.

Lastly, we may note that the ratio

$$\langle q \rangle = \frac{\langle E \rangle}{h\nu} = \frac{1}{\exp(h\nu/kT) - 1} \tag{2.19}$$

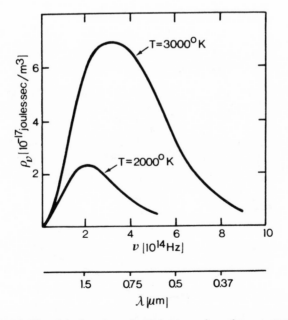

FIG. 2.3. Plot of the function $\rho_\nu(\nu, T)$ for two values of temperature T.

gives the average number of photons $\langle q \rangle$ for each mode. If we now
consider a frequency ν in the optical range (4×10^{14} Hz), we get $h\nu \simeq 1$ eV.
For $T \simeq 300°K$ we have $kT \simeq (1/40)$ eV, so that from equation (2.19),
$\langle q \rangle \simeq \exp(-40)$. We thus see that the average number of photons per
mode, for blackbody radiation at room temperature, is very much smaller
than unity. This value should be compared with the number of photons q_0
that can be obtained in a laser cavity in a single mode (see, for example,
Fig. 5.19).

Before ending this section it is of interest to derive the relationship
between energy density in a blackbody cavity and the intensity I emitted
by its walls. With reference to Fig. 2.4, let us calculate the energy density in
a small volume V inside the cavity due to emission from the cavity walls.
The cone of solid angle $d\Omega$ with apex at the elemental area dS of the
surface (at a distance r from V) will intersect V in a cylinder of cross
section ds and length l. According to (1.13), the energy emitted in unit time
by dS into the solid angle $d\Omega$ is $B \cos\theta \, dS \, d\Omega$, where B is the brightness of
the blackbody surface. Of this energy, the fraction l/c will be within the
volume V. Since $d\Omega = ds/r^2$, the energy in the volume V is $B \cos\theta \, dS$
$(l \, ds/r^2 c)$. To get the total contribution of dS to the volume V we must
integrate over all the solid angles which are coming from dS, which gives

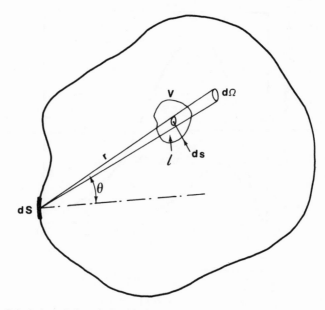

FIG. 2.4. Calculation of the relationship between the radiated intensity at the surface of a
blackbody cavity and its energy density.

$\int l\,ds = V$. Next, we must integrate the emission of the whole blackbody surface. We thus get the following expression for the energy density in the volume V:

$$\rho = \frac{B}{c} \int \frac{\cos\theta}{r^2} \, dS$$

Note now that $\cos\theta\, dS/r^2$ is equal to the solid angle $d\Omega'$ subtended by the surface dS as seen from any point of volume V (which is assumed to be very small). Thus

$$\rho = (B/c)\int d\Omega' = 4\pi B/c$$

The total intensity I emitted by dS, on the other hand, is

$$I = \int_{\theta=0}^{\pi/2} \int_{\phi=0}^{2\pi} B\cos\theta\, d\Omega = \pi B \qquad (2.19a)$$

With the help of this last expression, we finally get the result

$$\rho = (4/c)I \qquad (2.19b)$$

The same relationship obviously applies for the corresponding spectral densities, so that

$$\rho_\omega = (4/c)I_\omega \qquad (2.19c)$$

The substitution of (2.18a) in (2.19c) then gives an expression for the spectral intensity distribution of the light emitted by a blackbody surface.

2.2 ABSORPTION AND STIMULATED EMISSION

In this section, we will study in some detail the processes of absorption and stimulated emission induced in a two-level atomic system by a monochromatic electromagnetic wave. In particular our aims are: (i) to calculate the rates of absorption W_{12} and stimulated emission W_{21}, where W_{12} and W_{21} are defined by equations (1.5) and (1.3) respectively; (ii) to introduce and calculate the absorption and emission cross sections [see equations (1.4) and (1.6)]; (iii) to introduce two new parameters, the absorption and gain coefficients, which can often be directly measured by simple experiments.

2.2.1 Rates of Absorption and Stimulated Emission

The calculation which follows will make use of the so-called semiclassical treatment of the interaction of radiation and matter. In this treatment the atomic system is assumed to be quantized (and it is, therefore, described quantum mechanically) while the e.m. field of the incident wave is treated classically (i.e., by using Maxwell's equations).

We will first examine the phenomenon of absorption. We therefore consider the usual two-level system where we assume that, at time $t = 0$, the atom is in its ground state 1 and that a monochromatic e.m. wave at frequency ω is made to interact with it. Classically, the atom will acquire an additional energy H' when interacting with the e.m. wave. For instance this may be due to the interaction of the electric dipole moment of the atom μ_e with the electric field E of the e.m. wave ($H' = \mu_e \cdot E$). In this case we will talk about an electric dipole interaction. This is not the only type of interaction by which the transition can occur, however. For instance, the transition may result from the interaction of the magnetic dipole moment of the atom μ_m with the magnetic field B of the e.m. wave ($\mu_m \cdot B$, magnetic dipole interaction). To describe the time evolution of this two-level system, we must now resort to quantum mechanics. Thus, just as the classical B treatment involves an interaction energy H', so the quantum mechanical approach introduces an interaction Hamiltonian $\mathcal{H}'$. This Hamiltonian can be obtained from the classical expression for H' according to the well-known rules of quantum mechanics. The precise expression for $\mathcal{H}'$ need not concern us at this point, however. We only need to note that $\mathcal{H}'$ is a sinusoidal function of time with frequency ω equal to that of the incident wave. Accordingly we put

$$\mathcal{H}' = \mathcal{H}'^0 \sin \omega t \tag{2.20}$$

The total Hamiltonian $\mathcal{H}$ for the atom can then be written as

$$\mathcal{H} = \mathcal{H}_0 + \mathcal{H}' \tag{2.21}$$

where $\mathcal{H}_0$ is the atomic Hamiltonian in the absence of the e.m. wave. Once the total Hamiltonian $\mathcal{H}$ for $t > 0$ is known, the time evolution of the wave function ψ of the atom is obtained from the time-dependent Schrödinger equation

$$\mathcal{H}\psi = i\hbar \frac{\partial \psi}{\partial t} \tag{2.22}$$

To solve (2.22) for the function ψ we begin by introducing $\psi_1 = u_1 \exp[-(iE_1 t/\hbar)]$ and $\psi_2 = u_2 \exp[-(iE_2 t/\hbar)]$, the unperturbed eigenfunction of levels 1 and 2 respectively. Thus u_1 and u_2 satisfy the time-independent Schrödinger equation.

$$\mathcal{H}_0 u_i = E_i u_i \qquad (i = 1, 2) \tag{2.22a}$$

Under the influence of the e.m. wave, the wave function of the atom will be written as

$$\psi = a_1(t)\psi_1 + a_2(t)\psi_2 \tag{2.23}$$

where in general a_1 and a_2 will be time-dependent complex numbers. It is a well-known result from quantum mechanics that the coefficients $|a_1|^2$ and $|a_2|^2$ give, respectively, the probabilities that, at time t, the atom will be found in state 1 or 2, and they obey the following relation:

$$|a_1|^2 + |a_2|^2 = 1 \tag{2.24}$$

To calculate the transition probability W_{12} we must, therefore, calculate the quantity $|a_2(t)|^2$ (or $|a_1(t)|^2$). More generally, instead of (2.23), we will write

$$\psi = \sum_1^m a_k \psi_k = \sum_1^m a_k u_k \exp\left[-i(E_k/\hbar)t \right] \tag{2.25}$$

where m denotes the state of the atom. By substituting (2.25) into (2.22) we obtain

$$\sum_k (\mathcal{H}_0 + \mathcal{H}')a_k u_k \exp\left[-i(E_k/\hbar)t\right] = \sum_k \left\{ i\hbar \dot{a}_k u_k \exp\left[-i(E_k/\hbar)t \right] \right.$$
$$\left. + a_k u_k E_k \exp\left[-i(E_k/\hbar)t\right] \right\} \tag{2.25a}$$

This equation, with the help of (2.22a), reduces to

$$\sum i\hbar \dot{a}_k u_k \exp\left[-i(E_k/\hbar)t\right] = \sum a_k \mathcal{H}' u_k \exp\left[-i(E_k/\hbar)t\right] \tag{2.26}$$

By multiplying each side of this equation by the arbitrary eigenfunction u_n^* and then integrating over the whole of space, we get

$$\sum i\hbar \dot{a}_k \exp\left[-i(E_k/\hbar)t\right] \int u_k u_n^* \, dV = \sum a_k \exp\left[-i(E_k/\hbar)t\right] \int u_n^* \mathcal{H}' u_k \, dV \tag{2.26a}$$

Since the wave functions u_k are orthogonal (i.e., $\int u_n^* u_k \, dV = \delta_{kn}$), with the notation that

$$H_{nk}'(t) = \int u_n^* \mathcal{H}' u_k \, dV \tag{2.27}$$

(2.26a) reduces to

$$\dot{a}_n = \frac{1}{(i\hbar)} \sum_1^m {}_k H_{nk}' a_k \exp\left(-i\frac{(E_k - E_n)t}{\hbar} \right) \tag{2.28}$$

We therefore get m differential equations for the m variables $a_k(t)$, and these equations can be solved once the initial conditions are known. For a two-level system ($m = 2$), equation (2.28) gives the two equations

$$\dot{a}_1 = (1/i\hbar)\left\{ H_{11}'a_1 + H_{12}'a_2 \exp\left[-i(E_2 - E_1)t/\hbar\right] \right\}$$
$$\dot{a}_2 = (1/i\hbar)\left\{ H_{21}'a_1 \exp\left[-i(E_1 - E_2)t/\hbar\right] + H_{22}'a_2 \right\} \tag{2.29}$$

which must be solved with the initial condition $a_1(0) = 1$, $a_2(0) = 0$.

So far, no approximations have been made. Now, to simplify the solution of the equations (2.29), we will make use of a perturbation method.

We will assume that on the right-hand side of equation (2.29) we can make the approximations that $a_1(t) \simeq 1$ and $a_2(t) \simeq 0$. By solving (2.29) on this hypothesis, we shall get the first-order solutions for $a_1(t)$ and $a_2(t)$. For this reason, the theory that follows is known as first-order perturbation theory. The solutions $a_1(t)$ and $a_2(t)$ obtained in this way can then be substituted on the right-hand side of (2.29) to get a solution which will be a second-order approximation, and so on to higher orders. Correspondingly, one describes this as second-order perturbation theory, and so on. To first order, therefore, (2.29) gives

$$\dot{a}_1 = (1/i\hbar)H'_{11} \tag{2.30a}$$

$$\dot{a}_2 = (1/i\hbar)H'_{21}\exp(i\omega_0 t) \tag{2.30b}$$

where we have written $\omega_0 = (E_2 - E_1)/\hbar$ for the transition frequency of the atom. To calculate the transition probability, we need only to solve equation (2.30b). To this end, making use of (2.20) and (2.27), we can write

$$H'_{21} = H'^0_{21}\sin\omega t = \frac{H'^0_{21}\big[\exp(i\omega t) - \exp(-i\omega t)\big]}{2i} \tag{2.31}$$

where H'^0_{21} is given by

$$H'^0_{21} = \int u_2^* \mathcal{H}'^0 u_1 \, dV \tag{2.31a}$$

and is, in general, a complex constant. By substituting (2.31) into (2.30b) and integrating with the initial condition $a_2(0) = 0$, we get

$$a_2(t) = \frac{H'^0_{21}}{2i\hbar}\left[\frac{\exp\big[i(\omega_0 - \omega)t\big] - 1}{\omega_0 - \omega} - \frac{\exp\big[i(\omega_0 + \omega)t\big] - 1}{\omega_0 + \omega} \right] \tag{2.32}$$

If we now assume that $\omega \simeq \omega_0$, we see that the first term in the square brackets is much larger than the second. We can then write

$$a_2(t) \simeq -\frac{H'^0_{21}}{2i}\frac{\exp(-i\Delta\omega t) - 1}{\hbar\Delta\omega} \tag{2.33}$$

where $\Delta\omega = \omega - \omega_0$. It then follows that

$$|a_2(t)|^2 = \frac{|H'^0_{21}|^2}{\hbar^2}\left[\frac{\sin(\Delta\omega t/2)}{\Delta\omega} \right]^2 \tag{2.34}$$

The function $y = [\sin(\Delta\omega t/2)/\Delta\omega]^2$ is plotted in Fig. 2.5 as a function of $\Delta\omega$. We see that this function becomes higher and narrower as t increases. Since, furthermore, we can show that

$$\int_{-\infty}^{+\infty}\left[\frac{\sin(\Delta\omega t/2)}{\Delta\omega} \right]^2 d\Delta\omega = \frac{\pi t}{2} \tag{2.35}$$

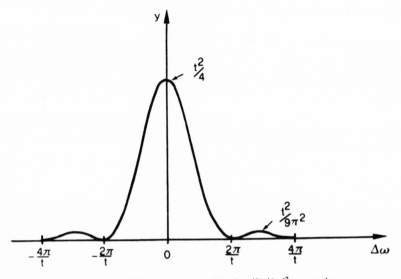

FIG. 2.5. Plot of the function $y = [\sin(\Delta\omega t/2)/\Delta\omega]^2$ versus $\Delta\omega$.

then, for large enough values of t we can put

$$\left[\frac{\sin(\Delta\omega t/2)}{\Delta\omega}\right]^2 \simeq \frac{\pi t}{2}\delta(\Delta\omega) \qquad (2.36)$$

where δ is the Dirac δ function. Therefore

$$|a_2(t)|^2 = \frac{|H_{21}'^{0}|^2}{\hbar^2}\frac{\pi}{2}t\delta(\Delta\omega) \qquad (2.37)$$

which shows that, after a long enough time, the probability $|a_2(t)|^2$ of finding the atom at time t in level two is proportional to the time itself. Therefore the transition probability W_{12} is given by

$$W_{12} = \frac{|a_2(t)|^2}{t} = \frac{\pi}{2}\frac{|H_{21}'^{0}|^2}{\hbar^2}\delta(\Delta\omega) \qquad (2.38)$$

To calculate W_{12} explicitly, we must calculate the quantity $|H_{21}'^{0}|^2$. To this end, if we assume that the interaction responsible for the transition occurs between the electric field of the e.m. wave and the electric dipole moment of the atom (electric dipole interaction) we can write [2]

$$\mathcal{H}' = e\mathbf{E}(\mathbf{r},t)\cdot\mathbf{r} \qquad (2.39)$$

In (2.39), e is the charge of the electron which undergoes the transition, the vector $\mathbf{r}$ gives its position, and $\mathbf{E}(\mathbf{r},t)$ is the electric field at point $\mathbf{r}$. For the sake of simplicity, we assume the origin ($\mathbf{r} = 0$) of our reference system to be coincident with the nucleus. Therefore, from (2.27) and (2.39), we have

$$H'_{21} = e \int u_2^* \mathbf{E} \cdot \mathbf{r} u_1 \, dV \qquad (2.40)$$

Let us now suppose that the wavelength of the e.m. wave is much greater than the atomic dimensions. This is satisfied very well for e.m. waves in the visible ($\lambda = 5000$ Å for green light, while the atomic dimensions are of the order of 1 Å). On this assumption, we can take $\mathbf{E}$ out of the integral in equation (2.40) and use its value at $\mathbf{r} = 0$, i.e., at the center of the nucleus (electric dipole approximation). Then, by defining

$$\mathbf{E}(0, t) = \mathbf{E}_0 \sin \omega t \qquad (2.41)$$

from (2.31), (2.40), and (2.41) we get

$$H'^0_{21} = \mathbf{E}_0 \cdot \boldsymbol{\mu}_{21} \qquad (2.42)$$

where $\boldsymbol{\mu}_{21}$ is given by

$$\boldsymbol{\mu}_{21} = e \int u_2^* \mathbf{r} u_1 \, dV \qquad (2.43)$$

and is called the matrix element of the electric dipole moment. Therefore, if we call θ the angle between $\boldsymbol{\mu}_{21}$ and $\mathbf{E}_0$, we have

$$|H'^0_{21}|^2 = E_0^2 |\mu_{21}|^2 \cos^2\theta \qquad (2.44)$$

where $|\mu_{21}|$ is the magnitude of the complex number μ_{21} (μ_{21} is in turn the magnitude of the vector $\boldsymbol{\mu}_{21}$). If we now assume that the e.m. wave interacts with several atoms whose vectors $\boldsymbol{\mu}_{21}$ are randomly oriented with respect to $\mathbf{E}_0$, the average value of $|H'^0_{21}|^2$ will be obtained by averaging (2.44) over all possible values of $\cos^2\theta$. If all angles θ are equally probable, it is known that the average value is $\langle \cos^2\theta \rangle = \frac{1}{3}$, so that

$$\langle |H'^0_{21}|^2 \rangle = \frac{1}{3} E_0^2 |\mu_{21}|^2 \qquad (2.45)$$

Instead of expressing $|H'^0_{21}|^2$ as a function of E_0, it is often more convenient to express it as a function of the energy density of the incident e.m. wave, which is given by

$$\rho = n^2 \epsilon_0 E_0^2 / 2 \qquad (2.46)$$

where n is the refractive index of the atomic system and ϵ_0 is the vacuum permittivity. From (2.38), (2.45), and (2.46) we get finally

$$W_{12} = \frac{\pi}{3 n^2 \epsilon_0 \hbar^2} |\mu_{21}|^2 \rho \delta(\Delta\omega) \qquad (2.47)$$

For a plane e.m. wave, it is sometimes useful to express W_{12} as a function of the intensity I of the incident e.m. radiation. Since $I = c_0 \rho / n$, where c_0 is the velocity of light *in vacuo*, from (2.47) we get

$$W_{12} = \frac{\pi}{3 n \epsilon_0 c_0 \hbar^2} |\mu_{21}|^2 I \delta(\Delta\omega) \tag{2.47a}$$

Equations (2.47) and (2.47a) summarize the results of our treatment so far. It should be noted again that while equation (2.47) is of general validity (within the approximations used, of course), equation (2.47a) only applies for a plane e.m. wave of uniform intensity. It is readily seen, however, that in their present form both expressions are physically unacceptable. In fact, the presence of the Dirac δ function implies that they would give $W_{12} = 0$ when $\omega \neq \omega_0$, and $W_{12} = \infty$ when $\omega = \omega_0$, i.e., when the frequency of the e.m. wave is exactly coincident with the frequency of the atomic transition. The reason for this nonphysical result can be traced back to the fact that we have let the time t go to infinity in (2.34), which means that we have assumed that the interaction between the e.m. wave and the system could continue coherently for an infinite time. Actually, there are a number of physical phenomena which prevent this from occurring. Although this point will be treated more extensively later on, it is worth giving an example here. Let us suppose that the collection of atoms with levels 1 and 2 (and subjected to the e.m. wave) is in gaseous form. In this case, collisions will occur between atoms. After a collision, the wavefunctions $u_1(\mathbf{r})$ and $u_2(\mathbf{r})$ of the atom will no longer have the same phase relative to that of the incident e.m. wave. Therefore, the interaction as described in the previous equations is valid only in the time interval between one collision and the next. After each collision, the initial conditions and in particular the relative phase between the atom's wavefunction and the electric field of the incoming wave undergoes a random jump. An equivalent way of treating this problem is to assume that it is just the phase of the electric field that undergoes a jump at each collision. The electric field will therefore no longer appear sinusoidal but will instead appear as shown in Fig. 2.6, where

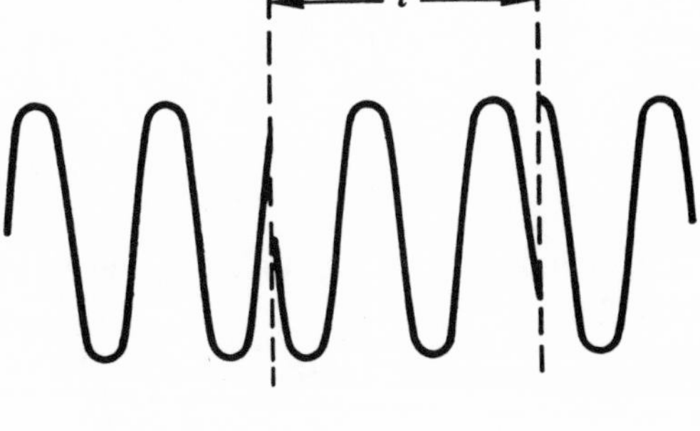

FIG. 2.6. Time behavior of the electric field of an e.m. wave as seen from an atom undergoing collisions.

each phase jump occurs at the time of a collision. It is therefore clear that, under these conditions, the atom no longer sees a monochromatic wave. In this case, if we write $d\rho = \rho_{\omega'} d\omega'$ for the energy density of the wave in the frequency interval between ω' and $\omega' + d\omega'$, we find from (2.47) that the transition probability is

$$W_{12} = \frac{\pi}{3n^2\epsilon_0\hbar^2} |\mu_{21}|^2 \int_{-\infty}^{+\infty} \rho_{\omega'}\delta(\omega' - \omega_0)\, d\omega' \qquad (2.48)$$

To calculate (2.48) explicitly, we must know $\rho_{\omega'}$, which is proportional to the square of the magnitude of the Fourier spectrum of the signal shown in Fig. 2.6. To find this, let us use the symbol τ (as indicated in Fig. 2.6) for the time interval between two collisions. This quantity will obviously be different for each collision. To specify this variability more precisely, we will assume that the distribution of values of τ can be described by a probability density

$$p_\tau = [\exp(-\tau/T_2)]/T_2 \qquad (2.49)$$

Here $p_\tau d\tau$ is the probability that the time interval between successive collisions lies between τ and $\tau + d\tau$. Note that T_2 has the physical meaning of the average time τ_c between two collisions. It is easy, in fact, to see that

$$\tau_c = \int_0^\infty \tau\, p_\tau\, d\tau = T_2 \qquad (2.50)$$

We can now show (see Section 2.5) that, for a signal as in Fig. 2.6 with a probability density as in (2.49), the spectral energy density $\rho_{\omega'}$ can be written as

$$\rho_{\omega'} = \rho g(\omega' - \omega) \qquad (2.51)$$

where ρ is the energy density of the wave,

$$g(\omega' - \omega) = \frac{T_2}{\pi} \frac{1}{\left[1 + (\omega' - \omega)^2 T_2^2\right]} \qquad (2.52)$$

and ω is the frequency of the sinusoidal signal of Fig. 2.6. We can verify that

$$\int_{-\infty}^{+\infty} g(\omega' - \omega)\, d\omega' = 1 \qquad (2.52a)$$

as it should be according to (2.51), since $\rho = \int \rho_{\omega'} d\omega'$. By substituting (2.51) into (2.48) we have

$$W_{12} = \frac{\pi}{3n^2\epsilon_0\hbar^2} |\mu_{12}|^2 \rho g(\Delta\omega) \qquad (2.53)$$

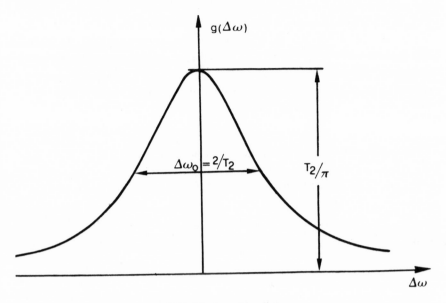

FIG. 2.7. Lorentzian line.

where $\Delta\omega = \omega - \omega_0$. Thus we now have an expression similar to (2.47) but with the function $\delta(\Delta\omega)$ replaced by the function $g(\Delta\omega)$. The function $g(\Delta\omega)$ is plotted in Fig. 2.7. Its maximum occurs for $\Delta\omega = 0$ (that is, $\omega = \omega_0$) and the corresponding value is T_2/π. The full width of the curve between points having half the maximum value (FWHM) is $\Delta\omega_0 = 2/T_2$. Such a curve is called *Lorentzian*.

Although equation (2.53) has been obtained for a particular case, it will be shown in Section 2.5 that W_{12} can always be expressed in a form similar to (2.53), namely

$$W_{12} = \frac{\pi}{3n^2\epsilon_0\hbar^2}|\mu_{21}|^2\rho g_t(\Delta\omega) \tag{2.53a}$$

where $g_t(\Delta\omega)$ is a normalized function (i.e., $\int g_t\, d\omega = 1$) whose particular form depends on the phenomenon responsible for the line broadening. Equation (2.53a) is therefore our final expression for the rate of absorption within the electric dipole approximation. For a plane e.m. wave of uniform intensity, equation (2.53a) can be recast in terms of the intensity I of the wave as

$$W_{12} = \frac{\pi}{3n\epsilon_0 c_0\hbar^2}|\mu_{21}|^2 I g_t(\Delta\omega) \tag{2.53b}$$

Having calculated the rate of absorption, we now go on to calculate the rate of stimulated emission. To do this, we should start again from (2.29), using now the initial conditions $a_1(0) = 0$ and $a_2(0) = 1$. We immediately see, however, that the equations in this case are readily obtainable from the corresponding equations [(2.29)–(2.53)] for the case of absorption simply by interchanging the indexes 1 and 2. Since it can easily be seen from (2.43) that $|\mu_{12}| = |\mu_{21}|$, it follows from (2.53a) that

$$W_{12} = W_{21} \tag{2.54}$$

which shows that the probabilities of absorption and stimulated emission are equal. From now on we will, therefore, set $W = W_{12} = W_{21}$ and $|\mu| = |\mu_{12}| = |\mu_{21}|$. Accordingly equations (2.53a) and (2.53b) become

$$W = \frac{\pi}{3n^2\epsilon_0\hbar^2} |\mu|^2 \rho g_t(\Delta\omega) \tag{2.53c}$$

and

$$W = \frac{\pi}{3n\epsilon_0 c_0 \hbar^2} |\mu|^2 I g_t(\Delta\omega) \tag{2.53d}$$

which are the final results of our calculations in this section.[†]

2.2.2 Allowed and Forbidden Transitions

Equation (2.53c) shows that $W = 0$ if $|\mu| = 0$. This occurs, for example [see equation (2.43)], when the eigenfunctions u_1 and u_2 are either both symmetric or both antisymmetric.[‡] In fact, in this case, the two contributions from the integrand of (2.43) at points $\mathbf{r}$ and $-\mathbf{r}$, respectively, are equal and opposite. It is therefore of interest to see when the wavefunctions $u(\mathbf{r})$ are either symmetric or antisymmetric.[(2)] This occurs when the Hamilto-

---

[†] It should be noted that all the expressions given so far have been derived under the assumption that the local (microscopic) field $E_{loc}(\mathbf{r}, t)$ experienced by the atom is equal to the total (average) field $E(\mathbf{r}, t)$. Although this may be taken to be true for a dilute medium, a suitable correction (the Lorentz local field correction factor) must be introduced for dense media. It can be shown that $E_{loc} = [(n^2 + 2)/3]E$, where n is the refractive index of the medium arising from all transitions except the transition under consideration. If E_{loc} is substituted for E in (2.39) we see that all previous expressions involving the transition probability W remain valid provided that we replace $|\mu_{21}|^2$ by $[(n^2 + 2)/3]^2|\mu_{21}|^2$.[(18)] The same correction should therefore be applied in all subsequent expressions concerning stimulated transitions [e.g., (2.61) and (2.65)]. A similar correction is perhaps also applicable to the expression for the rate of spontaneous emission [(2.86)] although the author is not aware of any treatment of spontaneous emission in which this point has been explicitly considered. The widely used relations between σ and τ_{sp} [equations (2.145) and (2.146)] thus remain (perhaps) valid since they are simply a re-expression of the (A/B) relation [see (2.83)] which was obtained by a thermodynamic argument.

[‡] It may be recalled here that a function $f(\mathbf{r})$ is symmetric (or of even parity) if $f(-\mathbf{r}) = f(\mathbf{r})$, while it is antisymmetric (or of odd parity) if $f(-\mathbf{r}) = -f(\mathbf{r})$.

nian $\mathcal{H}_0(\mathbf{r})$ of the system is unchanged by changing $\mathbf{r}$ to $-\mathbf{r}$, i.e., when[†]

$$\mathcal{H}_0(\mathbf{r}) = \mathcal{H}_0(-\mathbf{r}) \tag{2.55}$$

In this case, in fact, for any eigenfunction $u_n(\mathbf{r})$, one has

$$\mathcal{H}_0(\mathbf{r})u_n(\mathbf{r}) = E_n u_n(\mathbf{r}) \tag{2.56}$$

and changing $\mathbf{r}$ into $-\mathbf{r}$ and using (2.55) gives

$$\mathcal{H}_0(\mathbf{r})u_n(-\mathbf{r}) = E_n u_n(-\mathbf{r}) \tag{2.57}$$

Equations (2.56) and (2.57) show that $u_n(\mathbf{r})$ and $u_n(-\mathbf{r})$ are both eigenfunctions of the operator $\mathcal{H}_0$ with the same eigenvalue E_n. For nondegenerate energy levels, it is well known[3] that (apart from an arbitrary choice of sign) there is only one eigenfunction for each eigenvalue, so that

$$u_n(-\mathbf{r}) = \pm u_n(\mathbf{r}) \tag{2.58}$$

Therefore, if $\mathcal{H}_0(\mathbf{r})$ is symmetric, the eigenfunctions must be either symmetric or antisymmetric. In this case, it is usually said that the eigenfunctions have a well-defined parity.

It remains now to be seen when the Hamiltonian satisfies (2.55), i.e., when it is invariant under inversion. Obviously this occurs when the system has a center of symmetry. Another important case is that of an isolated atom. In this case, the potential energy of the kth electron of the atom is given by the sum of the potential energy due to the nucleus (which is symmetric) and that due to all other electrons. For the ith electron, this energy will depend on $|\mathbf{r}_i - \mathbf{r}_k|$, i.e., on the magnitude of the distance between the two electrons. Therefore, these terms will also be invariant under inversion. An important case where equation (2.55) is not valid is where an atom is placed in an external electric field (e.g., a crystal's electric field) which does not possess a center of inversion. In this case the wave functions u_n will not have a definite parity.

To sum up, we can say that electric dipole transitions only occur between states of opposite parity, and that the states have a well-defined parity if the Hamiltonian is invariant under inversion.

If $W = 0$, it is said that the transition is forbidden within the electric dipole approximation. This does not mean, however, that the atom cannot pass from level 1 to level 2 under the influence of the incident e.m. wave. In this case, the transition could occur, for instance, as a result of the interaction between the magnetic field of the e.m. wave and the magnetic dipole moment of the atom. For the sake of simplicity, we will not consider

---

[†]If the Hamiltonian $\mathcal{H}_0$ is a function of more than one coordinate $\mathbf{r}_1, \mathbf{r}_2, \ldots,$ the inversion operation must be simultaneously applied to all these coordinates.

this case any further (magnetic dipole interaction), but limit ourselves to observing that the analysis can be carried out in exactly the same manner as that used up to equation (2.38). Obviously, the value of $|H'^0_{12}|^2$ will now be different. We may also point out that a magnetic dipole transition is allowed between states of equal parity (even–even or odd–odd transitions). Therefore, a transition which is forbidden for electric dipole interaction is, however, allowed for magnetic dipole interaction and *vice versa*. It is also instructive to calculate the order of magnitude of the ratio of the electric dipole transition probability W_e to the magnetic dipole transition probability W_m. Obviously the calculation refers to two different transitions, one being allowed for electric dipole and the other for magnetic dipole interaction. We also assume that the intensity of the e.m. wave is the same for the two cases. For an allowed dipole transition, according to (2.38) and (2.39), we can write $W_e \propto (\mu_e E_0)^2 \simeq (eaE_0)^2$, where E_0 is the electric field amplitude and where the electric dipole moment of the atom μ_e has been approximated (for an allowed transition) by the product of the electron charge e times the radius a of the atom. In a similar way, we can write $W_m \propto (\mu_m B_0)^2 \simeq (\beta B_0)^2$, where B_0 is the magnetic field amplitude of the wave and where the magnetic dipole moment of the atom, μ_m, has been approximated (for an allowed transition) by the Bohr magneton β ($\beta = 9.27 \times 10^{-24}$A m^2). Thus we get

$$\left(\frac{W_e}{W_m} \right) = \left(\frac{eaE_0}{\beta B_0} \right)^2 = \left(\frac{eac}{\beta} \right)^2 \simeq 10^5 \qquad (2.59)$$

To obtain the numerical result of (2.59) we have made use of the fact that, for a plane wave, $E_0 = B_0 c$ (where c is the light velocity), and we have assumed that $a \simeq 0.5$Å. We see that the probability of an electric dipole transition is much greater than that of a magnetic dipole. This is essentially due to the fact that the interaction energy via an electric dipole ($\mu_e E_0$) is much greater than that due to a magnetic dipole ($\mu_m B_0$).

2.2.3 Transition Cross Section, Absorption and Gain Coefficient

Having calculated the transition rate W, we can now go on to define and calculate other parameters which are often used to describe the given transition.

The first of these parameters is the transition cross section σ which has already been briefly discussed in Chapter 1 [see (1.4) and (1.6)]. We have seen that, for a uniform plane wave, the transition rate is proportional to the intensity, and we can therefore define the cross section σ as

$$\sigma = W/F \qquad (2.60)$$

where $F = I/\hbar\omega$ is the photon flux of the incident e.m. wave. From $(2.53d)$ we then obtain the expression for σ as

$$\sigma = \frac{\pi}{3n\epsilon_0 c_0 \hbar} |\mu|^2 \omega g_t(\Delta\omega) \qquad (2.61)$$

We see from (2.61) that σ depends on the material parameters ($|\mu|^2$ and g_t) and on the frequency ω of the incident wave. A knowledge of σ as a function of ω is therefore all that is needed to describe the interaction process. The transition cross section σ is therefore a very important and widely used parameter for the transition. A physical interpretation of its significance can be obtained from equation (1.7). For simplicity, we assume all the atoms to be in the ground level so that $N_2 = 0$ and $N_1 = N_t$, where the symbol N_t denotes the total population of the system. From (1.7) we then get

$$dF = -\sigma N_t F \, dz \qquad (2.62)$$

Let us now suppose that we can associate with each atom an effective absorption cross section σ_a, in the sense that, if a photon enters this cross section, it will be absorbed by the atom (Fig. 2.8). If we call S the cross-sectional area of the e.m. wave in the material, the number of atoms in the element dz of the material irradiated by the wave (see also Fig. 1.2) is $N_t S \, dz$, thus giving a total absorption cross section of $\sigma_a N_t S \, dz$. The fractional change (dF/F) of photon flux in the element dz of the material is therefore

$$\frac{dF}{F} = -\frac{\sigma_a N_t S \, dz}{S} \qquad (2.63)$$

A comparison of (2.63) with (2.62) shows that $\sigma = \sigma_a$, so that the meaning

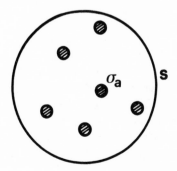

FIG. 2.8. Effective absorption cross section σ_a of atoms in a beam of cross section S.

we can attribute to σ is that of an effective absorption cross section as defined above.

Another way of describing the interaction of radiation with matter is obtained by defining a coefficient α as

$$\alpha = \sigma(N_1 - N_2) \tag{2.64}$$

If $N_1 > N_2$, the quantity α is referred to as the absorption coefficient of the material. Using (2.61), the following expression for α is obtained:

$$\alpha = \frac{\pi}{3n\epsilon_0 c_0\hbar} |\mu|^2 \omega(N_1 - N_2) g_t(\Delta\omega) \tag{2.65}$$

Since α depends on the population of the two levels, it is not the most suitable parameter to describe the interaction in a situation where the level populations are changing, such as, for example, in a laser. Its usefulness, however, lies in the fact that the absorption coefficient α can often be directly measured. From (1.7) and (2.64) we in fact get $dF = -\alpha F dz$. The ratio between the photon flux after traversing a length l of the material and the incident flux is therefore $F(l)/F(0) = \exp(-\alpha l)$. By experimentally measuring this ratio with a sufficiently monochromatic radiation, we can obtain the value of α for that particular incident wavelength. The corresponding transition cross section is then obtained from (2.64) once N_1 and N_2 are known. If the medium is in thermodynamic equilibrium, N_1 and N_2 can be obtained (given a knowledge of the total population $N_t = N_1 + N_2$) with the help of (1.8). The instrument used to measure the absorption coefficient α is known as an absorption spectrophotometer. We note, however, that an absorption measurement obviously cannot be performed for a transition in which level 1 is empty. This, for instance, occurs when level 1 is not the ground level and its energy above the ground level energy is much larger than kT. As a final observation we note that if $N_2 > N_1$, the absorption coefficient α, defined by (2.64), becomes negative and, of course, the wave gets amplified rather than absorbed in the material. In this case it is customary to define the new quantity

$$g = -\alpha = \sigma(N_2 - N_1) \tag{2.65a}$$

which is positive and is called the gain coefficient.

We now summarize the discussion of this section. Three transition coefficients, W, σ, and α have been introduced. They represent three different ways of describing the absorption and stimulated emission phenomena. The relative merits of the three parameters can be summarized as follows: (i) The transition rate W has a simple physical meaning [see equations (1.3) and (1.5)] and it is also directly obtained from the quantum mechanical calculation, (ii) The transition cross section σ depends only on the characteristics of the given material, (iii) The absorption coefficient α is a parameter that can often be experimentally measured in a straightforward way.

2.3 SPONTANEOUS EMISSION

The purpose of this section is to calculate the probability A of spontaneous emission, A being defined as in equation (1.2). Unfortunately, as discussed later in this section, the semiclassical treatment of the interaction between radiation and matter does not permit a correct prediction and understanding of the phenomenon of spontaneous emission. It is, nevertheless, instructive to first develop a treatment of this phenomenon based on semiclassical considerations. The results obtained will then be compared with the exact treatment, based on quantum electrodynamics, in which both atoms and radiation are quantized.

2.3.1 Semiclassical Approach

Let us first consider, from a purely classical viewpoint, an electric dipole oscillating at frequency ω_0. If the positive charge is assumed to be fixed, the instantaneous position $\mathbf{r}$ of the negative charge, with reference to a coordinate system whose origin is centered on the positive charge, can be written as

$$\mathbf{r} = \mathbf{r}_0 \cos(\omega_0 t + \varphi) = \mathrm{Re}\left[\mathbf{r}_0' \exp(i\omega_0 t)\right] \tag{2.66}$$

where Re stands for real part and $\mathbf{r}_0' = \mathbf{r}_0 \exp(i\phi)$. From Maxwell's equations we know that an accelerated electric charge radiates an e.m. wave whose power is proportional to the square of the acceleration. It can accordingly be shown[4] that the electron, during its oscillatory movement, radiates into the surrounding space a power P_r given by

$$P_r = \frac{n\mu^2\omega_0^4}{12\pi\epsilon_0 c_0^3} \tag{2.67}$$

where $\mu = er_0 = e|r_0'|$ is the amplitude of the electric dipole moment, n is the refractive index of the medium surrounding the dipole, and c_0 is the light velocity in vacuo. The average total energy of the oscillating electron is given by the sum of the average values of the kinetic and potential energies.

Since we know that these two quantities are equal, we can write

$$\langle E \rangle = \langle \text{kinetic energy} \rangle + \langle \text{potential energy} \rangle$$
$$= 2\langle \text{kinetic energy} \rangle$$

Therefore

$$\langle E \rangle = 2\frac{m}{2}\langle v^2 \rangle = \frac{1}{2}m\left(\frac{\mu\omega_0}{e}\right)^2 \tag{2.68}$$

where m is the electron mass and $\langle v^2 \rangle$ is the mean square velocity. In the elemental time dt the oscillator will therefore lose an energy dE given by $dE = P_r\,dt$. From (2.67) and (2.68) it then follows that we can write

$$dE = -\frac{E}{\tau_{cl}}dt \tag{2.69}$$

where

$$\tau_{cl} = \frac{E}{P_r} = \frac{6\pi\epsilon_0 mc_0^3}{ne^2\omega_0^2} \tag{2.70}$$

As a result of the dipole radiation, the amplitude r_0 of the oscillation and, therefore, μ will decrease in time. Since, however, τ_{cl} does not depend on μ, it will be independent of time. Equation (2.69) then shows that the energy E will decrease exponentially with a time constant equal to τ_{cl}. Classically, this was therefore called the lifetime of the oscillating dipole.

Let us now return to our problem and write $\psi_1(\mathbf{r}, t)$ and $\psi_2(\mathbf{r}, t)$ for the wave functions associated with levels 1 and 2, where $\mathbf{r} = PO$ is the vector connecting an arbitrary point P in the atom to the origin O. We recall that $\psi_j(\mathbf{r}, t) = u_j(\mathbf{r})\exp(-iE_jt/\hbar)$. Let us now suppose that the atom, which we assume to be initially in level 2, undergoes decay to the ground level 1. In a similar way to the analysis followed in the case of absorption, the wave function $\psi(\mathbf{r}, t)$ during the transition can be expressed as (2.23). Since the electric dipole moment is given by

$$\mathbf{M} = \int r e|\psi|^2\,dV \tag{2.71}$$

the substitution of (2.23) into (2.71) gives

$$\mathbf{M} = \int er|a_1|^2|u_1|^2\,dV + \int er|a_2|^2|u_2|^2\,dV$$

$$+ \int er\left[a_1a_2^*u_1u_2^*\exp(i\omega_0t) + a_1^*a_2u_1^*u_2\exp(-i\omega_0t)\right]dV \tag{2.72}$$

where * stands for complex conjugate and $\omega_0 = (E_2 - E_1)/\hbar$. Equation (2.72) shows that $\mathbf{M}$ has a term $\mathbf{M}_{21}$ oscillating at a frequency ω_0 which can be written as

$$\mathbf{M}_{21} = \operatorname{Re}\{[\exp(i\omega_0 t)]2a_1 a_2^* \mu_{21}\} \tag{2.73}$$

where use has been made of (2.43). By analogy with the previously considered case of a classical oscillator, we expect this term, which oscillates at frequency ω_0, to be responsible for the radiated power into the surrounding space and, therefore, to be responsible for the process of spontaneous emission. The rate of change per unit time of $|a_2|^2$ can then be obtained through a simple energy balance, *viz.*,

$$(\hbar\omega_0) \frac{d|a_2|^2}{dt} = -P_r \tag{2.74}$$

where the radiated power P_r can be obtained from (2.67) by taking into account that, according to (2.73), we now have $\mu = 2|a_1 a_2^* \mu_{21}|$. Equation (2.74) can then be written as

$$\frac{d|a_2|^2}{dt} = -\frac{1}{\tau_{sp}}|a_1|^2|a_2|^2 = -\frac{1}{\tau_{sp}}(1 - |a_2|^2)|a_2|^2 \tag{2.75}$$

where use has been made of (2.24) and where we have defined a characteristic time τ_{sp} as

$$\tau_{sp} = \frac{3\pi\hbar\epsilon_0 c_0^3}{\omega_0^3 n |\mu|^2} \tag{2.76}$$

which is called the spontaneous-emission (or radiative) lifetime of level 2. The solution of (2.75) is

$$|a_2|^2 = \frac{1}{2}\left[1 - \tanh\left(\frac{t - t_0}{2\tau_{sp}}\right)\right] \tag{2.76a}$$

where t_0 can be obtained from the initial condition (i.e., from the value of $|a_2(0)|^2$). As an example, Fig. 2.9 shows the time behavior of $|a_2(t)|^2$ for $|a_2(0)|^2 = 0.96$. The same figure also shows the normalized time behavior of the radiated power P_r. In the following, it is important to notice that the time behavior of $|a_2(t)|^2$ can be approximated by an exponential law, i.e.,

$$|a_2(t)|^2 = |a_2(0)|^2 \exp[-(t/\tau_{sp})] \tag{2.76b}$$

only when $|a_2(0)|^2 \ll 1$. In this case, in fact, we can put $|a_1|^2 \simeq 1$ into (2.75), thus getting (2.76b).

A particularly important case occurs when we have $|a_2(0)|^2 = 1$. In this case we get $t_0 = \infty$ and, according to semiclassical theory, the atom does not decay. Indeed, when $|a_2(0)|^2 = 1$, then $|a_1(0)|^2 = 0$, and from (2.75),

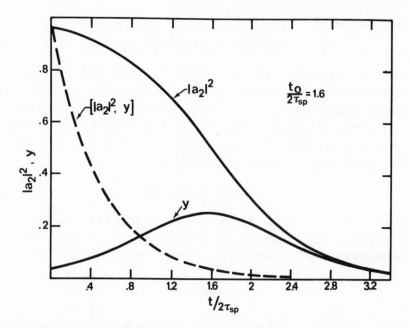

FIG. 2.9. Time behavior of the upper state occupation probability $|a_2|^2$ and of the normalized radiated power $y(y = \tau_{sp}P_r/\hbar\omega_0)$. Solid lines: semiclassical theory; dashed line: quantum electrodynamic theory.

$d|a_2|^2/dt = 0$. Another way of looking at this case is to observe that, when $|a_1(0)|^2 = 0$, $\mathbf{M}$ in (2.72) will reduce to $\mathbf{M}_{22} = \int e\mathbf{r}|a_2|^2|u_2|^2 dV$, which is constant in time ($\mathbf{M}_{22}$ is actually zero for states of well-defined parity). Since the atom does not have any oscillating dipole moment, it will not radiate and it will be in an equilibrium state. We want now to test the stability of this equilibrium state. To do this let us assume that we perturb the atom so as to produce $|a_1| \neq 0$ at $t = 0$. Physically, this means that, owing to this perturbation, we have a finite probability $|a_1|^2$ of finding the atom in level 1. Equation (2.72) then shows that a dipole moment oscillating at frequency ω_0 is now produced. This moment will radiate into the surrounding space, and the atom will tend to decay to level 1. This implies a decrease of $|a_2|^2$, as is also apparent from (2.75). The atom is, therefore, in unstable equilibrium.

Before proceeding, it is worth summarizing the more relevant results that we have obtained with this semiclassical approach: (i) The time behavior of $|a_2|^2$ can generally be described in terms of a hyperbolic tangent, equation (2.76a), but at very weak excitations (i.e., $|a_2(0)|^2 \ll 1$), it follows an approximately exponential law, equation (2.76b). (ii) When the atom is initially in the upper state (i.e., $|a_2(0)|^2 = 1$), the atom is in unstable equilibrium and no radiation occurs.

2.3.2 Quantum Electrodynamic Approach

Although a quantum electrodynamic approach is beyond the scope of this book, it is worthwhile to summarize the results obtained from such an approach and compare them with the semi-classical results. The most relevant results of the quantum electrodynamics approach can be summarized as follows[5,6]: (i) Unlike the semi-classical case, the time behavior of $|a_2|^2$ is now always described to a good approximation (Wigner–Weisskopf approximation) by an exponential law. This means that (2.76b) is now always true, no matter what the value of $|a_2(0)|^2$. (ii) The radiative lifetime is given, in this case too, by equation (2.76). What has just been said implies that an atom in the upper level is not in a state of unstable equilibrium. We see that the semiclassical and the quantum electro-dynamic approaches give completely different predictions for the phenomenon of spontaneous emission (see Fig. 2.9). On the basis of the available experimental results[†] we can, however, say that the quantum electrodynamic approach gives the correct answer to the problem. This approach is, therefore, the correct one, and strictly speaking it should also have been used in the previous section for the stimulated emission and absorption phenomena. Fortunately, however, the semiclassical and the quantum electrodynamic approaches give the same result in this case, so that the results obtained in the previous section still remain correct.

The physical reason for the disappearance of the unstable equilibrium behavior on passing from the classical to the quantum electrodynamic approach deserves some further discussion. In the semiclassical case, the atom, when in the upper level, is in an unstable equilibrium and therefore a very small perturbation is sufficient to remove the atom from this level. At first sight we may then be tempted to say that there will always be enough stray radiation around the material to upset the equilibrium. To be more specific, let us suppose that the material is contained in a blackbody cavity

---

[†]Of these we should like to mention the very accurate measurements of the so-called Lamb shift,[6] another phenomenon which occurs during spontaneous emission: The center frequency of the spontaneously emitted light does not occur at frequency ω_0 (the transition frequency) but at a slightly different value. Lamb-shift measurements for hydrogen are among the most careful measurements so far made in physics, and they have always exactly agreed (within experimental errors) with the predictions of the quantum electrodynamic approach.

whose walls are kept at temperature T. We might then think that the upset of equilibrium (i.e., the production of spontaneous emission) is due to the blackbody radiation within the cavity. This would be wrong, however, since the radiation produced in this way would actually be due to the phenomenon of stimulated emission, i.e., stimulated by the blackbody radiation. The perturbation element which is needed for spontaneous emission is provided by the quantum electrodynamic approach in which the e.m. field in the cavity is no longer treated classically (i.e., using Maxwell's equation) but rather is quantized. Again, referring elsewhere for details,[5,6] we limit ourselves to discussing a very important result. Let us consider a mode of the cavity at frequency ω. If the cavity is treated classically, we know that both the electric, E, and magnetic, H, field amplitudes can become zero (which occurs when $T = 0$). If the cavity is treated by quantum electrodynamics, however, both E^2 and H^2 turn out to have mean values different from zero even at $T = 0$. These limiting values are called zero-point field fluctuations. We may therefore consider these fluctuations as a perturbation which removes the unstable equilibrium predicted by the semiclassical treatment. Correspondingly we may think of the spontaneous emission process as originating from these zero-point fluctuations.

2.3.3 Einstein Thermodynamic Treatment

In this section we will discuss a rigorous calculation of A (due to Einstein) which does not depend on an explicit use of quantum electrodynamic calculations. In fact it was given by Einstein well before any development of the theory of quantum electrodynamics. The calculation makes use of an elegant thermodynamic argument. To this end, let us assume that the material is placed in a blackbody cavity whose walls are kept at a constant temperature T. Once thermodynamic equilibrium is reached, an e.m. energy density with a spectral distribution ρ_ω given by (2.18a) will be established and the material will be immersed in this radiation. In this material, both stimulated-emission and absorption processes will occur, in addition to the spontaneous-emission process. Since the system is in thermodynamic equilibrium, the number of transitions per second from level 1 to level 2 must be equal to the number of transitions from level 2 to level 1. We now set

$$W_{21} = B_{21}\rho_{\omega 0} \qquad (2.77)$$

$$W_{12} = B_{12}\rho_{\omega 0} \qquad (2.78)$$

where B_{21} and B_{12} are constant coefficients (the so-called Einstein B coefficients). If we write N_1^e and N_2^e for the equilibrium populations of level

1 and level 2, respectively, we can then write

$$AN_2^e + B_{21}\rho_{\omega 0}N_2^e = B_{12}\rho_{\omega 0}N_1^e \tag{2.79}$$

From Boltzmann statistics we also know that

$$N_2^e/N_1^e = \exp(-\hbar\omega_0/kT) \tag{2.80}$$

From (2.79) and (2.80) it then follows that

$$\rho_{\omega 0} = \frac{A}{B_{12}\exp(\hbar\omega_0/kT) - B_{21}} \tag{2.81}$$

A comparison of (2.81) with (2.18*a*) leads to the following relations:

$$B_{12} = B_{21} = B \tag{2.82}$$

$$\frac{A}{B} = \frac{\hbar\omega_0^3 n^3}{\pi^2 c_0^3} \tag{2.83}$$

Equation (2.82) shows that the probabilities of absorption and stimulated emission due to the blackbody radiation are equal. This relation is therefore analogous to that established in a completely different way in the case of monochromatic radiation [see (2.54)].

Equation (2.83) allows the calculation of A, once B, i.e., the coefficient for stimulated emission due to blackbody radiation, is known. This coefficient can easily be obtained from (2.53*c*). This equation is valid for perfectly monochromatic radiation. For blackbody radiation, let us write $\rho_{\omega'}d\omega'$ for the energy density of radiation whose frequency lies between ω' and $\omega' + d\omega'$. If we simulate this radiation by a monochromatic wave having the same power, the corresponding elemental transition probability dW is obtained from (2.53*c*) by substituting $\rho_\omega d\omega'$ for ρ. Upon integration of this equation with the assumption that $g_t(\Delta\omega)$ can be approximated by a Dirac δ function in comparison with $\rho_{\omega'}$ (see Fig. 2.3), we get

$$W = \frac{\pi}{3n^2\epsilon_0\hbar^2}|\mu|^2\rho_{\omega 0} \tag{2.84}$$

Comparing (2.84) with (2.77) or (2.78) then gives

$$B = \frac{\pi|\mu|^2}{3n^2\epsilon_0\hbar^2} \tag{2.85}$$

From (2.83) and (2.85) we finally get

$$A = \frac{n\omega_0^3|\mu|^2}{3\pi\hbar\epsilon_0 c_0^3} \tag{2.86}$$

It should be noted that the expression for A which we have just obtained is exactly the same as the result obtained from a quantum electrodynamic approach. Its calculation is in fact based on thermodynamics and the use of

Planck's law (which is quantum electrodynamically correct). Note also that, as already mentioned in Section 2.3.2, the spontaneous lifetime $\tau_{sp} = (1/A)$ obtained from (2.86) agrees with the semiclassical expression given by (2.76). Finally, we note that A increases as the cube of the frequency, so that the importance of the process of spontaneous emission increases rapidly with frequency. In fact spontaneous emission is often negligible in the middle- and far-infrared where nonradiative decay usually dominates. At a frequency corresponding to the middle of the visible range, one can find the order of magnitude of A by putting $\lambda = 2\pi c/\omega = 5 \times 10^{-5}$ cm and $|\mu| = ea$, where a is the atomic radius ($a \simeq 10^{-8}$ cm). We therefore get $A \simeq 10^8$ s^{-1} (i.e., $\tau_{sp} \simeq 10$ ns). For magnetic dipole transitions, A is approximately 10^5 times smaller, i.e., $A \simeq 10^3$ s^{-1}.

The thermodynamic argument of Einstein also allows one to investigate another important aspect of spontaneous emission, namely the spectrum of the emitted radiation. It can be shown in fact that, for any transition (i.e., for any line broadening mechanism), the spectrum of the emitted radiation is the same as that observed in absorption. To show this, let us define a spectral coefficient A_ω such that $N_2 A_\omega d\omega$ gives the number of atoms per unit time which, upon decay, give a photon of frequency between ω and $\omega + d\omega$. We will obviously have

$$A = \int A_\omega \, d\omega \tag{2.87}$$

Similarly, let us define a spectral coefficient B_ω in the following way: $B_\omega \rho_\omega d\omega$ gives the number of transitions per unit time (absorptions or stimulated emissions) induced by blackbody radiation with frequency between ω and $\omega + d\omega$. We can now easily see that $A_\omega/B_\omega = A/B$. Let us, in fact, assume that between the material under consideration and the walls of blackbody cavity an ideal e.m. filter is placed that only lets the frequencies lying between ω and $\omega + d\omega$ pass through. In this case also we can apply a thermodynamic argument [as in equation (2.79)] so that

$$A_\omega N_2^e d\omega + B_\omega \rho_\omega N_2^e d\omega = B_\omega \rho_\omega N_1^e d\omega \tag{2.88}$$

Using, as in the previous calculation, (2.80) and (2.18a), we get

$$\frac{A_\omega}{B_\omega} = \frac{A}{B} \tag{2.89}$$

On the other hand, the coefficient B_ω can be easily calculated from (2.53b) once it is noticed that $B_\omega \rho_\omega d\omega$ can be interpreted as a stimulated-emission coefficient for monochromatic radiation. From (2.53c) and (2.85) we then get

$$B_\omega = Bg_t(\Delta\omega) \tag{2.90}$$

and from (2.89),

$$A_\omega = Ag_t(\Delta\omega) \qquad (2.91)$$

Equation (2.91) shows that the spectrum of the radiated wave is again described by $g_t(\Delta\omega)$; in other words it is the same as for absorption or stimulated emission. From (2.91) we then get a new interpretation of $g_t(\Delta\omega)$: $g_t(\Delta\omega)d\omega$ gives the probability that a spontaneously emitted photon has a frequency lying between ω and $\omega + d\omega$.

2.3.4 Radiation Trapping, Superradiance, Superfluorescence, and Amplified Spontaneous Emission

So far the emission of an isolated atom has been considered. In a real situation, any atom will be surrounded by many other atoms, some in the ground and some in the excited state. In this case new phenomena may occur.

If the fraction of atoms which are initially in the upper state is very small, the phenomenon known as radiation trapping may play a significant role. A photon which is spontaneously emitted by one atom, instead of escaping from the medium, can be absorbed by another atom which thereby ends up in the excited state. The process therefore has the effect of slowing down the effective rate of spontaneous emission. A detailed discussion can be found elsewhere,[7] and we only wish to point out here that the lifetime increase due to radiation trapping depends on the atomic density, on the cross section of the transition involved, and on the geometry of the medium.

If the fraction of atoms initially in the upper state is such that population inversion occurs, emission may take place by a cooperative process in which the emission of one atom is influenced by the emission of the others. This leads to the phenomena of superradiance[8] and superfluorescence.[9] Again referring elsewhere for details[8,9] we point out here a few relevant properties of these phenomena: (i) A distinct threshold for the occurrence of the cooperative effect is observed. (ii) The length of the active material l must be smaller than some characteristic length l_c whose value depends on the initial inversion. (iii) The time evolution of the emitted light is no longer exponential. Instead it has the form of a bell-shaped curve, whose duration, for large initial inversion, can be much smaller than τ_{sp}. (iv) For a rod-shaped active material with diameter D, the light will be emitted into a solid angle corresponding to the diffraction angle $\theta_d = \lambda/D$. (v) The peak power of the emitted radiation now varies as N^2 (where N is the initial inversion) rather than N, as would be the case for the normal spontaneous emission process. The five characteristics cited above are typical of both superfluorescence and superradiance. The distinction between the two phenomena is rather subtle and it has to do with the way in which the initial inversion is prepared. If at time $t = 0$, the phases of the oscillating dipole moments $\mathbf{M}_{21}$ [see (2.73)] for each atom are random, the corresponding cooperative effect is called superfluorescence. The word superradiance is reserved for the situation where a phase correlation exists among the atomic dipole moments $\mathbf{M}_{21}$ so that a macroscopic dipole moment is present at $t = 0$. In the case of superfluorescence, since no macroscopic dipole moment is present at $t = 0$, the atoms start radiating independently as in normal fluorescence. The initial fluorescence intensity is thus proportional to N. A correlation then develops, this being induced by the field arising from spontaneous emission. The system then reaches a state in which the intensity is proportional to N^2. On the other hand, in the case of superradiance, the microscopic dipole moment already present at $t = 0$ gives, right from the start, a radiation field proportional to N and hence a radiation intensity proportional to N^2. The qualitative time behavior of superradiance and superfluorescence as compared to that of normal fluorescence is shown in Fig. 2.10.

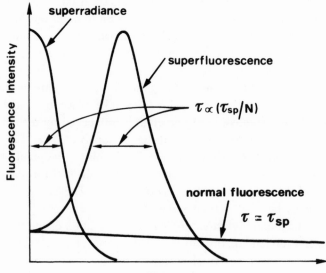

FIG. 2.10. Comparative time behavior of superradiance, superfluorescence, and normal fluorescence.

The phenomenon of superfluorescence should not be confused with the behavior occurring in many high-gain lasers, e.g., nitrogen or excimer lasers. In these lasers, if the inversion reaches a critical value, an intense emission within a solid angle Ω around the axis of the active material is observed even when no mirrors (or perhaps only one mirror) are used. In fact, in this case also, just as for superfluorescence, the time evolution of the emitted light is given by a bell-shaped curve with duration much smaller than τ_{sp}. However, the following differences distinguish this amplified spontaneous emission from superfluorescence: (i) The length l of the active material is much longer than l_c. (ii) The solid angle of emission Ω no longer corresponds to the diffraction angle but is simply given by geometrical considerations. If no mirrors are used we thus get

$$\Omega = \frac{\pi D^2 n^2}{4 l^2} \tag{2.91a}$$

where D is the diameter, l is the length, and n is the refractive index of the active material (Fig. 2.11a). (iii) The peak intensity of the radiation field is no longer proportional to N^2. To conclude this discussion of amplified spontaneous emission, we note that there is a very simple expression for the critical inversion $N_c = N_2 - N_1$ required for this process. It can in fact be shown[10] that, if we call $G = \exp(\sigma N_c l)$ the single-pass gain of the active material, the following condition must be satisfied:

$$\frac{[\ln G]^{1/2}}{G} = \frac{\Omega}{4} \tag{2.91b}$$

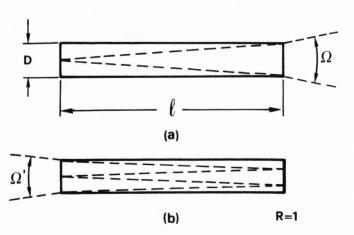

(a)

(b)   **R=1**

FIG. 2.11. Solid angle of emission in the case of amplified spontaneous emission: (a) active material without any end mirror; (b) active material with one end mirror.

If a mirror with 100% reflectivity is placed at one end of the material (Fig. 2.11*b*), equations (2.91*a*) and (2.91*b*) are modified as follows:

$$\Omega' = \frac{\pi D^2 n^2}{16 l^2} \tag{2.91c}$$

$$\frac{\left[\ln G^2\right]^{1/2}}{G^2} = \frac{\Omega'}{4} \tag{2.91d}$$

2.4 NONRADIATIVE DECAY[11]

Besides decaying in a radiative way, an atom can also undergo nonradiative decay from level 2 to level 1. In this case the energy difference $E_2 - E_1$ is given to the surrounding molecules in the form of translational, rotational, vibrational, or electronic energy. In the case of a gas, this energy can also be released by means of collisions with the walls of the container. In the case of a gas discharge, a collision between an electron and the excited atom in which the atom gives up its energy to the electron (collision of second kind) may also occur. Thus, in the case of a gas or liquid, nonradiative transitions can occur as a result of inelastic collisions. This is not always the case, however, as nonradiative decay can also take place in an isolated molecule (unimolecular processes). If, for example, levels 1 and 2 correspond to a given vibrational mode of a molecule, the corresponding energy can be given to another vibrational mode of the molecule or it can be used to dissociate the molecule (predissociation). In the case of ionic crystals, nonradiative decay usually occurs through interaction with the lattice vibrations. For a semiconductor, in which there are electrons in the upper (conduction) band and holes in the lower (valence) band, nonradiative decay occurs through electron–hole recombination at deep traps (which arise from dislocations, vacancies, or impurities).

From what has been said above, it is apparent that the mechanisms for nonradiative decay are complicated. Despite this, however, the upper state population change due to nonradiative decay can, in general, be written as $(dN_2/dt)_{nr} = -N_2/\tau_{nr}$ where τ_{nr} is a characteristic time constant known as the nonradiative lifetime. Its value will depend very much on the type of atom or molecule involved and on the nature of the surrounding material. As a result of the simultaneous occurrence of both radiative decay and nonradiative decay, the time variation of the upper state population N_2 can be written as

$$\frac{dN_2}{dt} = -\left(\frac{N_2}{\tau_{sp}} + \frac{N_2}{\tau_{nr}}\right) \tag{2.92}$$

Equation (2.92) shows that we can define an overall lifetime τ given by

$$\frac{1}{\tau} = \frac{1}{\tau_{sp}} + \frac{1}{\tau_{nr}} \tag{2.93}$$

We will call this quantity the lifetime of the upper state 2. It can easily be measured by monitoring the time behavior of the spontaneously emitted light. Let us assume in fact that at time $t = 0$ there are $N_2(0)$ atoms in the upper level, and let V be the volume of the material. According to (2.92) the power emitted by spontaneous emission will be

$$P(t) = \frac{N_2(t)\hbar\omega_0 V}{\tau_{sp}} \tag{2.94}$$

The population $N_2(t)$ at time t is obtained by integrating (2.92), i.e., $N_2(t) = N_2(0)\exp(-t/\tau)$, so that

$$P(t) = \frac{N_2(0)\hbar\omega_0 V}{\tau_{sp}} \exp(-t/\tau) \tag{2.95}$$

Note that the light emitted as a result of spontaneous emission decays exponentially with a time constant equal to τ rather than to τ_{sp}.

It is customary to define the fluorescence quantum yield ϕ as the ratio of the number of emitted photons to the number of atoms initially raised to level 2. Using (2.95), we therefore have

$$\phi = \frac{\int \dfrac{P(t)}{\hbar\omega_0} dt}{N_2(0)V} = \frac{\tau}{\tau_{sp}} \tag{2.96}$$

From a measurement of the quantum yield ϕ and lifetime τ it is therefore possible to find both τ_{sp} and τ_{nr}.

2.5 LINE BROADENING MECHANISMS

In this section we intend to discuss briefly the various line broadening mechanisms and the corresponding behavior of the function $g(\Delta\omega)$. Note that, according to what has been said in Section 2.3.3, the frequency spectrum [and therefore $g(\Delta\omega)$] is the same for the processes of spontaneous emission, stimulated emission, and absorption. Therefore, in what follows, when we discuss the various line-shape functions, we will study whichever process is the most convenient to analyze.

There is an important distinction, which it is convenient to introduce immediately, between homogeneous and inhomogeneous line broadening mechanisms. A line broadening mechanism will be called *homogeneous* when it broadens the line of each individual atom, and therefore of the whole system, in the same way. Conversely, a line broadening mechanism will be called *inhomogeneous* when it distributes the resonance frequencies of the atoms over a given band and therefore gives a broadened line for the system as a whole without broadening the line of individual atoms.

2.5.1 Homogeneous Broadening

The first homogeneous line broadening mechanism we shall consider is one due to collisions and is known as collision broadening. In a gas, it is due to the collision of an atom with other atoms, ions, free electrons, or the walls of the container. In a solid it is due to the interaction of the atom with the lattice. These collisions interrupt the process of coherent interaction between the atom and the incident e.m. wave. As already mentioned [see equation (2.52)], this mechanism gives a Lorentzian line shape function whose width depends on the average time between collisions $\tau_c(\Delta\omega_0 = 2/\tau_c)$.

Let us now see how to derive the expression (2.52) for the line shape function $g(\Delta\omega)$. Apart from a proportionality factor, this function will be given by the power spectrum $W(\omega)$ of the signal $E(t)$ shown in Fig. 2.6. For this proportionality factor to be unity, we require that $W(\omega)$ be such that $\int W(\omega)\,d\omega = 1$. The implication of this condition is readily obtained from Parseval's theorem which gives

$$\int_{-\infty}^{+\infty} W(\omega)\,d\omega = \lim_{T\to\infty} \frac{\pi}{T} \int_{-T}^{+T} E^2(t)\,dt = \pi E_0^2 \qquad (2.97)$$

where we have called E_0 the amplitude of the signal in Fig. 2.6. To have $\int W\,d\omega = 1$, we require that $E_0^2 = 1/\pi$. The function $g(\omega' - \omega)$ will therefore be given by the power spectrum of a signal $E(t)$ of the type shown in Fig. 2.6 and of amplitude $E_0 = (\pi)^{-1/2}$. This spectrum can then be obtained as the Fourier transform of the signal autocorrelation function (Wiener–Kintchine theorem). If the function in Fig. 2.6 were a perfect sine wave of frequency ω, its correlation function would be $(E_0^2/2)\cos\omega\tau = (1/2\pi)\cos\omega\tau$. The wave of Fig. 2.6 undergoes interruption, however, with a probability density p_τ given by (2.49). It then follows that the quantity $p(\tau) = \int_\tau^\infty p_\tau\,d\tau = \exp(-\tau/\tau_c)$ represents the probability that a collision occurs after a time τ from the previous one. Accordingly, if we consider two points of Fig. 2.6 separated in time by τ, the probability that they are correlated (i.e., they are on the same uninterrupted portion of a sinusoidal wave) is $p(\tau)$, while the probability that they are not correlated owing to an intervening interruption is $1 - p(\tau)$. Therefore, the desired correlation function is $[p(\tau)\cos\omega\tau]/2\pi$. What we have calculated so far, only applies for $\tau > 0$. To get the correlation function for times $\tau < 0$, it is enough to remember that the correlation function is symmetrical $[G(-\tau) = G(\tau)]$. Therefore the desired correlation function is

$$G(\tau) = \frac{1}{2\pi} \exp(-|\tau|/\tau_c)\cos\omega\tau \qquad (2.98)$$

From the Wiener–Kintchine theorem it therefore follows that

$$g(\omega' - \omega) = \frac{1}{2\pi}\left[\frac{\tau_c}{1 + (\omega' - \omega)^2\tau_c^2} + \frac{\tau_c}{1 + (\omega' + \omega)^2\tau_c^2}\right] \qquad (2.99)$$

The two terms in the brackets give two spectra centered around $+\omega$ and $-\omega$, respectively. If we limit ourselves to considering only positive values of ω, we can neglect the second term in (2.99) provided we multiply the first term by a factor of 2. Thus we get (2.52). Actually, Fig. 2.6 represents a very rough description of the physical phenomenon that is occurring. There it was assumed that the jumps of phase were instantaneous, which in turn implies an instantaneous duration of the collision. What actually happens is that, upon collision, the atom (or molecule) will be subjected to a potential energy of either an attractive (Fig. 2.22) or repulsive (Fig. 6.21) type. Due to this potential energy, the levels 1 and 2 will shift in energy by $\Delta E_1(R)$ and $\Delta E_2(R)$, respectively, where R is the distance between the two colliding atoms. The corresponding change of transition frequency will be given by

$$\Delta\omega_0(t) = \frac{\Delta E_2(R) - \Delta E_1(R)}{\hbar} \qquad (2.100)$$

where $\Delta\omega_0$ is a function of time since the distance R is itself a function of time. Again we can look at this problem in an equivalent way in which, instead of the transition frequency undergoing a change upon collision, we treat it as though the frequency of the incoming wave undergoes the change $\Delta\omega(t) = [\Delta E_2 - \Delta E_1]/\hbar$ (Fig. 2.12). A more realistic theory of collision

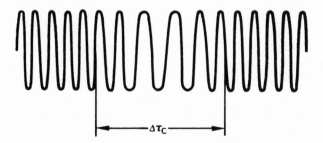

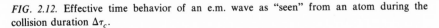

FIG. 2.12. Effective time behavior of an e.m. wave as "seen" from an atom during the collision duration $\Delta\tau_c$.

broadening should therefore take into account the finite collision duration $\Delta\tau_c$ and the phenomena occurring during this time. It can be shown, however, that, if $\Delta\tau_c \ll \tau_c$, the function $g(\omega' - \omega)$ is again accurately described by a Lorentzian curve at least up to frequencies $|\omega' - \omega| \leqslant 1/\Delta\tau_c$. An order-of-magnitude estimate for $\Delta\tau_c$ can be obtained from the relationship

$$\Delta\tau_c \simeq \frac{a}{v_{\text{th}}} \tag{2.101}$$

where a is the interatomic distance at which atoms (or molecules) appreciably influence one another and v_{th} is the average velocity due to the thermal movement of the molecules. In practice, a is approximately equal to the molecular dimensions and is therefore of the order of 1 Å (see, for example, Fig. 6.19). The average thermal velocity can be obtained from the relationship

$$v_{\text{th}} = \left(\frac{8kT}{\pi M} \right)^{1/2} \tag{2.102}$$

where M is the mass of the molecule. If, for example, we consider a Ne atom at room temperature, we get from (2.102) and (2.101),

$$\Delta\tau_c \simeq 10^{-13} \text{ s} \tag{2.103}$$

Note that during this time, for an optical wave ($\nu \simeq 5 \times 10^{14}$ Hz) several cycles will occur. On the other hand, an order-of-magnitude estimate for the time interval τ_c between two collisions is given by the ratio of the mean free path to the mean velocity v_{th}. Therefore, we have

$$\tau_c \simeq \frac{(MkT)^{1/2}}{16\pi^{1/2}pa^2} \tag{2.104}$$

where p is the pressure of the gas and a is the radius of the molecule. For Ne atoms at a pressure $p \simeq 0.5$ Torr (typical pressure in a He–Ne gas laser, see Chapter 6) and at room temperature, we get

$$\tau_c \simeq 0.5 \times 10^{-6} \text{ s} \tag{2.105}$$

This shows that $\Delta\tau_c \ll \tau_c$. The corresponding linewidth (see Fig. 2.7) is

$$\Delta\nu_0 = \frac{1}{\pi\tau_c} = 0.64 \text{ MHz} \tag{2.105a}$$

Note that τ_c is inversely proportional to the pressure p, so that the linewidth $\Delta\nu_0$ is proportional to p. As a rough "rule of thumb" we can say that, for any atom, collisions contribute to the line broadening by an amount $(\Delta\nu_0/p) \simeq 1$ MHz/Torr, comparable to that shown for the case of Ne atoms. To conclude this discussion on collision broadening we wish to point out a fundamental difference between the collision time τ_c referred to in this section and the nonradiative decay time τ_{nr} defined in Section 2.4. Obviously, for a gas, they are both due to collisions. However, nonradiative decay requires an inelastic collision, and the atom decays by delivering its energy to the surrounding molecules. On the other hand, line broadening may be due either to inelastic collisions or (as we have seen in this section) to elastic collisions, which produce jumps of phase of the incident e.m. wave with respect to the phase of the atom's wave function.

A second homogeneous line broadening mechanism is that due to spontaneous emission itself, and is called natural or intrinsic broadening. Since the energy emitted by the atom during the spontaneous emission decays in time according to $\exp(-t/\tau_{sp})$,† it follows that its Fourier spectrum covers a frequency range of the order of $1/\tau_{sp}$. More precisely, it can be shown that the line-shape function $g(\Delta\omega)$ associated with this phenomenon is again Lorentzian and can be obtained from (2.52) by substituting $2\tau_{sp}$ for T_2. It is, however, obvious that since the phenomenon of spontaneous emission cannot be explained by semiclassical considerations, this is also true for the corresponding line-shape function $g(\Delta\omega)$.[12,13] We can, nevertheless, use a heuristic argument by assuming that the electric field due to the spontaneous emission decays in time according to the relationship $E(t) = \exp(-t/2\tau_{sp})\cos\omega t$. In this case, in fact, the decay of emitted intensity [which is proportional to $\langle E^2(t)\rangle$] would show the correct temporal behavior, namely, $\exp(-t/\tau_{sp})$. We can easily calculate the power spectrum corresponding to such a field $E(t)$ and verify the previously mentioned result for $g(\Delta\omega)$. Natural broadening is always present for any line, but it is usually much less important than other mechanisms (e.g., Doppler broadening). For example, for an allowed transition, we have seen that $\tau_{sp} \simeq 10^{-8}$ s, so that (see Fig. 2.7)

$$\Delta\nu_{nat} = \frac{\Delta\omega_0}{2\pi} = \frac{1}{2\pi\tau_{sp}} \simeq 20 \text{ MHz} \qquad (2.106)$$

This broadening would be observable in a material whose atoms were rigidly fixed and did not interact with other atoms.

2.5.2 Inhomogeneous Broadening

Let us suppose that a line broadening mechanism distributes the resonance frequencies of the atoms over a given band centered at the frequency ω_0, and call $g^*(\omega_0' - \omega_0)$ the relative distribution density. This

---

†Nonradiative decay is not considered here since its effect is included in the collision broadening already discussed.

means that $g^*(\omega_0' - \omega_0) d\omega_0'$ gives the probability that an atom has its resonance frequency between ω_0' and $\omega_0' + d\omega_0'$. The average value of the stimulated emission or absorption coefficients can then be obtained from (2.47), or more generally from (2.53) if some other broadening mechanism (e.g., collision broadening) is also present. We get

$$W = \frac{\pi}{3n^2\epsilon_0\hbar^2} |\mu|^2\rho \int_{-\infty}^{+\infty} g^*(\omega_0' - \omega_0)g(\omega - \omega_0') d\omega_0'$$

$$= \frac{\pi}{3n^2\epsilon_0\hbar^2} |\mu|^2\rho g_t(\omega - \omega_0) \tag{2.107}$$

where we have written $g_t(\omega - \omega_0)$ for the function

$$g_t = \int_{-\infty}^{+\infty} g^*(x)g[(\omega - \omega_0) - x] dx \tag{2.107a}$$

and where we have put $x = \omega_0' - \omega_0$. We see, therefore, that when two broadening mechanisms, one homogeneous (g) and the other inhomogeneous (g^*) are present, equation (2.53a) is indeed obtained, with g_t being given by the convolution of the two line-shape functions. If the homogeneous broadening $g(\omega - \omega_0')$ is much smaller than the inhomogeneous broadening $g^*(\omega_0' - \omega_0)$, then $g(\omega - \omega_0')$ can be approximated by a Dirac δ function so that

$$W = \frac{\pi}{3n^2\epsilon_0\hbar^2} |\mu|^2\rho g^*(\omega - \omega_0) \tag{2.108}$$

This case is sometimes described as being purely inhomogeneous.

An inhomogeneous broadening mechanism which is typical of gases is that due to atomic motion and is called Doppler broadening. To illustrate this type of broadening, let us consider the case of a molecule that is moving while it is subjected to e.m. radiation of frequency ω (where this is the frequency measured in a coordinate system fixed relative to the laboratory). If we call v the velocity component of the molecule (measured relative to these laboratory coordinates) in the propagation direction of the e.m. wave, the frequency of this wave, as seen from the atom, is (Doppler effect) $\omega' = \omega[1 \pm (v/c)]$ where the sign $-$ or $+$ applies according to whether the velocity is the same or opposite direction to that of the wave: If the molecule is in fact moving in the opposite direction to that of the wave, it is well known that the frequency ω' observed by the atom is higher than the value ω observed in the laboratory coordinate system. Of course, then, absorption will occur only when the apparent frequency ω' of the e.m. wave as seen from the atom is equal to the atomic transition frequency ω_0, i.e., when

$$\omega[1 \pm (v/c)] = \omega_0 \tag{2.109}$$

If we take the bracketed expression of (2.109) to the right-hand side of the equation, we arrive at a different interpretation of the process: As regards the interaction of the e.m. radiation with the atom, the results would be the same if the atom were not moving but on the other hand had a resonant frequency ω_0' given by

$$\omega_0' = \frac{\omega_0}{1 \pm v/c} \qquad (2.109a)$$

where ω_0 is the transition frequency of the atom. When looked at in this way, one can then say that this broadening mechanism indeed belongs to the inhomogeneous category as defined at the beginning of this section. To calculate the corresponding line shape $g^*(\omega_0' - \omega_0)$ it is sufficient to remember that the probability $p_v \, dv$ that an atom of mass M in a gas at temperature T has a velocity component between v and $v + dv$ is given by the Maxwell distribution

$$p_v \, dv = \left(\frac{M}{2\pi kT} \right)^{1/2} \exp\left(-\frac{Mv^2}{2kT} \right) dv \qquad (2.110)$$

Since it follows from (2.109a) that

$$v = \frac{c(\omega_0' - \omega_0)}{\omega_0'} \simeq \frac{c(\omega_0' - \omega_0)}{\omega_0} \qquad (2.111)$$

then from (2.110) and (2.111) we get the desired distribution by establishing the condition that $g^*(\omega_0' - \omega_0) \, d\omega_0' = p_v \, dv$. Thus we get

$$g^*(\omega_0' - \omega_0) = \frac{c}{\omega_0} \left(\frac{M}{2\pi kT} \right)^{1/2} \exp\left(-\frac{Mc^2}{2kT} \frac{(\omega_0' - \omega_0)^2}{\omega_0^2} \right) \qquad (2.112)$$

The shape of the curve corresponding to this equation is called Gaussian. Its full width at half-maximum (FWHM) is

$$\Delta\omega_0^* = 2\omega_0 \left(\frac{2kT}{Mc^2} \ln 2 \right)^{1/2} \qquad (2.113)$$

The Doppler linewidth $\Delta\nu_0^* = \Delta\omega_0^*/2\pi$ for the Ne line at the wavelength $\lambda = 0.6328 \ \mu m$ (one of the lines that exhibit laser action in Ne, see Chapter 6) is (for $T = 300°K$)

$$\Delta\nu_0^* = 1.7 \ \text{GHz} \qquad (2.114)$$

A comparison of (2.114) with (2.106) and (2.105a) shows that, in the example considered, Doppler broadening is much greater than natural broadening and this, in turn, is much greater than collision broadening. This is not always true, however, since collision broadening dominates Doppler broadening at sufficiently high gas pressures (e.g., CO_2 laser at atmospheric pressure, see Chapter 6).

Just as in the Doppler effect, so in general a Gaussian broadening will result from any mechanism that produces a random distribution of the transition frequencies of the atoms. For example, if the crystalline electric field in a solid has local inhomogeneities due, for instance, to lattice imperfections, then local variations of the energy levels of the atoms, due to the Stark effect, will be produced. This will lead to a line broadening, which is often of a Gaussian form. In this case, $\Delta\omega_0^*$ will depend on the average value of the inhomogeneities of the crystalline electrical field.

2.5.3 Combined Effects of Line Broadening Mechanisms

Before proceeding it is worth summarizing the results on the broadening mechanisms which we have obtained so far. We have seen that the function $g(\omega - \omega_0)$ can be either of Lorentzian form, in which case it can be written as

$$g(\omega - \omega_0) = \frac{2}{\pi\Delta\omega_0} \frac{1}{1 + \left(\dfrac{\omega - \omega_0}{\Delta\omega_0/2}\right)^2} \qquad (2.115)$$

or of Gaussian form, in which case it can be written as

$$g(\omega - \omega_0) = \frac{2}{\Delta\omega_0}\left(\frac{\ln 2}{\pi}\right)^{1/2}\exp\left[-\left(\frac{\omega - \omega_0}{\Delta\omega_0/2}\right)^2\ln 2\right] \qquad (2.116)$$

In both equations (2.115) and (2.116), $\Delta\omega_0$ stands for the full width at half-maximum, and specific expressions for this have already been given for various cases. Figure 2.13 shows the plots of the dimensionless function $g\Delta\omega_0$ vs. dimensionless frequency offset $2(\omega - \omega_0)/\Delta\omega_0$ for the two cases. Note that the Gaussian curve is much sharper than the Lorentzian. The peak value of $g(\omega - \omega_0)$ is indeed

$$g(0) = \frac{2}{\pi\Delta\omega_0} = \frac{0.637}{\Delta\omega_0} \qquad (2.117)$$

for the Lorentzian and

$$g(0) = \frac{1}{\Delta\omega_0}\left(\frac{\ln 2}{\pi}\right)^{1/2} = \frac{0.939}{\Delta\omega_0} \qquad (2.118)$$

for the Gaussian. We have also seen that, in general, a Lorentzian line is a homogeneous line while a Gaussian line is an inhomogeneous line.

Let us see now what happens when the overall broadening is due to the combined effect of more than one of the previously mentioned broadening mechanisms. For any two broadening processes which are simultaneously present but independent (i.e., not correlated with each other) it can be shown that the overall line shape is given by a convolution [of the type

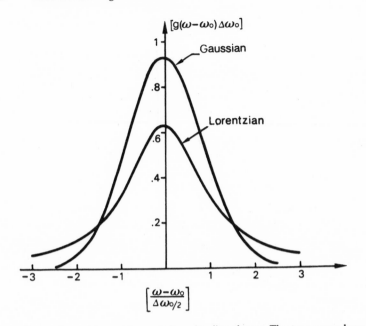

FIG. 2.13. Comparison of Lorentzian and Gaussian line shapes. The two curves have been drawn with the same width at half-power points.

indicated in (2.107*a*)] between the two processes. It can be shown that the convolution of a Lorentzian line of width $\Delta\omega_1$ with another Lorentzian line of width $\Delta\omega_2$ gives again a Lorentzian line whose width is $\Delta\omega = \Delta\omega_1 + \Delta\omega_2$. The convolution of a Gaussian line of width $\Delta\omega_1$ with another Gaussian line of width $\Delta\omega_2$ is again a Gaussian line of width $\Delta\omega = (\Delta\omega_1^2 + \Delta\omega_2^2)^{1/2}$. It is, therefore, always possible to reduce the problem to the convolution of a single Lorentzian line with a single Gaussian line, and this integral [which is known as the Voigt integral[14]] is tabulated. Sometimes, however, (e.g., as in the previously discussed cases for Ne) one mechanism is predominant: in this case, it is then possible to talk about either a Lorentzian or a Gaussian line.

As examples of combined effects of homogeneous and inhomogeneous broadening, Fig. 2.14 shows the behavior of the laser linewidth versus temperature for a ruby and a Nd^{3+}:YAG crystal. Ruby is a crystal of Al_2O_3 doped with Cr^{3+} ions which substitute for some of the Al^{3+} ions in the lattice (fraction of Al^{3+} replaced by Cr^{3+} ions is $\sim 0.5\%$). The Nd^{3+}:YAG crystal consists of YAG (an acronym for yttrium aluminum garnet, $Y_3Al_5O_{12}$) doped with Nd^{3+} ions which substitute for some of the Y^{3+} ions in the lattice (fraction of Nd^{3+} ions is $\sim 1\%$). The laser transition is one of the Cr^{3+} transitions in ruby ($\lambda = 694.3$ nm) and one of the Nd^{3+}

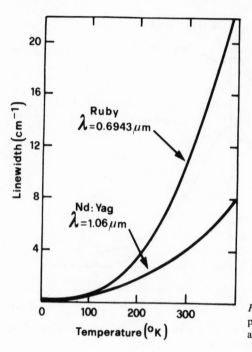

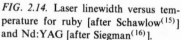

FIG. 2.14. Laser linewidth versus temperature for ruby [after Schawlow[15]] and Nd:YAG [after Siegman[16]].

transitions in Nd^{3+}:YAG ($\lambda = 1.06\mu m$). In each case the laser linewidth is predominantly due to collisions of the ions with the lattice phonons. This explains the rapid increase of linewidth with temperature. The residual linewidth observed as $T \to 0$ (which can be barely seen in Fig. 2.14) arises from inhomogeneous broadening due to the inhomogeneities of the crystal field around each of the Cr^{3+} or Nd^{3+} ions.

2.6 SATURATION

The purpose of this section is to investigate the behavior of a two-level transition (frequency ω_0) of a medium in the presence of a strong monochromatic e.m. wave of intensity I and frequency $\omega \simeq \omega_0$. In general this wave will have the effect of tending to equalize the populations N_1 and N_2 of the two levels. In fact, if N_1 is initially greater than N_2 the absorption processes (WN_1) will dominate the stimulated emission processes (WN_2), i.e., there are more atoms undergoing the $1 \to 2$ transition than those undergoing the $2 \to 1$ transition. At sufficiently high values of I the two populations will thus tend to equalize. This phenomenon is called *saturation*.

2.6.1 Saturation of Absorption: Homogeneous Line

We will consider first an absorbing transition ($N_1 > N_2$) and suppose that the line is homogeneously broadened. Taking account of both spontaneous emission and stimulated emission induced by the incident wave (Fig. 2.15) we can write the following two equations for the populations N_1 and N_2 of the two levels:

$$N_1 + N_2 = N_t \qquad (2.119a)$$

$$\frac{dN_2}{dt} = -W(N_2 - N_1) - \frac{N_2}{\tau} \qquad (2.119b)$$

In equation (2.119a) N_t is the total population of the given material. If we write

$$\Delta N = N_1 - N_2 \qquad (2.120)$$

the two equations (2.119) can be reduced to a single differential equation:

$$\Delta \dot{N} = -\Delta N \left(\frac{1}{\tau} + 2W \right) + \frac{1}{\tau} N_t \qquad (2.121)$$

When $\Delta \dot{N} = 0$, i.e., at steady state, we get

$$\Delta N = \frac{N_t}{1 + 2W\tau} \qquad (2.122)$$

The population difference ΔN between the two levels therefore depends on τ and W, i.e., on the upper state decay time (which is a characteristic of the material) and on the intensity I of the incident radiation. When I increases, W increases, the population difference ΔN will decrease, and when $W\tau \gg 1$ we have $\Delta N \simeq 0$, i.e., $N_1 \simeq N_2 \simeq N_t/2$. The populations of the two levels thus tend to become equal.

To maintain a given population difference ΔN, the material needs to absorb from the incident radiation a power per unit volume (dP/dV) given

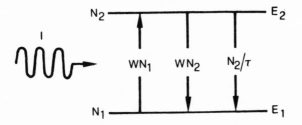

FIG. 2.15. Two-level system interacting with an e.m. wave of intensity I.

by

$$\frac{dP}{dV} = (\hbar\omega)W\Delta N = (\hbar\omega)\frac{N_t W}{1 + 2W\tau} \tag{2.123}$$

which, at saturation, i.e., for $W\tau \gg 1$, becomes

$$\left(\frac{dP}{dV}\right)_s = \frac{(\hbar\omega)N_t}{2\tau} \tag{2.124}$$

Equation (2.124) shows that the power $(dP/dV)_s$ which must be absorbed by the system to keep it in saturation is (as expected) equal to the power lost by the material due to the decay of the upper state.

It is sometimes useful to have (2.122) and (2.123) rewritten in a more convenient form. To do this we first notice that according to (2.60), W can be expressed as

$$W = \sigma I/\hbar\omega \tag{2.125}$$

where σ is the absorption cross section of the transition. Equations (2.122) and (2.123), with the help of (2.125), can be now recast in the following forms:

$$\frac{\Delta N}{N_t} = \frac{1}{1 + (I/I_s)} \tag{2.126}$$

$$\frac{dP/dV}{(dP/dV)_s} = \frac{I/I_s}{1 + I/I_s} \tag{2.127}$$

where

$$I_s = \frac{\hbar\omega}{2\sigma\tau} \tag{2.128}$$

is a parameter which depends on the given material and on the frequency of the incident wave. Its physical meaning is obvious from (2.126). In fact, for $I = I_s$ we get $\Delta N = N_t/2$. When $\omega = \omega_0$, the quantity I_s has a value which depends only on the parameters of the transition: This quantity is called the *saturation intensity*.

Let us now see how the shape of an absorption line changes with increasing value of the intensity I of the saturating beam. To do this, let us consider the idealized experimental situation shown in Fig. 2.16, where the absorption measurements are made using a probe beam of variable frequency ω' and whose intensity I' is small enough so as not to perturb the system appreciably. In practice the beams need to be more or less collinear to ensure that the probe beam interacts only with the saturated region. Under these conditions, the absorption coefficient as seen by the probe beam will be given by (2.65) where $g_t(\Delta\omega) = g(\omega' - \omega_0)$ and where $N_1 - N_2$

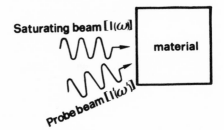

Saturating beam [I(ω)]

material

Probe beam [I'(ω')]

FIG. 2.16. Measurement of the absorption or gain coefficient at frequency ω' [$I(\omega)$ $\gg I'(\omega')$].

$= \Delta N$ is given by (2.126). We can, therefore, write

$$\alpha = \frac{\alpha_0}{1 + (I/I_s)} \tag{2.129}$$

where $\alpha_0 = g(\omega' - \omega_0)$ is the absorption coefficient when the saturating wave at frequency ω is absent (i.e., $I = 0$) and is given by

$$\alpha_0 = \frac{\pi}{3n\epsilon_0 c_0 \hbar} |\mu|^2 \omega' N_t g(\omega' - \omega_0) \tag{2.130}$$

Equations (2.129) and (2.130) show that when the intensity I of the saturating beam is increased, the absorption coefficient is reduced. The line shape, however, remains the same since it is always described by the function $g(\omega' - \omega_0)$. Figure 2.17 shows three plots of the absorption coefficient α versus ω' at three different values of I/I_s.

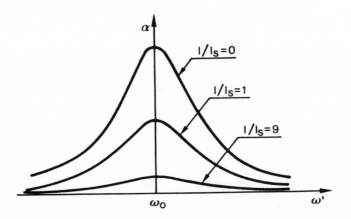

FIG. 2.17. Saturation behavior of homogeneous line absorption coefficient versus frequency ω' is shown for increasing values of the intensity.

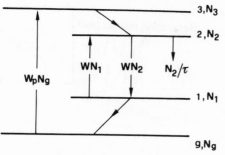

FIG. 2.18. Gain saturation in a four-level
laser.

2.6.2 Gain Saturation: Homogeneous Line

We now consider the case where the transition $2 \to 1$ exhibits net gain rather than net absorption. We assume that the medium behaves as a four-level system (Fig. 2.18) and that the inversion between levels 2 and 1 is produced by some suitable pumping process. We will further assume that the $3 \to 2$ and $1 \to g$ transitions are so fast that we can take $N_3 \simeq N_1 \simeq 0$. With these simplifying assumptions we can write the following rate equation for the population of level 2:

$$\frac{dN_2}{dt} = W_p(N_t - N_2) - WN_2 - \frac{N_2}{\tau} \qquad (2.131)$$

where W_p is the pumping rate and N_t is the total population. In the steady state (i.e., for $dN_2/dt = 0$) we find from (2.131)

$$N_2 = \frac{W_p N_t \tau}{1 + W\tau} \qquad (2.132)$$

In deriving (2.132) we have assumed $W_p\tau \ll 1$, a condition which is commonly met in laser materials. With the help of (2.125), equation (2.132) can be rewritten as

$$N_2 = \frac{N_{2o}}{1 + (I/I_s)} \qquad (2.133)$$

where $N_{2o} = W_p N_t \tau$ is the population of level 2 in the absence of the saturating beam (i.e., for $I = 0$) and

$$I_s = \frac{\hbar\omega}{\sigma\tau} \qquad (2.134)$$

A comparison of (2.134) with (2.128) shows that, for the same values of $\hbar\omega, \sigma$, and τ, the expression for the saturation intensity I_s of a four-level system is twice that of the two-level system of Fig. 2.15.

In an experiment such as that shown in Fig. 2.16, the probe beam at frequency ω' will now measure gain rather than absorption. According to

(2.65a) and (2.133) the resulting gain coefficient can be written as

$$g = \frac{g_o}{1 + (I/I_s)} \tag{2.135}$$

where $g_o = \sigma N_{2o}$ is the gain coefficient when the saturating beam is absent (unsaturated gain coefficient). Using (2.61), an expression for g_o can be found:

$$g_o = \frac{\pi}{3n\epsilon_0 c_0 \hbar} |\mu|^2 \omega' N_{2o} g(\omega' - \omega_0) \tag{2.136}$$

From (2.135) and (2.136) we see that, just as in the case of absorption considered in the previous section, when I increases, the gain g decreases, but the shape of the line remains unchanged.

2.6.3 Inhomogeneously Broadened Line

When the line is inhomogeneously broadened the saturation phenomenon is more complicated, and we will limit ourselves to just a qualitative description (see Problems 2.21 and 2.22 for further details). For the sake of generality, we will assume that the line is broadened both by homogeneous and inhomogeneous mechanisms so that its shape is expressed as in (2.107a). The overall line $g_t(\omega - \omega_0)$ is given by the convolution of the homogeneous contributions $g(\Delta\omega)$ of the various atoms. Thus, in the case of absorption, the resulting absorption coefficient can be visualized as shown in Fig. 2.19. In this case, for an experiment such as that envisaged in Fig.

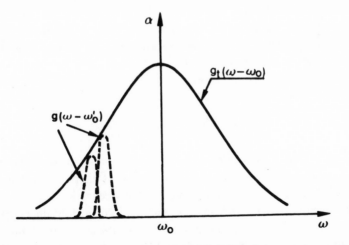

FIG. 2.19. Line shape of a transition which is broadened by both homogeneous and inhomogeneous mechanisms. The corresponding $g_t(\omega - \omega_0)$ is obtainable as a convolution [see equation (2.107a)] of the lines $g(\omega - \omega_0')$ of the individual atoms.

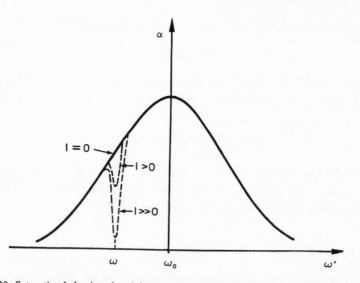

FIG. 2.20. Saturation behavior of an inhomogeneously broadened line. The plot of absorption coefficient versus frequency shows a hole of increasing depth for increasing intensity $I(\omega)$.

2.16, the intensity $I(\omega)$ will interact only with those atoms whose resonant frequency is in the neighborhood of ω and accordingly only these atoms will undergo saturation when $I(\omega)$ becomes sufficiently intense. The modified shape of the absorption line for various values of $I(\omega)$ will then be as shown in Fig. 2.20. In this case, as $I(\omega)$ is increased, a hole will be produced in the absorption line at frequency ω. The width of this hole is of the same order as the width of each of the dashed absorption profiles of Fig. 2.19, i.e., the width of the homogeneous line. A similar argument applies if a transition with net gain rather than absorption is considered. The effect of the saturating beam will, in this case, be to burn holes in the gain profile rather than in the absorption profile.

2.7 DEGENERATE LEVELS

So far we have considered only the simplest case, in which both levels 1 and 2 are nondegenerate. Let us briefly see what happens when the levels are degenerate, a situation which very often occurs in practice. This is depicted in Fig. 2.21, where level 1 is assumed to be g_1-fold degenerate and level 2, g_2-fold degenerate. Let us write N_1 for the total population of the manifold of degenerate lower levels and N_2 for the higher levels. We will use N_{2j} and N_{1i} to indicate the population of a particular level of the upper or lower manifolds, respectively.

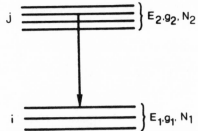

FIG. 2.21. Two-level system with degeneracy g_1 and g_2 corresponding to each level.

Let us now see how equations (2.119*a*) and (2.119*b*) are modified in the case of degenerate levels. Obviously, (2.119*a*) is still valid. The rate of change of the total population N_2 of the upper level must now include all possible transitions between levels i and j. Therefore,

$$\left(\frac{dN_2}{dt}\right) = -\sum_{i=1}^{g_1}\sum_{j=1}^{g_2}\left(W_{ji}N_{2j} - W_{ij}N_{1i} + \frac{N_{2j}}{\tau_{ji}}\right) \tag{2.137}$$

On the other hand, from the previous assumptions, all the upper levels will be equally populated, and similarly for all the lower levels i. Therefore,

$$N_{2j} = N_2/g_2 \tag{2.138a}$$

$$N_{1i} = N_1/g_1 \tag{2.138b}$$

Upon substitution of (2.138) into (2.137), we get

$$\dot{N}_2 = -W\left(\frac{N_2}{g_2} - \frac{N_1}{g_1}\right) - \frac{N_2}{\tau} \tag{2.139}$$

where we have defined

$$W = \sum_{i=1}^{g_1}\sum_{j=1}^{g_2} W_{ij} = \sum_{i=1}^{g_1}\sum_{j=1}^{g_2} W_{ji} \tag{2.140}$$

and

$$\frac{1}{\tau} = \frac{\displaystyle\sum_{i=1}^{g_1}\sum_{j=1}^{g_2}\frac{1}{\tau_{ji}}}{g_2} \tag{2.141}$$

The change in photon flux dF when the beam travels a distance dz in the material (see Fig. 1.2) can now be written, from (2.139), as

$$dF = W\left(\frac{N_2}{g_2} - \frac{N_1}{g_1}\right)dz \tag{2.142a}$$

We can then define a stimulated emission cross section, σ_{21}, and absorption cross section, σ_{12}, as [compare with (2.60)]

$$\sigma_{21} = W/(g_2 F) \tag{2.142b}$$

$$\sigma_{12} = W/(g_1 F) \tag{2.142c}$$

from which we obviously have

$$g_2 \sigma_{21} = g_1 \sigma_{12} \tag{2.142d}$$

When $N_1 > N_2$, equation (2.142a) with the help of (2.142c) can be put in the familiar form $dF = -\alpha F\,dz$ provided we define the absorption coefficient α as

$$\alpha = \sigma_{12}\left(N_1 - N_2\frac{g_1}{g_2}\right) \tag{2.142e}$$

Similarly, when $N_2 > N_1$, equation (2.142a) with the help of (2.142b) can be put in the familiar form $dF = gF\,dz$ provided we define the gain coefficient g as

$$g = \sigma_{21}\left(N_2 - N_1\frac{g_2}{g_1}\right) \tag{2.142f}$$

The reason for defining σ_{21} and σ_{12} respectively as in (2.142b) and (2.142c) is now apparent. When in fact $N_1 \gg N_2$ (as usually applies to absorption measurements involving optical transitions) (2.142e) simply reduces to $\alpha = \sigma_{12} N_1$. Conversely, when $N_2 \gg N_1$ (as applies in a four-level laser), (2.142f) simply reduces to $g = \sigma_{21} N_2$.

Another case of interest is where the upper (2) or lower (1) levels actually consist of sublevels (themselves degenerate) differing in energy, but with a relaxation between sublevels that can be considered instantaneous. In this case, thermalization between each of the sublevels of levels 1 and 2 will occur and, instead of (2.138), we write

$$N_{2j} = z_{2j} N_2 \tag{2.142g}$$

$$N_{1i} = z_{1i} N_1 \tag{2.142h}$$

where $z_{2j}(z_{1i})$ is the fraction of total population of level 2 (level 1) which, according to Boltzmann's statistics, is found in sublevel $j(i)$ (the level partition function). With the further assumption that the stimulated transition is from a given sublevel (say l) of level 2 to a given sublevel (say m) of level 1, equation (2.137), with the help of (2.142g) and (2.142h), can be written as

$$\frac{dN_2}{dt} = -W_{21}N_2 + W_{12}N_1 - \frac{N_2}{\tau} \tag{2.142i}$$

where we have defined the effective rates of stimulated emission, W_{21}, stimulated absorption, W_{12}, and spontaneous decay $(1/\tau)$ respectively as

$$W_{21} = z_{2l} W_{lm} \tag{2.142j}$$

$$W_{12} = z_{1m} W_{ml} \tag{2.142k}$$

$$\frac{1}{\tau} = \sum_{ji} \frac{z_{2j}}{\tau_{ji}} \tag{2.142l}$$

Note that, according to (2.142j) and (2.142k), we can define an effective stimulated emission cross section, σ_{21}, and an effective absorption cross section, σ_{12}, as

$$\sigma_{21} = z_{2l}\sigma \tag{2.142m}$$

$$\sigma_{12} = z_{1m}\sigma \tag{2.142n}$$

where $\sigma = \sigma_{lm} = W_{lm}/F$ is the actual cross section of the given transition.

Finally, it is interesting to obtain the ratio between the populations of the degenerate levels in thermal equilibrium. Obviously, we have

$$N_{2j}^e = N_{1i}^e \exp\left[-(E_2 - E_1)/kT \right] \tag{2.143}$$

and therefore, using (2.138), we obtain

$$\frac{N_2^e}{g_2} = \frac{N_1^e}{g_1} \exp\left[-(E_2 - E_1)/kT \right] \tag{2.144}$$

2.8 RELATION BETWEEN CROSS SECTION AND SPONTANEOUS RADIATIVE LIFETIME

From (2.61) and (2.86) it is seen that the cross section and the Einstein coefficient A are both proportional to $|\mu|^2$. For any transition, a simple expression relating σ and $\tau_{sp} = 1/A$, and independent of the dipole moment $|\mu|$, can thus be established. From (2.61) and (2.86) we get in fact

$$\sigma = \left(\frac{\lambda}{2} \right)^2 \frac{g_t(\Delta\omega)}{\tau_{sp}} \tag{2.145}$$

where $\lambda = 2\pi c_0/n\omega_0$ is the wavelength (in the medium) of an e.m. wave whose frequency corresponds to the center of the line. Equation (2.145) can be used either to obtain the value of σ, when τ_{sp} is known, or the value of τ_{sp} when σ is known.

Let us assume first that σ cannot be easily measured. This, for instance, would be the case when level 1 is not the ground state and its energy above the ground state value is much greater than kT. In this case, at thermal equilibrium, level 1 will be essentially empty and the absorption due to the transition $1 \rightarrow 2$ would be too weak for measurement. To calculate σ from (2.145) we need to know both τ_{sp} and $g_t(\Delta\omega)$. The spontaneous radiative lifetime τ_{sp} can be obtained from (2.96) if the lifetime τ of the emitted radiation [see (2.95)] and the fluorescence quantum yield ϕ have both been measured. The function $g_t(\Delta\omega)$ can be obtained from the measured shape of the line $S(\Delta\omega)$ as observed in emission. We have in fact $g_t(\Delta\omega) = S(\Delta\omega)/\int_{-\infty}^{+\infty} S(\Delta\omega)\,d\omega$.

We now consider the situation where σ can be measured (as is the case when level 1 is the ground level). To calculate τ_{sp} from (2.145) we multiply both sides by $d\omega$ and integrate. Since $\int g_t(\Delta\omega)\,d\omega = 1$, we get

$$\tau_{sp} = \left(\frac{\lambda}{2}\right)^2 \frac{1}{\int \sigma\,d\omega} \tag{2.146}$$

The spontaneous radiative lifetime is seen to be related in a simple way to the integrated cross section of the transition. Equation (2.146) is particularly useful if the upper state lifetime τ is so short (e.g., in the picosecond range) that its value and hence the value of τ_{sp} cannot be directly measured.

2.9 MOLECULAR SYSTEMS

The purpose of this section is to specialize some of the results and considerations of previous sections to the particular case of molecular media, since these play a very important role in the laser field. Here, again we will limit our description to the coarse features of the complex phenomena which occur. Our discussion does, however, provide the basis for a deeper understanding of the laser physics of systems such as molecular gas or dye lasers. For a more extensive treatment of this subject the reader is referred to specialized texts [17,18].

2.9.1 Energy Levels of a Molecule

The total energy of a molecule consists generally of a sum of four contributions: (i) electronic energy E_e, due to the motion of electrons about the nuclei: (ii) vibrational energy E_v, due to the motion (vibrations) of the nuclei; (iii) rotational energy E_r, due to the rotational motion of the molecule; and (iv) translational energy. We will not consider the translational energy any further since it is not usually quantized. The other types of energy, however, are quantized. Before going into any detailed discussion, it is instructive to derive, from simple arguments, the order of magnitude of the energy difference between electronic levels (ΔE_e), vibrational levels (ΔE_v), and rotational levels (ΔE_r). The order of magnitude of ΔE_e is given by

$$\Delta E_e \simeq \frac{\hbar^2}{ma^2} \qquad (2.147)$$

where m is the mass of the electron and a is the size of molecule. In fact, if we consider an outer electron of the molecule, the uncertainty in its position is of the order of a, the uncertainty in momentum is then $\hbar/a$, and the minimum kinetic energy is therefore $\hbar^2/ma^2$. For a diatomic molecule the energy difference ΔE_v between two vibrational levels turns out to be approximately equal to

$$\Delta E_v = \hbar\omega_v \simeq \hbar\left(\frac{K_0}{M}\right)^{1/2} \qquad (2.148)$$

where M is the mass of the atom and K_0 is the elastic constant for the attraction between the two atoms. We expect that a separation of the two atoms by an amount equal to the size a of the molecule would produce an energy change of about ΔE_e since this separation would produce a considerable distortion of the electronic wave functions. We can therefore put $K_0 = \Delta E_e/a^2$. Then from (2.147) and (2.148) we get

$$\Delta E_v = \left(\frac{m}{M}\right)^{1/2}\Delta E_e \qquad (2.149)$$

The rotational energy is of the order of $\hbar^2 J(J+1)/2Ma^2$, where J is a positive integer (rotational quantum number). Therefore, the difference ΔE_r in rotational energy between the $J = 0$ and $J = 1$ levels is given by

$$\Delta E_r \simeq \frac{\hbar^2}{Ma^2} \simeq \left(\frac{m}{M}\right)^{1/2}\Delta E_v \qquad (2.150)$$

where equations (2.147) and (2.149) have been used. Since $m/M \simeq 10^{-4}$, it follows that the separation of rotational levels is about one-hundredth that of the vibrational levels. The spacing of the vibrational levels is, in turn, about one-hundredth of ΔE_e. Bearing these facts in mind, we can see that the order of magnitude of the frequency $\nu_v = \Delta E_v/h$ is about 1000 cm^{-1} (i.e., $\nu_v \simeq 3 \times 10^{13}$ Hz).

After these preliminary considerations, we will now consider in some detail the simplest case: a molecule consisting of two identical atoms. Following the Born–Oppenheimer approximation, we first consider the two atoms as fixed at a separation R. By solving Schrödinger's equation for this situation it is then possible to find the dependence of the electronic energy levels on the separation R. Even without actually solving the equation (which is usually very complicated), it is easy to see that this dependence of energy on R must be of the form shown in Fig. 2.22, where the ground level 1 and first excited state 2 are shown as examples. If the atomic separation is very large ($R \to \infty$), the levels will obviously be the same as those of the single atom. If the separation R is finite, then because of the interaction between the atoms, the energy levels will be displaced. Since the derivative of energy with respect to R represents the force exerted by the atoms on each other, it can be seen that this force is at first attractive at large separations and then becomes repulsive for small separations. The force is zero for the position corresponding to the minimum (e.g., R_0) of each curve: This is, therefore, the separation that the atoms tend to take up (in the absence of oscillation). One notes that the curve for the excited state is shifted to the right relative to the ground-state curve, which indicates that for the excited state the equilibrium interatomic distance is somewhat greater than for the ground state.

What has been said so far refers to the case in which the two atoms are held fixed at a separation R. If we now suppose that the atoms are released at some value of R (where $R \neq R_0$), then they would tend to oscillate around the equilibrium position R_0. In this case, the total energy is the sum of the energy already discussed above, plus the vibrational energy. The latter can be calculated after noting that the curves in Fig. 2.22 also represent (except for the addition of an arbitrary constant) the variation of potential energy of one atom in the field produced by the other. The problem is, therefore, related to that of a single atom bound to the position R_0 by a potential energy of the form given in curve 1 (a similar reasoning can also be applied for the molecule in its excited state 2). For small oscillations about the position R_0, curve 1 can be approximated by a parabola (elastic restoring force). In this case the problem has a well-known solution (harmonic oscillator). The energy levels are equally spaced by an amount $\hbar \omega_v$ given by (2.148) where the elastic force constant K_0 is equal to the curvature of the parabola. Therefore, when the vibrations are taken into account, it is seen that the energy levels (for each of the two states) are given by levels $0, 1, 2, \ldots$ of Fig. 2.22. We note that the $v = 0$ level does not coincide with the minimum of the curve because of the well-known zero-point energy ($\hbar \omega_v / 2$) of a harmonic oscillator. Curves 1 and 2 now no longer represent the energy of the system since the atoms are no longer fixed. Therefore, instead of using the representation of Fig. 2.22, sometimes the simpler representation of Fig. 2.23a is used.

However, the representation of Fig. 2.22 is, in fact, more meaningful than that of Fig. 2.23. Suppose, for example, that the system is in the $v'' = 3$ vibrational level of the ground level 1. From Fig. 2.22 it is readily seen that the nuclear distance R oscillates between values corresponding to the points

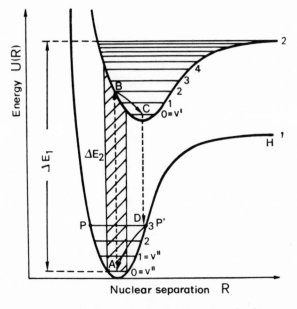

FIG. 2.22. Energy levels of a diatomic molecule.

P and P' shown in the figure. Finally we note that for large oscillations about the equilibrium position R_0, the potential energy variation cannot be approximated by a parabola. Consequently, the higher vibrational levels are no longer equally spaced. Also we note that, for polyatomic molecules, it will be necessary to use the representation of Fig. 2.23 since obviously the representation of Fig. 2.22 is then no longer appropriate in general.

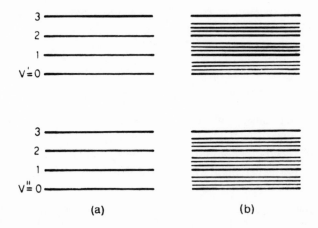

FIG. 2.23. (a) Vibrational levels and (b) rotational-vibrational levels of a molecule.

The description given so far still does not give a complete picture of the molecular system since we have ignored the fact that the molecule can also rotate. Thus, the total energy of the system is given by the sum of the electronic energy plus vibrational energy and rotational energy. Since, as we have seen, the separation between the rotational levels is much smaller than for vibrational levels, the complete picture is as shown in Fig. 2.23b.

2.9.2 Level Occupation at Thermal Equilibrium

At thermodynamic equilibrium the population of a given rotational-vibrational level belonging to a given electronic state can be written as

$$N(E_e, E_v, E_r) \propto g_e g_v g_r \exp - \left[(E_e + E_v + E_r)/kT \right] \quad (2.151)$$

where E_e, E_v and E_r are the electronic, vibrational, and rotational energies of the level and g_e, g_v, g_r are the corresponding level degeneracies [see (2.144)]. According to the estimates of the previous section, the order of magnitude of E_v/hc is 1000 cm^{-1} while E_e/hc is more than an order of magnitude larger. Since $kT/hc \simeq 209$ cm$^{-1}(T = 300°$K), it then follows that both E_e and E_v are appreciably larger than kT. Accordingly we can say that, at thermal equilibrium, a molecule lies in the lowest vibrational level[†] of the ground electronic state. The probability of occupation of a given rotational state of this lowest vibrational level can then be written, according to (2.151), as

$$N_j \propto (2J + 1) \exp - \left[BJ(J + 1)/kT \right] \quad (2.152)$$

where $B = \hbar^2/2I$ is the rotational constant (I is the moment of inertia of the molecule about the rotation axis). The factor $(2J + 1)$ accounts for level degeneracy: a rotational level of quantum number J is in fact $(2J + 1)$-fold degenerate. On account of this factor, the most heavily populated level is not the ground (i.e., $J = 0$) level but the level having rotational number J such that $(2J + 1) = (2kT/B)^{1/2}$, as can be readily shown from (2.152).

2.9.3 Radiative and Nonradiative Transitions[(17)]

Let us first consider what happens when the molecule is subjected to e.m. radiation (see Fig. 2.22). If the photon energy is greater than ΔE_1, the molecule dissociates (photolysis) upon absorption. If the incident photon energy ΔE_2 is smaller than ΔE_1, and of suitable value, the molecule can

---

[†]While this conclusion is generally true for diatomic molecules, it is not generally true for polyatomic molecules. In the latter case (e.g., SF$_6$ molecule) the spacing between vibrational levels is often appreciably smaller than 1000 cm^{-1} (down to 100 cm^{-1}) and many vibrational levels of the ground electronic state may have a significant population at room temperature.

undergo a transition from the lowest vibrational level of the ground state[†] to one of the vibrational levels (e.g., B) of the electronic excited state. If we assume that the electronic transition occurs in a time much shorter than the vibrational period, we arrive at the so-called Franck–Condon principle: The nuclear separation does not change during the process of absorption and the transition therefore occurs vertically in Fig. 2.22. Thus, if the molecule is initially in the $v'' = 0$ level of the ground electronic state, transitions will take place predominantly in the hatched region of Fig. 2.22. More precisely, the transition probability to a given v' level of the upper electronic state can be obtained from the general expression for W given by (2.53c) once the appropriate value for $|\mu|^2$ is known. To find an expression for $|\mu|^2$ we begin by remembering that, according to the Born–Oppenheimer approximation, the molecular wavefunction $\psi(\mathbf{r}_i, \mathbf{R}_J)$, which is a function of both the electron coordinates $\mathbf{r}_i$ and nuclear coordinates $\mathbf{R}_J$, can be written as

$$\psi(\mathbf{r}_i, \mathbf{R}_J) = u(\mathbf{r}_i, \mathbf{R}_J)w(\mathbf{R}_J) \tag{2.153}$$

In equation (2.153), $u(\mathbf{r}_i, \mathbf{R}_J)$ and $w(\mathbf{R}_J)$ are the electronic and nuclear wavefunctions respectively. The electronic wavefunction is obtained from the time-independent Schrödinger equation for the electrons with the nuclear coordinates $\mathbf{R}_J$ held fixed. The nuclear wavefunction $w(\mathbf{R}_J)$ is obtained from the time-independent Schrödinger equation for the nuclei in which the potential energy term is the electronic energy calculated for a given nuclear separation, i.e., the $U(\mathbf{R}_J)$ referred to in our discussion of the diatomic molecule (see Fig. 2.24). If the function $U(\mathbf{R}_J)$ is approximated by a parabola (elastic restoring force) the wavefunctions $w(\mathbf{R}_J)$ will be given by the well-known harmonic oscillator functions, which are the product of Hermite polynomials with a Gaussian function. Some of these functions are indicated in Fig. 2.24 for a diatomic molecule. Once the total wavefunction $\psi(\mathbf{r}_i, \mathbf{R}_J)$ is obtained, the dipole moment μ can be calculated as in (2.43), giving,

$$\mu_{21} = e\sum_1^n {}_i\sum_1^N {}_J\int \psi_2^*\mathbf{r}_i\psi_1 \, d\mathbf{r}_i \, d\mathbf{R}_J \tag{2.154}$$

where the sums are taken over all electrons, n, and over all nuclei, N, of the molecule. With the help of (2.153) we find

$$\mu_{21} = \left(\sum_1^N {}_J\int w_{v'}^* w_{v''} \, d\mathbf{R}_J\right)\left(e\sum_1^n {}_i\int u_2^*\mathbf{r}_i u_1 \, d\mathbf{r}_i\right) \tag{2.155}$$

[†]When many vibrational levels of the ground electronic state are occupied, transitions may occur starting from any of these levels. Absorption bands originating from $v'' > 0$ are referred to as hot bands.

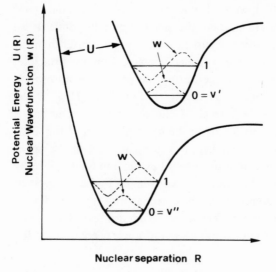

FIG. 2.24. Potential energy $U(R)$ and nuclear wavefunctions $w(R)$ for a diatomic molecule.

Nuclear separation R

where v'' and v' are the vibrational quantum numbers of the vibrational levels in the ground and excited electronic states respectively (see Fig. 2.24). The quantity $|\mu|^2$ is thus seen to be proportional to $|\sum_J \int w_{v'}^* w_{v''} \, dR_J|^2$. This quantity is called the Franck–Condon factor. For a diatomic molecule this factor reduces to $|\int w_{v'}^*(R)w_{v''}(R) \, dR|^2$, where R is the nuclear separation. Once $|\mu|^2$ is calculated, the transition probability W is obtained from (2.53c). This probability is therefore proportional to the corresponding Franck–Condon factor.

So far we have considered radiative transitions between two vibrational levels belonging to two different electronic states. The case of radiative transitions between vibrational levels of the same electronic state [e.g., $(v'' = 0) \rightarrow (v'' = 1)$ transition of Fig. 2.24] can be treated in a similar way. According to (2.148) we now have

$$\mu_{21} = \left(\sum_{1^J}^{N} \int w_{v'=1}^* w_{v''=0} \, d\mathbf{R} \right)\left(e\sum_{1^i}^{n} \int u_1^* \mathbf{r}_i u_1 \, d\mathbf{r}_i \right) \qquad (2.156)$$

The transition probability is therefore again proportional to the Franck–Condon factor involving the two vibrational states. Note that, if the Hamiltonian of the molecule is invariant under inversion, the second factor in the expression (2.156) is zero and the transition probability is therefore zero. In the case of a diatomic molecule this applies when the two atoms are identical (e.g., a N_2 molecule involving the same isotopic species). Indeed, in this case, on symmetry grounds, the molecule cannot have a net dipole moment.

In the discussion above, we have so far ignored the fact that each vibrational level actually represents a whole set of closely spaced rotational levels. If this is taken into account, we realize that the absorption takes place between a rotational level of the lower $v'' = 0$ state to some rotational level of the upper $v'' = 1$ state. For diatomic or linear triatomic molecules the selection rules usually require that $\Delta J = \pm 1$ ($\Delta J = J'' - J'$, where J'' and J' are the rotational numbers of the lower and upper vibrational states). Therefore, a given transition (e.g., $v'' = 0 \to v'' = 1$ of Fig. 2.24) which, in the absence of rotation, would consist of just a single frequency ω_0, is in fact made up of two sets of lines (Fig. 2.25). The first set, having the lower frequencies, is called the P branch and corresponds to the $\Delta J = 1$ transition. The transition frequencies of this branch are lower than ω_0 because the rotational energy of the upper level is smaller than that of the lower level. The second set, having the higher frequencies, is called the R branch and corresponds to $\Delta J = -1$. Finally we note that, for more complex molecules, the selection rule $\Delta J = 0$ also holds. When this rule applies, the transitions from all the rotational levels of a given vibrational state give a single line centered at frequency ω_0 (Q branch).

Let us now consider again a transition from the $v'' = 0$ level of the ground electronic state to some vibrational level (v') of the upper electronic state (Fig. 2.22). The transition probability will be proportional to the corresponding Franck–Condon factor. On account of the form of the nuclear wavefunction of the ground vibrational state ($w_{v''=0}$, see Fig. 2.24), the transition will take place predominantly in the hatched region of Fig. 2.22 (Franck–Condon principle). After absorption the molecule is thus left in an excited vibrational state. This vibration often decays rapidly in a

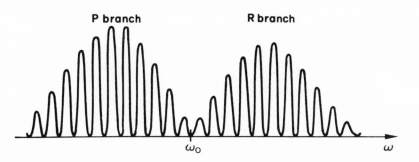

FIG. 2.25. Transitions between two vibrational levels with account taken of the rotational splitting. This transition, which, in the absence of rotational energy, would consist of a single line centered at ω_0, actually consists of two groups of lines: the so-called P-branch, which corresponds to a jump in rotational quantum number of $\Delta J = +1$, and the so-called R-branch, which corresponds to a jump in rotational quantum number of $\Delta J = -1$.

nonradiative way, usually by collisions with the surrounding molecules. This results in the molecule falling to the lowest vibrational level of the upper electronic state.[†] From there it usually decays by spontaneous emission (fluorescence) to one of the vibrational levels of the ground state. We recall that the probability of spontaneous emission A is also proportional to $|\mu|^2$. The transition probability to a given v'' state of the ground electronic level will therefore be proportional to the corresponding Franck–Condon factor $|\int w_{v''}^* w_{v'=0}\, dR|^2$. Again we see that the molecule is left in an excited vibrational state. Finally, by collisions, the molecule rapidly returns to the $v'' = 0$ level of the ground electronic state (or more precisely, thermal equilibrium is again established in the ground electronic state). It is now clear from Fig. 2.22 why the wavelength of fluorescence is longer than that of absorption (referred to as the Stokes law).

PROBLEMS

2.1 For a cavity of volume $V = 1$ cm^3 calculate the number of modes which fall within a bandwidth $\Delta\lambda = 100$ Å centered at $\lambda = 600$ nm.

2.2 The wavelength λ_M at which the maximum occurs for the distribution in Fig. 2.3 satisfies the relation $\lambda_M T = 2.9 \times 10^{-3}$ m°K. (Wien's law). Calculate λ_M for $T = 6000$°K. What is the color corresponding to this wavelength?

2.3 The R_1 laser transition of ruby has, to a good approximation, a Lorentzian shape of width (FWHM) 330 GHz at room temperature (see Fig. 2.14). The measured peak transition cross section is $\sigma = 2.5 \times 10^{-20}$cm^2. Calculate the radiative lifetime (the refractive index is $n = 1.76$). Since the observed room temperature lifetime is 3 ms, what is the fluorescence quantum yield?

2.4 Nd:YAG, a typical active laser material, is a crystal of $Y_3Al_5O_{12}$ (yttrium aluminum garnet, YAG) in which some of the Y^{3+} ions are substituted by Nd^{3+} ions. The typical Nd^{3+} atomic concentration used is 1%, i.e., 1% of Y^{3+} ions are replaced by Nd^{3+}. The YAG density is 4.56 g/cm^3. Calculate the Nd^{3+} concentration in the ground ($^4I_{9/2}$) level. This level is actually made up of five (doubly degenerate) levels, the four higher levels being spaced from the lowest level by 134, 197, 311, and 848 cm^{-1}, respectively. Calculate the Nd^{3+} concentration in the lowest level of the $^4I_{9/2}$ state.

2.5 The Nd:YAG laser transition has, to a good approximation, a Lorentzian shape of width (FWHM)~195 GHz at room temperature (see Fig. 2.14). The

---

[†]Actually, this rapid decay results in a "thermalization" of the molecule in the upper electronic state. The probability of occupation of a given vibrational level of this state is thus given by (2.151). For simple molecules, it is therefore the lowest vibrational level that will be predominantly populated.

upper state lifetime is $\tau = 230$ μs, the fluorescence quantum yield of the laser transition is ~ 0.42, and the YAG refractive index is 1.82. Calculate the peak transition cross section.

2.6 The neon laser transition at $\lambda = 1.15$ μm is predominantly Doppler broadened to $\Delta\nu_0 = 9 \times 10^8$ Hz. The upper state lifetime is $\sim 10^{-7}$s. Calculate the peak cross section assuming that the laser transition lifetime is equal to the total upper state lifetime.

2.7 The quantum yield of the $S_1 \rightarrow S_0$ transition (see Fig. 6.24) for Rhodamine 6G is 0.87, and the corresponding lifetime is ~ 5 ns. Calculate the radiative and nonradiative lifetimes (τ_{sp} and τ_{nr}, respectively) of the S_1 level.

2.8 Calculate the Doppler linewidth of a CO_2 molecule for a transition at $\lambda = 10.6$ μm ($T = 400°$K). Since collision broadening of this CO_2 laser transition is ~ 6.5 MHz/Torr, calculate the CO_2 pressure at which the two broadening mechanisms give equal contributions to the linewidth.

2.9 Calculate the total homogeneous linewidth of the 0.633 μm transition of Ne knowing that $\Delta\nu_{nat} \simeq 20$ MHz and $\Delta\nu_c = 0.64$ MHz [see (2.105a)]. What is the shape of the overall line?

2.10 A cylindrical rod of Nd:YAG with a diameter of 6.3 mm and a length of 7.5 cm is pumped very hard by a suitable flashlamp. The peak cross section for the 1.06 μm laser transition is $\sigma = 3.5 \times 10^{-19}$cm^2 and the refractive index of YAG is $n = 1.82$. Calculate the critical inversion for the onset of the amplified spontaneous emission (ASE) process (the two rod end faces are assumed to be perfectly antireflection coated, i.e., nonreflecting). Also calculate the maximum energy which can be stored in the rod if the ASE process is to be avoided.

2.11 A solution of cryptocyanine (1, 1'-diethyl-4, 4'-carbocyanine iodide) in methanol is often used simultaneously to Q-switch and mode lock (see Chapter 5) a ruby laser. The absorption cross section of cryptocyanine for ruby laser radiation (0.6943 μm) is 8.1×10^{-16} cm^2. The upper state lifetime is $\tau = 22 \times 10^{-12}$ sec. Calculate the saturation intensity at this wavelength.

2.12 The Nd:YAG laser ($\lambda = 1.06$ μm) operates as a four-level laser with a peak transition cross section of $\sigma_p = 3.5 \times 10^{-19}$ cm^2 and a lifetime $\lambda = 0.23$ ms. Calculate the gain saturation intensity.

2.13 At thermal equilibrium ($T = 400°$K), the most heavily populated level of a CO_2 molecule in the (0, 0, 1) level (see Fig. 6.13) is that corresponding to the rotational quantum number $J' = 21$. Calculate the rotational constant B for the CO_2 molecule in its (0, 0, 1) level.

2.14 Using the result of the previous problem, calculate the frequency spacing between the rotational lines (Fig. 2.25) of the CO_2 laser transition at $\lambda = 10.6$ μm (assume that the rotational constant B is the same for upper and lower levels and take into account that, due to selection rules, stimulated emissions arising from odd values of J only occur in a CO_2 molecule).

2.15 What CO_2 gas pressure would be needed to cause all the rotational lines to merge together? What would the width of the gain curve be for such a pressure?

2.16 Instead of ρ_ν, a spectral energy density ρ_λ can also be defined, ρ_λ being such that $\rho_\lambda d\lambda$ gives the energy density for the e.m. waves of wavelength lying between λ and $\lambda + d\lambda$. Find the relationship between ρ_λ and ρ_ν.

2.17 For blackbody radiation find the maximum of ρ_λ versus λ. Show in this way that the wavelength λ_M at which the maximum occurs satisfies the relationship $\lambda_M T = hc/ky$ (Wien's law), where the quantity y satisfies the equation $5[1 - \exp(-y)] = y$. From this equation find an approximate value of y ($y = 4.965$).

2.18 Find the relationship between the intensity I and the corresponding energy density for a plane e.m. wave [compare with (2.19b) and explain the difference].

2.19 Instead of observing saturation as in Fig. 2.16, we can also do this by using only the beam $I(\omega)$ and measuring the absorption coefficient for this beam. For a homogeneous line, show that the absorption coefficient is, in this case,

$$\alpha(\omega - \omega_0) = \frac{\alpha_0(0)}{1 + (\omega - \omega_0)^2 T_2^2 + I/I_{s0}}$$

where $\alpha_0(0)$ is the weak-signal ($I \ll I_{s0}$) absorption coefficient at $\omega = \omega_0$ and I_{s0} is the saturation intensity as defined by (2.128) at $\omega = \omega_0$. Hint: begin by showing that

$$\alpha = \frac{\alpha_0(0)}{1 + (\omega - \omega_0)^2 T_2^2} \frac{1}{1 + I/I_s}$$

$$= \frac{\alpha_0(0)}{1 + (\omega - \omega_0)^2 T_2^2} \frac{1}{1 + \dfrac{I}{I_{s0}} \dfrac{1}{1 + (\omega - \omega_0)^2 T_2^2}}$$

2.20 From the previously derived expression, find the behavior of the peak absorption coefficient and the linewidth versus I. How would you then measure the saturation intensity I_{s0}?

2.21 Show that, for an inhomogeneous line with line shape function g, the saturated absorption coefficient for an experiment as in Fig. 2.16 can be written as

$$\alpha = \left(\frac{\pi}{3n\epsilon_0 c_0 \hbar} \right) |\mu|^2 N \int \frac{T_2}{\pi} \frac{\omega'' g^*(\omega'' - \omega_0)}{1 + (\omega' - \omega'')^2 T_2^2} \frac{1}{1 + \dfrac{I}{I_{s0}} \dfrac{1}{1 + (\omega - \omega'')^2 T_2^2}} d\omega''$$

where the homogeneous contribution is accounted for by a Lorentzian line. [Hint: begin by calculating the elemental contribution $d\alpha$ due to the fraction $g^*(\omega'' - \omega_0)d\omega''$ of atoms whose resonant frequencies lie between ω'' and $\omega'' + d\omega''$.]

2.22 Under the assumption that (i) the homogeneous linewidth is much smaller than the inhomogeneous linewidth and (ii) that $I \ll I_{s0}$, show that the previous expression for α can be approximately written as

$$\alpha \simeq \frac{\pi}{3n\epsilon_0 c_0 \hbar} |\mu|^2 N \omega' g^*(\omega' - \omega_0)$$

$$\times \left(1 - \frac{T_2}{\pi} \frac{I}{I_{s0}} \int \frac{d\omega''}{\left[1 + (\omega' - \omega'')^2 T_2^2 \right]\left[1 + (\omega - \omega'')^2 T_2^2 \right]} \right)$$

Since the integral is now the convolution of two Lorentzian lines, what is the width of the hole in Fig. 2.20?

2.23 At short wavelengths (vuv, soft x-rays) the predominant line broadening mechanism is natural broadening. In this case, from (2.145), show that the peak cross section is $\sigma_0 = \lambda^2/2\pi$.

REFERENCES

1. R. Reif, *Fundamentals of Statistical and Thermal Physics* (McGraw-Hill Book Co., New York, 1965), Chap. 9.
2. R. H. Pantell and H. E. Puthoff, *Fundamentals of Quantum Electronics* (John Wiley and Sons, New York, 1969), Chap. 2, Secs. 2.1–2.3.
3. A. Messiah, *Quantum Mechanics* (North–Holland Publishing Co., Amsterdam, 1961), Vol. 1, pp. 112, 113.
4. J. A. Stratton, *Electromagnetic Theory*, 1st ed. (McGraw–Hill Book Co., New York, 1941), pp. 431–438.
5. R. H. Pantell and H. E. Puthoff, *Fundamentals of Quantum Electronics*, (John Wiley and Sons, New York, 1969), Chap. 6.
6. W. Louisell, *Radiation and Noise in Quantum Electronics*, (McGraw–Hill Book Co., New York, 1964), Chap. 4.
7. T. Holstein, *Phys. Rev.* **72**, 1212 (1947).
8. R. H. Dicke, *Phys. Rev.* **93**, 99 (1954).
9. R. Bonifacio and L. Lugiato, *Phys. Rev. A* **11**, 1507 (1975).
10. G. J. Linford and L. W. Hill, *Appl. Opt.* **13**, 1387 (1974).
11. *Radiationless Transitions*, ed. by F. J. Fong (Springer–Verlag, Berlin, 1976).
12. W. Louisell, *Radiation and Noise in Quantum Electronics* (McGraw–Hill Book Co., New York, 1964), Chap. 5.
13. W. Heitler, *The Quantum Theory of Radiation*, 3rd ed. (Oxford University Press, London, 1953), pp. 181–189.
14. H. G. Kuhn, *Atomic Spectra*, 2nd ed. (Longmans, Green and Co., London, 1969). Chap. VII.
15. A. L. Schawlow, in *Advances in Quantum Electronics*, ed. by J. R. Singer (Columbia University Press, New York, 1961), pp. 50–62.
16. A. E. Siegman, *An Introduction to Lasers and Masers*, (McGraw–Hill Book Co., New York, 1971), p. 362.
17. A. S. Davydov, *Quantum Mechanics*, (Pergamon Press, Oxford, 1966), pp. 513–520.
18. R. H. Pantell and H. E. Puthoff, *Fundamentals of Quantum Electronics*, (John Wiley and Sons, New York, 1969), pp. 40–41, 60, 62, and Appendix 4.

3
Pumping Processes

3.1 INTRODUCTION

We have seen in Chapter 1 that the process by which atoms are raised from level 1 to level 3 (for a three-level laser, Fig. 1.4a) or from level 0 to level 3 (for a four-level laser, Fig. 1.4b) is called the pumping process. Usually it is performed in one of the following two ways: optically or electrically. In *optical pumping* the light from a powerful source is absorbed by the active material and the atoms are thereby pumped into the upper pump level. This method is particularly suited to solid state (e.g., ruby or neodymium) and liquid (e.g., dye) lasers. The line broadening mechanisms in solids and liquids produce a very considerable broadening, so that usually one is dealing with pump bands rather than levels. These bands can, therefore, absorb a sizable fraction of the (usually broad-band) light emitted by the pumping lamp. *Electrical pumping* is accomplished by means of a sufficiently intense electrical discharge and is particularly suited to gas and semiconductor lasers. Gas lasers, in particular, do not usually lend themselves so readily to optical pumping because of the small widths of their absorption lines. On the other hand, semiconductor lasers can be optically pumped quite effectively, although electrical pumping is much more convenient. The two pumping processes mentioned above are not the only ones available for pumping lasers: For instance, pumping can also be achieved by a suitable chemical reaction (chemical pumping) or by a means of a supersonic gas expansion (gas-dynamic pumping). It should also be noted that, increasingly, lasers are being used for optical pumping of other lasers (solid-state, dye, and gas lasers). These latter pumping processes will not be treated any further here and we refer the reader to Chapter 6 for further details.

If the pump level (or bands) are empty, the rate at which the upper state becomes populated by the pumping, $(dN_2/dt)_p$, is given by (1.10), where W_p is the pump rate. The purpose of this chapter is to give specific expressions for W_p, for both optical and electrical pumping.

3.2 OPTICAL PUMPING

Figure 3.1 is a schematic illustration of a quite general optical pumping system. The light from a powerful incoherent lamp is conveyed, by a suitable optical system, to the active material. The following two cases will be considered: (i) Pulsed lasers. In this case medium to high pressure (450–1500 Torr) Xe or Kr flashlamps are used. (ii) Continuous wave (cw) lasers. In this case high pressure (4000–8000 Torr) Kr or tungsten–iodine lamps are most often used. In case (i), the electrical energy stored in a capacitor bank is discharged into the flashlamp. The discharge is usually initiated by a high-voltage trigger pulse to an auxiliary electrode, and this pulse pre-ionizes the gas. The lamp then produces an intense flash of light whose duration (given by the product of storage capacitance and the lamp resistance) ranges from a few microseconds up to a few hundred microseconds. In both cases (i) and (ii), the active material is usually in the form of a cylindrical rod with a diameter ranging from a few millimeters up to a few centimeters and a length ranging from a few centimeters up to a few tens of centimeters.

Figure 3.2 shows three configurations which are particularly important examples of the general system sketched in Fig. 3.1.[1, 2] In Figure 3.2a the lamp (usually a flashlamp) has a helical form, and the light reaches the active material either directly or after reflection at the specular cylindrical surface 1. This system was used for the first ruby laser, and it is still widely used for pulsed lasers. In Fig. 3.2b the lamp is in the form of a cylinder (linear lamp) of radius and length equal to those of the active rod. The

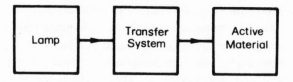

FIG. 3.1. General scheme of an optical pumping system.

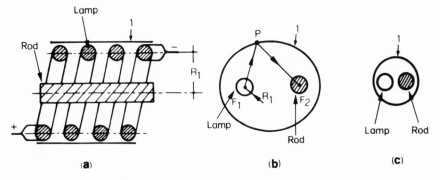

FIG. 3.2. Most commonly used optical pumping systems.

lamp is placed along one of the two focal axes, F_1, of a specularly reflecting elliptical cylinder (labeled 1 in the figure).[2] The laser rod is placed along the second focal axis F_2. A well-known property of an ellipse is that a ray F_1P leaving the first focus F_1 passes, after reflection by the elliptical surface, through the second focus F_2 (ray PF_2). This means that a large fraction of the light emitted by the lamp is conveyed, by reflection from the elliptical cylinder, to the active rod. Figure 3.2c shows what is called the close-coupling configuration. The rod and the linear lamp are placed as close as possible and are surrounded by a close-coupled cylindrical reflector (labeled 1 in the figure). The efficiency of the close-coupling configuration is usually not much lower than that of an elliptical cylinder. Note that cylinders made of diffusely reflecting material (such as compressed MgO or $BaSO_4$ powders or white ceramic) are sometimes used instead of the specular reflectors shown in Fig. 3.2. Although diffusive surfaces somewhat reduce the pump transfer efficiency, they have the advantage of providing a more uniform pumping of the active material. Multiple configurations using more than one elliptical cylinder or several lamps in close-coupling configurations have also been used. Figure 3.3 gives just two possible

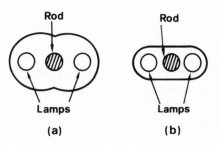

FIG. 3.3. Pump configurations using two lamps: (a) double-ellipse; (b) close-coupling.

examples. The efficiency of these multiple configurations is lower than for the corresponding single configurations of Fig. 3.2b and c. Nevertheless, they are often used in high-power (or high-energy) systems.

3.2.1. Pumping Efficiency

The overall pumping efficiency can be split up into three types of efficiency:

(i) *Transfer Efficiency* η_t, which is defined as the ratio of the pump power (or energy) actually entering the rod to the power (or energy) emitted by the lamp.

(ii) *Lamp Radiative Efficiency* η_r, which gives the efficiency of conversion of electrical input to light output in the wavelength range λ_1 to λ_2 in which the effective pump bands of the laser medium lie (e.g., 0.3 to 0.9 μm for Nd^{3+}:YAG). The lamp radiative efficiency is therefore given by

$$\eta_r = \frac{(2\pi Rl)\int_{\lambda_1}^{\lambda_2} I_\lambda \, d\lambda}{P} \tag{3.1}$$

where R is the radius and l the length of the lamp, I_λ is its spectral intensity, and P is the electrical power delivered to the lamp. Notice that, according to (3.1), I_λ can be written as

$$I_\lambda = \frac{P}{2\pi Rl} \eta_r g_\lambda \tag{3.2}$$

where g_λ is a normalized spectral intensity distribution (i.e., $\int_{\lambda_1}^{\lambda_2} g_\lambda \, d\lambda = 1$). Equation (3.2) allows one to calibrate I_λ once the uncalibrated spectrum of the emitted light and the lamp radiative efficiency η_r are known.

(iii) *Pump Quantum Efficiency* η_q, which accounts for the fact that not all of the atoms raised to the pump bands subsequently decay to the upper laser level. Some of these atoms can in fact decay from the pump bands straight back to the ground state or perhaps to other levels which are not useful. We will define the pump quantum efficiency $\eta_q(\lambda)$ as the ratio of the number of atoms which decay to the upper laser level to the number of atoms which are raised to the pump band by a monochromatic pump at wavelength λ.

The problem of improving the radiative efficiency is a challenging technical one for a lamp manufacturer. What are needed are lamps whose emission spectrum is a good match to the absorption spectrum of the pump bands. The quantum efficiency, on the other hand, is a quantity over which one can have little control since it depends on the properties of the given material. The transfer efficiency, however, depends a great deal on the optical system chosen to convey the pump light to the laser rod. Its calculation is, therefore, important if one is to provide the optimum transfer conditions. The remainder of this section is devoted to this topic.

Before dealing with the calculation of the transfer efficiency, let us begin by finding a unified approach for analyzing the two pump configurations of Fig. 3.2a and b. Thus we shall assume that the pitch of the helix of Fig. 3.2a is very small. The presence of the reflecting cylindrical surface 1 allows us to represent the helical pumping system as shown schematically in Fig. 3.4a, where the shaded rod (lateral surface labeled S_2) is the laser rod and where the lamp is represented by the cylindrical surface S_1 having the same radius R_1 as the lamp radius (see Fig. 3.2a). In the case of Fig. 3.2b, all the rays emitted by the lamp tangentially to its surface S_1 will be transformed, upon reflection at the surface of the elliptical cylinder, to a bundle of rays around the second focal line F_2. The envelope of these rays is a surface S_1', this being the lamp image as formed by the elliptical cylinder. In Fig. 3.4b the particular rays which bound the S_1' surface both horizontally and vertically have been indicated. It is apparent that the image S_1' is elongated in the direction of the minor axis of the elliptical mirror. It can be shown that this image is itself an ellipse. The major and minor axes of this ellipse, R_M and R_m respectively, can then be obtained from Fig. 3.4b by simple geometrical considerations. If we assume that the radius R_L of the lamp is much smaller than the minor axis of the elliptical mirror, we get

$$R_M = R_L\left(\frac{1+e}{1-e}\right) \tag{3.2a}$$

$$R_m = R_L\left(\frac{1-e^2}{1+e^2}\right) \tag{3.2b}$$

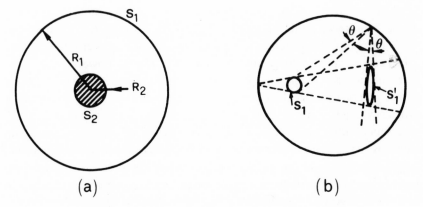

(a)   **(b)**

FIG. 3.4. Reduction of the two systems of Fig. 3.2a and b to a single system.

where e is the eccentricity of the elliptical mirror. Now, if this eccentricity is very small, the image S_1' is again a circle of the same radius as that of the lamp. In this case the system of Fig. 3.4b reduces to that of Fig. 3.4a and the surface S_1 of Fig. 3.4a is the surface S_1' of Fig. 3.4b.

Now that the two systems of Fig. 3.2a and b have been reduced to the single system shown in Fig. 3.4a, we can proceed to calculate the fraction of the power emitted by the surface S_1 of Fig. 3.4a which actually enters the surface S_2 of the active rod. To do this we will assume that S_1 can be considered as a blackbody surface at temperature T. According to the Stefan–Boltzmann law, the total power emitted by the lamp is given by

$$P_1 = \sigma_{SB} T^4 S_1 \tag{3.3}$$

where σ_{SB} is the Stefan–Boltzmann constant. The calculation of the power entering the rod then follows from a simple thermodynamic argument.[3] Let us suppose the laser rod is replaced by a blackbody cylinder having the same dimensions as the rod. Obviously, the power P_{2i} entering the surface S_2 will remain the same. Now, if the blackbody cylinder is kept at the same temperature T as the lamp, then, according to the second law of thermodynamics, there will be no net exchange of power between the two blackbody surfaces (lamp and rod). This means that the incident power P_{2i} must equal the power P_{2e} emitted by the rod. Since P_{2e} is given by $P_{2e} = \sigma_{SB} T^4 S_2$, we get

$$P_{2i} = P_{2e} = \sigma_{SB} T^4 S_2 \tag{3.4}$$

Then we readily find from equations (3.3) and (3.4) that the value of the transfer efficiency η_t is given by

$$\eta_t = \frac{P_{2i}}{P_1} = \frac{S_2}{S_1} = \frac{R_2}{R_1} \tag{3.5}$$

where rod and lamp have been assumed to have the same length. The above expression holds provided $R_2 < R_1$. If $R_2 \geqslant R_1$ (a situation that obviously can only be achieved with the system of Fig. 3.2b), the transfer efficiency is expected to be always equal to 1. This is, however, rigorously true only when the elliptical pump cavity has zero eccentricity. For finite values of eccentricity, there are calculations available giving the transfer efficiency as a function of the ratio between the lamp and rod diameters.[19] One should also take account of the fact that the reflectivity of the pump cavity is never 100%. In practice, the transfer efficiency of an optimized elliptical cylinder can be as high as 80%. Since the radius R_1 of a helical lamp is usually at least twice the rod radius R_2, the efficiency of a helical lamp is appreciably smaller than that of a linear lamp in an elliptical reflector. On the other hand, helical lamps provide a more uniform pumping of the laser rod (see the next section) and are often used in high-energy systems when laser beam uniformity is more important than laser efficiency.

3.2.2 Pump Light Distribution

In the previous section we have calculated the fraction of pump light reaching the rod. In this section we want to calculate, for a few representative cases, the distribution of this light within the active rod.

As a first example, we consider the case of a helical flashlamp or, equivalently, that of a very low eccentricity elliptical reflector with lamp diameter larger than the rod diameter. In both of these cases the configuration of Fig. 3.4a applies. We further assume that the lateral surface of the rod is polished. Then, since the refractive index of the rod is usually larger than that of the surrounding medium, the pump light tends to be concentrated toward the rod axis. This can be understood with the help of Fig. 3.5, which shows a rod of radius R and refractive index n surrounded by a medium of unit refractive index. The lamp is not shown. We recall, however, that its radius has been assumed to be equal to (or larger than) R. In this case, the rays falling on point P of the rod surface can arrive from any direction within the angle π shown in Fig. 3.5. The two extreme rays 2 and 3 are indicated in the figure. Upon entering the rod, these rays are refracted and become the rays 2′ and 3′, where the angle θ is the critical

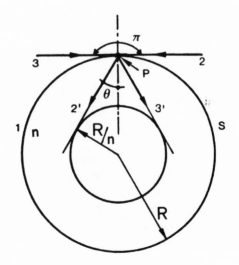

FIG. 3.5. Concentration of rays in a central core of a cylindrical rod, due to refraction.

angle ($\sin\theta = 1/n$). Therefore, all rays arriving from the lamp will be refracted within the angle 2θ between the rays 2' and 3'. Applying the same argument to all points P of surface S, we arrive at the conclusion that the central core (radius R/n) of the rod is more heavily pumped than the outer part of the rod. The calculation of pump energy density ρ in the rod becomes particularly simple if we make the following assumptions: (i) We only consider light entering the rod in a plane orthogonal to the rod axis and (ii) we neglect the attenuation of this light as it propagates into the rod. In this case the energy density ρ_n at a point within the rod at a distance r from the axis is[4]

$$\rho_n = n^2\rho \qquad\qquad (0 < r < R/n) \qquad\qquad (3.6a)$$

$$\rho_n = \frac{2n^2}{\pi}\rho \sin^{-1}\left(\frac{R}{nr}\right) \qquad (R/n < r < R) \qquad\qquad (3.6b)$$

where ρ is the value that the energy density would have at that same point of the rod if its refractive index were unity. This density is related to the intensity of the light emitted by the lamp by equation (2.19b). If the simplifying assumptions (i) and (ii) are not made, the expression for ρ_n is much more complicated.[5] In Fig. 3.6, computed plots of the dimensionless quantity

$$f(\alpha R, r/R) = \rho_n/n^2\rho \qquad\qquad (3.7)$$

are shown for several values of αR, where α is the absorption coefficient at the pump wavelength (the pump light is assumed to be monochromatic). The same figure also shows the predictions of equation (3.6), indicated by a dashed line. Note the difference between the dashed curve and the solid curve corresponding to $\alpha R = 0$. While both curves refer to the case where there is no absorption in the rod, the solid curve, unlike the dashed one, takes account of the fact that light can enter the rod from any direction. Note also that, when $\alpha R \neq 0$, the attenuation of the pump light as it propagates inward from the rod surface tends to smooth out the distribution ρ_n. From the data of Fig. 3.6 it is seen that, at the center of the rod ($r = 0$), the quantity $f(\alpha R, 0)$ can be closely approximated by the expression $f = \exp(-1.1\alpha R)$.

The fact that, for very small values of αR, the energy density in the central region of the rod is $n^2\rho$ deserves some further consideration. Let us assume that the lamp has the same radius as the rod and is placed along the focal axis F_1 of Fig. 3.2b. Since the rays 2 and 3 of Fig. 3.5 are tangent to S, they must have come from two rays which are tangent to the lamp surface. After refraction, rays 2 and 3 become rays 2' and 3' which are

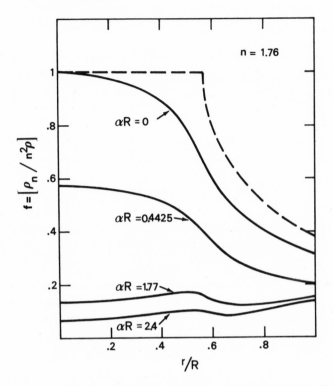

FIG. 3.6. Radial variation of the pump energy density ρ_n for several values of the pump absorption coefficient α (monochromatic pump). The data have been taken from Reference 5.

tangent to a circle of radius R/n. We can, therefore, say that the rod acts like a cylindrical lens in producing an image of the lamp at its center which is n times smaller than the lamp itself. Since the volume of this image is n^2 times smaller than that of the lamp, we can now understand why the corresponding energy density ρ_n is increased by n^2.

We have seen that, for very small values of αR, the pump energy density is uniform only for $r < R/n$, while it is nonuniform outside this central core. A nonuniform energy density is certainly not desirable for an active material. This inconvenience can be overcome[5] by surrounding the active rod by a cylindrical cladding of transparent material with the same refractive index as the rod (Fig. 3.7). In this case, if the radii of both the cladding and the lamp are made equal to nR, we can repeat the argument of Fig. 3.5, starting from a point P of the cladding. The refracted rays $2'$ and $3'$ will, in this case, be tangent to the surface of the active material, and all the incoming light will be concentrated into the active material. In the

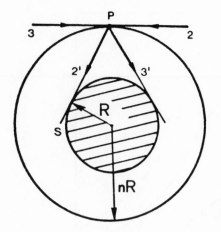

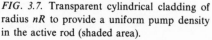

FIG. 3.7. Transparent cylindrical cladding of radius nR to provide a uniform pump density in the active rod (shaded area).

case of $\alpha R = 0$ and for light entering only in the plane of Fig. 3.7, the energy density will now be uniform in the active material and be given by equation (3.6a). Another way to provide a more uniform pumping is by grinding the lateral surface of the rod. In this case the pump light, upon entering the rod, will be diffused, and the light concentration shown in Fig. 3.5 will not occur. In Fig. 3.8 computed plots of the dimensionless quantity

$$f_1(\alpha R, r/R) = \rho_n / n\rho \tag{3.8}$$

for several values of αR are shown for this case.[1] Here again α is the absorption coefficient at the pump wavelength (for monochromatic pump light). Note that for $\alpha R = 0$ we have $\rho_n = n\rho$. The factor n arises in this case simply from the fact that the light velocity in the rod is n times smaller than its vacuum value. For a given lamp emission the energy density ρ_n is thus expected to be n times larger than the value ρ which a rod of unit refractive index would produce. From the data of Fig. 3.8 it is seen that, at the center of the rod, $f_1(\alpha R, 0)$ can be closely approximated by the expression $f_1 \simeq \exp(-1.27\alpha R)$. A comparison of (3.8) with (3.7), at $r = 0$, then shows that, apart from the relatively small difference between f and f_1, the pump energy density at the center of the rod is actually reduced by an amount n as a result of roughening the lateral surface. However, the whole cross section of the rod, rather than just the central core of radius R/n, is now more or less uniformly illuminated. In fact it can be shown from Figs. 3.6 and 3.8 that the integrated pump energy density over the rod cross section is approximately the same in the two cases.

We now consider the case where the lamp radius R_L is smaller than the rod radius R_R. We assume the pumping geometry to be that of Fig. 3.2b. If

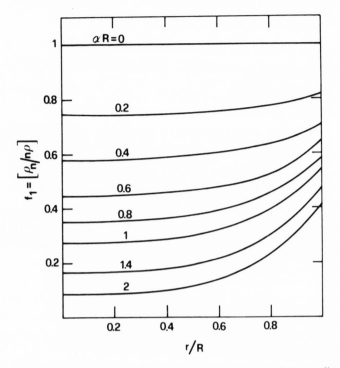

FIG. 3.8. Rod with rough-ground lateral surface. Radial variation of the normalized pump energy density $(\rho_n/n\rho)$ versus normalized radius (r/R) for several values of the pump absorption coefficient α (after Reference 1).

the rod lateral surface is polished, an elliptical image of the lamp will be formed in the rod (see Fig. 3.4b). Due to refraction at the rod surface, the major and minor axes of this image are both reduced by a factor n from the values given in expressions (3.2a) and (3.2b). To avoid this nonuniform pump distribution, the rod lateral surface can again be rough ground.

So far, the discussion applies to monochromatic pump radiation. For polychromatic radiation, the same relations as in (3.6) to (3.8) apply and so also do the curves of Figs. 3.6 and 3.8 provided, however, that ρ_n and ρ are replaced by the spectral quantities $\rho_{n\lambda}$ and ρ_λ.

3.2.3 Pumping Rate[6]

Let us first consider a monochromatic pump of frequency ω. The pump power absorbed per unit volume of the rod, dP/dV, is then given by

$$\frac{dP}{dV} = WN_g\hbar\omega \tag{3.9}$$

where W is the absorption rate and the upper pump level has been assumed empty. With the help of (2.53c) and (2.61), equation (3.9) can be written as

$$\frac{dP}{dV} = \frac{c_0}{n} \sigma N_g \rho_n \tag{3.10}$$

where ρ_n is the pump energy density at the point in question. For polychromatic pump radiation, (3.10) can be written in terms of the corresponding spectrally dependent variables, *viz.*,

$$\frac{dP_\lambda}{dV} = \frac{c_0}{n} \sigma N_g \rho_{n\lambda} \tag{3.10a}$$

where P_λ is such that $(dP_\lambda/dV)\,d\lambda$ is the power absorbed per unit volume from pump radiation with wavelength lying between λ and $\lambda + d\lambda$.

As a particularly relevant example, we now consider the case in which the lateral surface of the rod is rough ground. With the help of (3.8) and (2.19c), equation (3.10a) can be written as

$$\frac{dP_\lambda}{dV} = 4\eta_t \sigma N_g f_1 I_\lambda \tag{3.11}$$

where η_t is the transfer efficiency of the given pump configuration. The rate at which the upper state becomes populated by the pumping process is then

$$\frac{dN_2}{dt} = \int \eta_q \frac{1}{\hbar\omega} \frac{dP_\lambda}{dV} \, d\lambda$$

$$= 4\eta_t N_g \int \frac{\eta_q \sigma f_1}{\hbar\omega} I_\lambda \, d\lambda \tag{3.12}$$

where $\eta_q = \eta_q(\lambda)$ is the pump quantum efficiency. A comparison of (3.12) with (1.10) then gives

$$W_p = 4\eta_t \int \frac{\eta_q \sigma f_1}{\hbar\omega} I_\lambda \, d\lambda \tag{3.13}$$

With the help of (3.2), we can re-express Eq. (3.13) more conveniently as

$$W_p = 4\eta_t \eta_r \frac{P}{2\pi R l} \int \frac{\eta_q \sigma f_1}{\hbar\omega} g_\lambda \, d\lambda \tag{3.14}$$

Note that, according to (3.7), the right-hand side of (3.13) and (3.14) should be multiplied by n, and f_1 replaced by f, in the case of a rod with a polished surface.

Equations (3.13) and (3.14) are the desired expressions for the pump rate. They depend on the properties of the active material [quantum efficiency $\eta_q(\lambda)$ and absorption cross section $\sigma(\lambda)$ of the pump bands] and on the spectral emission of the lamp (I_λ or g_λ). Since $f_1 = f_1(\alpha R, r/R)$, it follows that W_p will also depend on the concentration of active ions, on the rod radius R, and on the normalized radial coordinate r/R. A calculation

TABLE 3.1. Efficiency Terms for Optical Pumping (%)

Case	η_t	η_r	η_a	η_{pq}	η_p
1	30–40	25	30–60	50	1.1–3
2	80	50	16	40	2.6

of W_p therefore requires knowledge of all these quantities. To simplify matters, an overall pumping efficiency η_p is sometimes introduced. This is defined as the ratio of the minimum possible power required to produce a given pumping in the rod (i.e., $\langle W_p \rangle N_g V \hbar \omega_0$, where $\langle W_p \rangle$ is the average value of W_p over the rod volume V, and ω_0 is the frequency of laser transition) to the actual electrical power input, P, to the lamp required to produce this pumping. Therefore we can write

$$\langle W_p \rangle = \eta_p \frac{P}{VN_g \hbar \omega_0} \qquad (3.15)$$

The pump efficiency η_p can be split up into the product of four terms[7]: (i) the transfer efficiency η_t; (ii) the lamp radiative efficiency η_r; (iii) the absorption efficiency η_a which gives the fraction of the useful radiation which is actually absorbed by the rod; (iv) the power quantum efficiency η_{pq}, which is that fraction of the absorbed power which leads to population of the laser level. Notice that this last quantity is similar to the pump quantum efficiency η_q defined previously. Estimates for these efficiency factors defined above are available in the literature.[7] Table 3.1 gives these values for a 6.3-mm-diameter ruby rod pumped by a xenon helical flash-tube (Case 1) and for a 6.3-mm-diameter Nd^{3+}:YAG rod pumped by a Kr lamp (Case 2). It must be stressed, however, that the values given in the table are only rough estimates, and an accurate calculation of W_p at each point of the rod can only be obtained through Eq. (3.14).

3.3 ELECTRICAL PUMPING[8,9]

This type of pumping is used for gas and semiconductor lasers. We will limit ourselves here to a discussion of the electrical pumping of gas lasers. In this case pumping is achieved by allowing a current of suitable value to pass through the gas. Ions and free electrons are produced, and since they are accelerated by the electric field, they acquire additional kinetic energy and are able to excite a neutral atom by collision. For this impact excitation, the movement of the ions is usually less important than that of the electrons. For a low-pressure gas, in fact, the average electron energy is

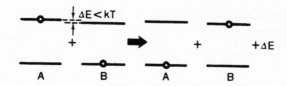

FIG. 3.9. Near-resonant energy transfer between two atoms (or molecules) A and B.

much greater than the corresponding ion energy. After a short time, an equilibrium condition is established among the electrons, and this can be described by an effective electron temperature T_e.

Electrical pumping of a gas usually occurs via one of the following two processes: (i) For a gas consisting of only one species, the excitation can only be produced by electron impact, i.e., according to the process

$$e + X \rightarrow X^* + e \qquad (3.16)$$

where X and X^* represent the atom in the ground and excited states, respectively. Such a process is called a collision of the *first kind*. (ii) For a gas consisting of two species (say A and B), excitation can also occur as a result of collisions between atoms of different species through a process known as *resonant energy transfer*. With reference to Fig. 3.9, let us assume that species A is in the excited state and species B in the ground state. We will also assume that the energy difference ΔE between the two transitions is less than kT. In this case, there is an appreciable probability that, after the collision, species A will be found in its ground state and species B in its excited state. The process can be denoted by

$$A^* + B \rightarrow A + B^* + \Delta E \qquad (3.17)$$

where the energy difference ΔE will be added to or subtracted from the translational energy, depending on its sign. This process provides a particularly attractive way of pumping species B, if the upper state of A is metastable (forbidden transition). In this case, once A is excited to its upper level by electron impact, it will remain there for a long time, thus constituting an energy reservoir for excitation of the B species. A process of the type indicated in (3.17) is called a collision of the *second kind*.[†]

[†]Collisions of the *first kind* involve conversion of the kinetic energy of one species into potential energy of another species. In collisions of the *second kind*, potential energy is converted into some other form of energy (other than radiation) such as kinetic energy, or is transferred as potential energy (in the form of electronic, vibrational, or rotational energy) to another like or unlike species. Collisions of the second kind therefore include not only the reverse of collisions of the first kind (e.g., $e + X^* \rightarrow e + X$) but also, for instance, the conversion of excitation energy into chemical energy.

3.3.1 Electron Impact Excitation[10]

Electron impacts can involve both elastic and inelastic collisions. In an inelastic collision, the atom may either be excited to a higher state or be ionized. All three of these phenomena take place in an electrical discharge and influence its behavior in a complicated way.

For the sake of simplicity, let us first consider the case of impact excitation by a beam of collimated monoenergetic electrons. If F_e is the electron flux (electrons/cm²sec), a total collision cross section σ_e can be defined in a similar way to the case of a photon flux [see equation (2.62)], namely,

$$dF_e = -\sigma_e N_g F_e \, dz \qquad (3.18)$$

Here dF_e is the change of flux which takes place when the beam propagates a distance dz in the material. Collisions which produce electronic excitation will only account for some fraction of this total cross section. If we let σ_{e2} be the cross section for electronic excitation from the ground level to the upper laser level, then, according to (3.18), the rate of population of the upper state due to the pumping process is

$$(dN_2/dt)_p = \sigma_{e2} N_g F_e = N_g N_e v \sigma_{e2} \qquad (3.19)$$

where v is the electron velocity and N_e is the electron density. A calculation of the pump rate requires a knowledge of the σ_{e2} value in addition to information about the e-beam parameters. This quantity σ_{e2} is in turn a function of the e-beam energy E (i.e., of v), and its qualitative behavior is sketched in Fig. 3.10. Note that there is a threshhold E_{th} for the process to occur and that this threshhold is approximately equal to the energy which is required for the $0 \rightarrow 2$ atomic transition. The cross section σ then reaches a maximum value (at an energy which may be a few electron volts higher than E_{th}) and decreases thereafter. The peak value of σ and the width of the $\sigma = \sigma(E)$ curve depend on the type of transition. The simplest calculation for electron-impact cross section is made using the Born approximation. The basic assumption here is that there is only a weak electrostatic interaction between the incident electron [which is described by the wave function $\exp(i\mathbf{k}_0 \cdot \mathbf{r})$] and the electrons of the atom, so that the chance of a transition occurring in the atom during impact is very small and the chance of two such transitions may be neglected. In this case the Schrödinger equation for the problem can be linearized. The transition cross section involves a factor of the form $\int u_n^* \exp i[(\mathbf{k}_0 - \mathbf{k}_n) \cdot \mathbf{r}] u_0 \, dV$, where u_0 and u_n are the wavefunctions of the ground and excited states respectively, and $\mathbf{k}_n$ is the wavevector of the scattered electron. It is further assumed that the electron wavelength $\lambda = 2\pi/k_0$ is appreciably larger than the size of the

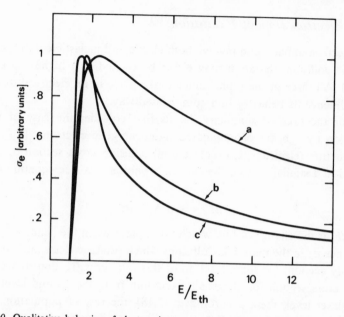

FIG. 3.10. Qualitative behavior of electron-impact excitation cross section versus the energy of the incident electron: (a) optically allowed transition; (b) optically forbidden transition involving no change of multiplicity; (c) optically forbidden transition involving a change of multiplicity. Curves a, b, and c have been derived from those given in Reference 12 for $2p$ and $2s$ transitions in H and 2^3S transition in He, respectively.

atom $[\lambda = (12.26/V)\mathring{A}$, where V is the electron energy in electron volts]. In this case the factor $\exp i[(\mathbf{k}_0 - \mathbf{k}_n) \cdot \mathbf{r}]$ appearing in the above integral can be expanded in a power series about the atom position. One can distinguish three general types of electron impact depending on the type of transition involved: (i) optically allowed transitions; (ii) optically forbidden transitions involving no change of multiplicity; (iii) transitions involving a change of multiplicity.

For optically allowed transitions, we retain only the first nonvanishing term in the expression of $\exp i(\mathbf{k} \cdot \mathbf{r})$ (i.e., $i\mathbf{k} \cdot \mathbf{r}$, where $\mathbf{k} = \mathbf{k}_0 - \mathbf{k}_n$) and this leads to a cross section of the form

$$\sigma_e \propto |\mu|^2 g(E) \tag{3.20}$$

where $|\mu|^2$ is given by (2.43) and $g(E)$ is a function of the electron energy. For an optically allowed transition, the electron impact cross section σ_e is seen to depend on the same matrix element $|\mu|$ which occurs in the expression for the photon absorption cross section. The transition probability for electron impact is thus proportional to the corresponding photon absorption probability. The quantity $g(E)$ turns out to be a relatively slowly varying function of E. The decreasing part of the corresponding $\sigma(E)$ curve in Fig. 3.10 varies as $E^{-1}\ln E$ and the width of the curve may be typically 10 times larger than the threshold energy E_{th} (Fig. 3.10a). The peak value of σ is typically 10^{-16}cm^2.

For optically forbidden transitions involving no change in multiplicity ($\Delta S = 0$, e.g., $1^1S \rightarrow 2^1S$ transition in He, see Fig. 6.4), the Born approximation gives a nonvanishing cross section for the next-higher-order term in the expansion of $\exp i(\mathbf{k} \cdot \mathbf{r})$. The corresponding cross section σ_e can again be expressed as in (3.20). The quantity $|\mu|^2$ is now given by $|e\int u_2^* x^2 u_1\,dx|^2$ rather than $|e\int u_2^* x u_1\,dx|^2$, which of course is zero. The rate of fall of the $g(E)$ curve is somewhat faster than in the previous case, the curve behaving as E^{-1} rather than $E^{-1}\ln E$. Peak values of σ are typically of the order of 10^{-19} cm^2 and the width of the curve now may be only 3–4 times larger than the threshold energy E_{th} (Fig. 3.10b).

When a change of multiplicity is involved (e.g., $1^1S \rightarrow 2^3S$ in He, Fig. 6.4) the Born approximation gives a zero cross section in any order of expansion of $\exp(i\mathbf{k} \cdot \mathbf{r})$. In fact, such a transition involves a spin change while, within the Born approximation, the incoming electron only couples to the orbital motion of the atom.[†] It must be remembered, however, that it is the total spin of the atom plus the incident electron which must be conserved, not necessarily that of the atom alone. Transitions may, therefore, occur via electron *exchange* collisions, wherein the incoming electron replaces the electron of the atom involved in the transition and this electron is in turn ejected by the atom (during the collision, however, the two electrons are quantum mechanically indistinguishable). To conserve spin, the incoming electron must have its spin opposite to that of the ejected one. The peak cross section for this type of transition is usually fairly high ($\sim 10^{-16}$ cm^2). The cross section rises very sharply at threshold and falls off rapidly thereafter. The width of the curve may now be typically equal to or smaller than the value of the threshold energy (Fig. 3.10c).

The discussion so far applies to a monoenergetic beam of electrons. In a gas discharge, however, the electrons will not be monoenergetic but will instead have some particular energy distribution $f(E)$ [$f(E)\,dE$ is the probability for an electron to have its energy lying between E and $E + dE$]. In this case the rate of population of the upper state is obtained from (3.19) by averaging over this distribution, *viz.*,

$$\left(\frac{dN_2}{dt}\right)_p = N_g N_e \langle v\sigma_{e2} \rangle \qquad (3.21)$$

[†]This assumes a negligible spin–orbit coupling, which is true for light atoms (e.g., He, Ne) while it is not true for heavy atoms like Hg.

where

$$\langle v\sigma \rangle = \int v\sigma(E)f(E)\,dE \tag{3.22}$$

If a Maxwellian energy distribution is assumed then $f(E) \propto E^{1/2}$ · $\exp[-(E/kT_e)]$, and the only quantity which needs to be known is the electron temperature T_e. This can be related to the applied electric field $\mathcal{E}$ provided we make the assumption that, at each collision, some fraction δ of the kinetic energy of the electron is lost. If v_{th} is the average thermal electron velocity, the average kinetic energy is about $mv_{th}^2/2$. The collision rate is v_{th}/l, where l is the electron mean free path. The power lost by the electron is therefore $\delta(v_{th}/l)(mv_{th}^2/2)$, and this must be equal to the power delivered by the electric field, which is $e\mathcal{E}v_{drift}$. Since the drift velocity v_{drift} is in turn given by $el\mathcal{E}/mv_{th}$, the power delivered by the electric field is $e^2l\mathcal{E}^2/mv_{th}$. By equating the two above expressions we finally get the following expression for the electron temperature ($T_e = mv_{th}^2/2k$), viz.,

$$T_e = \frac{e}{(2\delta)^{1/2}k}(\mathcal{E}l) \tag{3.23}$$

Since the mean free path l is inversely proportional to the gas pressure p, (3.23) shows that, for a given gas, T_e depends solely on the $\mathcal{E}/p$ ratio. This ratio is the fundamental quantity involved in establishing a given electron temperature, and it is often used in practice as a useful parameter for specifying the discharge conditions. For a given gas mixture there generally exists some value of the $\mathcal{E}/p$ ratio which maximizes the pump rate. Too low a value of $\mathcal{E}/p$ results in too low a value of the electron temperature T_e to excite the laser pump levels effectively. Conversely, too high a value of $\mathcal{E}/p$ (i.e., of T_e) leads to excitation of higher levels of the gas mixture (which may not be so strongly coupled to the laser transition) and to excessive ionization of the gas mixture (which may result in a discharge instability, i.e., a transition from a glow discharge to an arc).

According to (1.10) and (3.21) the pump rate W_p is

$$W_p = N_e\langle v\sigma \rangle \tag{3.24}$$

where $\langle v\sigma \rangle$ is given by (3.22), the electron temperature being expressed through (3.23) as a function of the applied field $\mathcal{E}$. The electron density N_e can then be expressed as a function of the current density J and the drift velocity v_{drift} as

$$N_e = J/ev_{drift} \tag{3.24a}$$

According to the previous calculation v_{drift} can be written as

$$v_{drift} = \frac{el\mathcal{E}}{mv_{th}} = \left(\frac{\delta}{2}\right)^{1/4}\left(\frac{el\mathcal{E}}{m}\right)^{1/2} \tag{3.24b}$$

From (3.24), with the help of (3.24a) and (3.24b) we obtain

$$W_p = \frac{J}{c}\left[\langle v\sigma\rangle \left(\frac{2}{\delta}\right)^{1/4}\left(\frac{m}{el\mathcal{E}}\right)^{1/2}\right] \qquad (3.24c)$$

and the expression in the square brackets depends only on the product $l\mathcal{E}$, i.e., on the $\mathcal{E}/p$ ratio. Since this ratio is generally kept at its optimum value, a change in pump rate is achieved by changing the current density J in the gas discharge.

The calculation given above is a rather crude one since it is based on the assumption of a Maxwellian distribution which in fact is not found to hold in practice.[11] For neutral atom and ion gas lasers, however, the departure from a Maxwellian distribution is not so great and this distribution is therefore often used. However, in molecular gas lasers oscillating on vibrational transitions, the gas is usually weakly ionized and the mean electron energy is low ($E \simeq 1$ eV, since only vibrational states need to be excited) in comparison with that (10–30 eV) needed for neutral atom and ion gas lasers. Accordingly, the assumption of a Maxwellian distribution is expected to be inadequate for molecular lasers. One needs in this case to carry out an *ab initio* calculation to obtain the electron energy distribution $f(E)$. This is done using the appropriate electron transport equation (the Boltzmann equation), and it requires a knowledge of all possible electron collision processes leading to excitation (or de-excitation) of the vibrational or electronic levels of all gas species in the discharge. The calculation therefore gets quite involved, and sometimes it may be impracticable because of the lack of appropriate data on electron collision cross sections. Detailed computer calculations have therefore only been performed for gas mixtures of particular importance such as the CO_2–N_2–He mixture used in high-power CO_2 lasers.[12, 13] These calculations do indeed show a considerable departure from a Maxwellian distribution. However, the average electron temperature and the overall excitation rates still turn out to be, for a given gas mixture, a function only of the ($\mathcal{E}/p$) ratio, as indeed indicated by our crude calculation.

3.3.2 Spatial Distribution of the Pump Rate

In the positive column of a glow discharge the dc electric field, hence the drift velocity v_{drift}, are independent of the discharge current J. It then follows that the spatial dependence of the electron density N_e[see (3.24a)], hence of the pump rate W_p [see (3.24)], is the same as that of the current density J.

In the situation where the gas is contained in a cylindrical tube and the discharge current is flowing along the tube, the radial dependence of J can be analytically specified.[14, 15] For both neutral atom and ion gas lasers, electron–ion recombination can be assumed to occur only at the walls.† Thus, if the ion mean free path is much shorter than the tube radius R, recombination occurs by ambipolar diffusion to the walls. In this case one can apply the Schottky theory for a gas positive column, and the radial distribution of the discharge electrons is predicted to vary as $J_0(2.4r/R)$ where J_0 is the Bessel function of zeroth order. This function is plotted in Fig. 3.11. Note that the electron density drops to zero at the tube walls. Note also that an ion balance equation can be obtained using the condition that the rate of electron–ion pair production must equal the rate of electron–ion pair recombination at the tube walls. This equation leads to a relation between the electron temperature‡ T_e (whose value establishes the ionization rate) and the product pR (whose value, via diffusion, establishes the recombination rate). Accordingly, for a given gas, T_e turns out to be a function of pR only. The ion balance equation thus leads to a relation between T_e and pR in much the same way as the energy balance equation leads to a relation between T_e and $\mathcal{E}/p$ [see (3.23)]. Experimental results have shown that the Schottky theory does hold for noble gas lasers involving neutral atoms and for high-pressure noble gas ion lasers. It is also interesting to note that a Bessel-like radial behavior of the electron density in the discharge has also been used to give accurate predictions of the radial distribution of inversion in a CO_2 laser.[16]

When the ion mean free path becomes comparable with the tube radius (as happens in the relatively low-pressure ion gas lasers), electrons and ions reach the walls by free flight rather than by diffusion. In this case one should use the "free-fall" model of Tonks–Langmuir for the plasma discharge.[17] In this case the radial distribution of the discharge electrons, although no longer given by a Bessel function, still has a bell-shaped form (Fig. 3.11). Note also that the ion balance equation again leads to a relation between the electron temperature and the pR product.

When the gas is excited by a current flowing transversely to the resonator axis (as, for example, with two electrodes placed along the resonator axis, see Fig. 6.15) a reliable prediction of the spatial distribution of the pump rate becomes difficult. In fact the distribution is affected by

---

†Ion–electron $(e + A_i)$ recombination within the discharge volume is an unlikely process since it would require the recombination energy to be removed (radiatively) within the short duration of the collision. A three-body process $e + A_i + M$ where the excess energy is given to the third partner, M, is also unlikely at the gas pressure used (a few Torr).
‡A Maxwellian distribution is assumed in the Schottky theory.

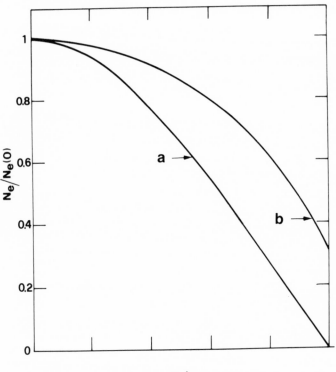

FIG. 3.11. Radial behavior of the electron density for a gas contained in a cylindrical tube (longitudinal discharge): (a) Schottky theory (high-pressure gas); (b) Tonks–Langmuir theory (low-pressure gas).

the shape of the electrodes, by the type and geometry of the auxiliary ionizing sources which are sometimes used, and by the flow conditions of the gas mixture in the discharge chamber. Experimental measurements of the resulting population inversion have indicated a rather nonuniform and asymmetric pump distribution for this type of discharge (a 50% variation of pump rate from center to periphery of the discharge channel may typically be observed).

3.3.3 Pumping Efficiency

As we have seen from the preceding discussion, the electrical pumping of gas lasers is a very complicated process, and a closed expression for the pump rate (such as that obtained for the optical pumping case) cannot, in general, be given. Just as in the optical pumping case, however, we can again define an overall pumping efficiency η_p as the ratio between the minimum power required to produce a given inversion (i.e., $\langle W_p \rangle N_g V \hbar \omega_p$, where $\langle W_p \rangle$ is the average value of W_p over the discharge volume V, and $\hbar \omega_p$ is the energy of the upper laser level) and the electrical power input P to the discharge. Therefore we can write

$$\langle W_p \rangle = \eta_p \frac{P}{VN_g \hbar \omega_p} \tag{3.25}$$

Note that it has been assumed that only one pump level (of energy $\hbar \omega_p$) plays a role, and η_p has therefore been defined in a slightly different way from that for the case of optical pumping [compare (3.25) with (3.15)]. Calculations of η_p are available in the literature for a few gas mixtures of notable interest. In particular, for a $CO_2:N_2:He$ (1:1:8) gas mixture and for an average electron energy of 1 eV, η_p may be as high as 70%.[12, 18]

3.3.4 Excitation by (Near) Resonant Energy Transfer[8,10]

In this case, too, the phenomenon can be described by a suitable collision cross section σ_{AB}:

$$\left(\frac{dN}{dt} \right)_{AB} = N_A N_B v \sigma_{AB} \tag{3.26}$$

where $(dN/dt)_{AB}$ is the number of transitions per unit volume per unit time for the process (3.17), N_A is the upper state population of atoms A, N_B is the lower state population of atoms B, and v is the (relative) velocity of the two atoms. For a gas at temperature T, the quantity $v\sigma_{AB}$ must be averaged over the velocity distribution.

The behavior of σ_{AB} versus the energy defect ΔE between the two levels deserves some comment. Since we are dealing with a resonant process we would expect $\sigma_{AB}(\Delta E)$ to be a sharply peaked function of ΔE, with the maximum obviously occurring at $\Delta E = 0$. In this excitation process, what actually occurs physically is as follows: When atom A approaches atom B, the latter will be subjected to a potential energy of either attractive (see Fig. 2.22) or repulsive (see Fig. 6.21) type. We shall denote this potential by $U(\mathbf{r}, \mathbf{R})$ where $\mathbf{r}$ refers to the electron coordinates and $\mathbf{R}$ the nuclear coordinates of the two-atom system (see also section 2.9.3). The relative motion of the two atoms [i.e., $\mathbf{R} = \mathbf{R}(t)$] therefore produces a time-varying potential $U(\mathbf{r}, t)$.[†] This term will act as a time-dependent Hamiltonian $\mathcal{H}_u(\mathbf{r}, t)$ which couples together the translational and internal motions of

---

[†]Note that, when the two colliding species are atoms, the only nuclear coordinate of interest is the internuclear distance. When, however, the colliding species are molecules, the interaction potential will also depend on the mutual orientation of the two molecules.

the two-atom system. A time-dependent perturbation analysis shows[8] that the transition cross section σ_{AB} can be written as

$$\sigma_{AB} \propto \left| \int_{-\infty}^{+\infty} H_u'(t) \exp(i\omega_{if}t)\, dt \right|^2 \tag{3.27}$$

where $H_u'(t) = \int \psi_f^*(\mathbf{r})\mathcal{K}_u(\mathbf{r},t)\psi_i(\mathbf{r})\, d\mathbf{r}$ is the transition matrix element between the initial state ψ_i [species A in the excited state and B in the gound state] and the final state ψ_f [species A in the ground state and B in the excited state]. In equation (3.27) ω_{if} is given by $\omega_{if} = \Delta E/\hbar$, where ΔE is the energy defect of the resonant process (see Fig. 3.9). The energy transfer cross section σ_{AB} is thus proportional to the power spectrum $|H_u(\omega_{if})|^2$ of the matrix element $H_u'(t)$ at the frequency $\Delta E/\hbar$. We can therefore say that σ_{AB} is established by the Fourier transform $U(\mathbf{r},\omega)$ of the time-varying potential $U(\mathbf{r},t)$ at the frequency ω_{if} required to induce the transition. Since $U(\mathbf{r},t)$ is expected to be nonzero only for a time of the order of the collision duration $\Delta\tau_c$ [given by (2.101)], its Fourier transform is expected to have a bandwidth of the order of $1/\Delta\tau_c$. More precisely, for binary collisions, the frequency behavior of both $|H_u'(\nu)|^2$ and σ_{AB} can be shown to be of the form $\exp(-\nu\Delta\tau_c)$. Thus σ_{AB} is resonantly large over a range ΔE_r of the energy defect ΔE, given by

$$\Delta E_r = \frac{h}{\Delta\tau_c} \tag{3.28}$$

In the case of Ne one has $\Delta\tau_c \simeq 10^{-13}$ s [see (2.103)], and from (3.28) we find $\Delta E_r = 0.006$ eV. Note that this value is appreciably smaller than $kT (\simeq 0.025$ eV at room temperature). For an energy defect ΔE less than ΔE_r, σ_{AB} may be as large as 10^{-14} cm^2. Therefore near-resonant collisions provide a very convenient way of selectively populating a given transition.

PROBLEMS

3.1 A ruby rod with 6.3 mm diameter is pumped by a helical flashlamp of $\sim$2 cm diameter. Calculate the pump transfer efficiency.

3.2 A laser rod in an elliptical pumping chamber has rough ground sides in order to achieve a uniform pump distribution. The flashlamp and rod diameters are assumed equal. Let I_λ be the lamp spectral intensity, S the lateral surface, and V the volume of the active material. By considering only radially propagating rays, show that the (average) pump rate is

$$W_p = \frac{\eta_t}{N_g V} \int \eta_q (1 - e^{-2\alpha R}) \frac{SI_\lambda\, d\lambda}{\hbar\omega}$$

$$= \frac{S\eta_t}{N_g V} \int \eta_q [e^{\alpha R} - e^{-\alpha R}] e^{-\alpha R} \frac{I_\lambda\, d\lambda}{\hbar\omega}$$

Show that, if we assume $\exp(\alpha R) - \exp(-\alpha R) \simeq 2\alpha R$ and $\exp(-\alpha R) \simeq f_1$, the above expression reduces to (3.13).

3.3 Using (3.14) and (3.15) show that $\eta_p = 2\eta_t\eta_r\int \eta_q \alpha R \langle f_1\rangle (\lambda/\lambda_0) g_\lambda \, d\lambda$, where $\langle f_1\rangle$ is the average of f_1 over the rod cross section.

3.4 Show that the power quantum efficiency η_{pq} is given by

$$\eta_{pq} = \frac{\int W_p N_g \hbar\omega_0 \, dV}{\int (dP_\lambda/dV) \, d\lambda \, dV}$$

where the volume integrals are taken over the rod volume. With the help of (3.11), (3.14), and (3.2) show that

$$\eta_{pq} = \frac{\int\int \eta_q \sigma \langle f_1\rangle (\lambda/\lambda_0) g_\lambda \, d\lambda}{\int \sigma \langle f_1\rangle g_\lambda \, d\lambda}$$

where $\langle f_1\rangle$ is the average of f_1 over the rod cross section.

3.5 Using the results obtained in Problems 3.3 and 3.4, show that $\eta_p = \eta_r\eta_t\eta_{pq}\eta_a$, where the absorption efficiency η_a is given by

$$\eta_a = 2\int \alpha R \langle f_1\rangle g_\lambda \, d\lambda$$

3.6 For radially propagating rays, using the expression for W_p given in Problem 3.2, show that $\eta_{pq} = \int \eta_q h(\lambda)(\lambda/\lambda_0) g_\lambda \, d\lambda / \int h(\lambda) g_\lambda \, d\lambda$ and $\eta_a = \int h(\lambda) g_\lambda \, d\lambda$, where $h(\lambda) = 1 - \exp(-2\alpha R)$.

3.7 With the help of Fig. 3.8 calculate $\langle f_1\rangle$ for each value of αR.

REFERENCES

1. W. Koechner, *Solid-State Laser Engineering* (Springer-Verlag, New York, 1976), Chapter 6, Springer Series in Optical Sciences, Vol. 1.
2. D. Ross, *Lasers, Light Amplifiers, and Oscillators* (Academic Press, New York, 1969), Chapter 14.
3. O. Svelto, *Appl. Opt.* **1**, 745 (1962).
4. G. E. Devlin, J. McKenna, A. D. May, and A. L. Schawlow, *Appl. Opt.* **1**, 11 (1962).
5. C. H. Cooke, J. McKenna, and J. R. Skinner, *Appl. Opt.* **3**, 957 (1964).
6. W. Kaiser, C. G. B. Garret, and D. L. Wood, *Phys. Rev.* **123**, 766 (1961).
7. W. Koechner, *Solid-State Laser Engineering* (Springer-Verlag, New York, 1976), pp. 85, 105, and 324.
8. C. K. Rhodes and A. Szoke, "Gaseous Lasers: Atomic, Molecular and Ionic," in *Laser Handbook*, ed. by F. T. Arecchi and E. O. Schultz-Dubois (North-Holland Publishing Co., Amsterdam, 1972), Vol. I, pp. 265–324.
9. C. S. Willett, *An Introduction to Gas Lasers: Population Inversion Mechanisms* (Pergamon Press, Oxford, 1974).
10. H. S. W. Massey and E. H. S. Burhop, *Electronic and Ionic Impact Phenomena* (Oxford University Press, London, 1969), Vols. I and II.

11. C. S. Willett, *An Introduction to Gas Lasers*: *Population Inversion Mechanisms* (Pergamon Press, Oxford, 1974), pp. 84, 280, and 327.
12. W. L. Nighan, *Phys. Rev. A* **2**, 1989 (1970).
13. K. Smith and R. M. Thomson, *Computer Modeling of Gas Lasers* (Plenum Press, New York, 1978).
14. C. E. Webb, in *High-Power Gas Lasers*, ed. by E. R. Pike (The Institute of Physics, Bristol, 1976), pp. 1–28.
15. C. S. Willet, *An Introduction to Gas Lasers*: *Population Inversion Mechanisms* (Pergamon Press, Oxford, 1974), Section 3.2.2.
16. P. K. Cheo, in *Lasers*, ed. by A. K. Levine and A. J. DeMaria (Marcel Dekker, New York, 1971), Chapter 2.
17. C. C. Davis and T. A. King, in *Advances in Quantum Electronics*, ed. by D. W. Goodwin (Academic Press, New York, 1975), pp. 170–437.
18. A. J. DeMaria, in *Principles of Laser Plasmas*, ed. by G. Bekefi (Wiley-Interscience, Inc., New York, 1976), Chapter 8.
19. C. Bowness, *Appl. Opt.* **4**, 103 (1965).

4
Passive Optical Resonators

4.1 INTRODUCTION

This chapter deals with the theory of passive optical resonators. What we mean by a passive optical resonator is a cavity consisting of reflecting surfaces and containing a homogeneous, isotropic, and passive dielectric medium. We recall that a mode of a resonator was defined in Section 2.1 as a stationary e.m. field configuration which satisfies both Maxwell's equations and the boundary conditions. The electric field of this configuration can then be written as

$$\mathbf{E}(\mathbf{r}, t) = E_0 \mathbf{u}(\mathbf{r}) \exp(i\omega t) \tag{4.1}$$

where $\omega/2\pi$ is the mode frequency.

The resonators used in the laser field differ from those used in the microwave field in two main aspects: (i) Laser resonators are usually open, i.e., no lateral surface is used. (ii) The resonator dimensions are much greater than the laser wavelength. Since this wavelength usually ranges from a fraction of a micron to a few tens of microns, a laser cavity with dimensions comparable to these wavelengths would, in fact, have too low a gain to allow laser oscillation. The above-mentioned properties (i) and (ii) have a considerable effect on the performance of an optical resonator. For example, the fact that the resonator is open means that, for any cavity mode, there will inevitably be some losses. These losses are due to diffraction of the e.m. field, which leads to some fraction of the energy leaving the sides of the cavity. They are, therefore, known as *diffraction losses*. Strictly speaking, therefore, the mode definition (4.1) cannot be applied to an open optical resonator, and true modes (i.e., stationary configurations) do not exist for such a resonator. In what follows, however, we shall see that e.m. standing-wave configurations having very small losses do exist in open resonators. We will therefore define as a mode (sometimes called a quasi-mode) an e.m. configuration whose electric field can be written as

$$E(\mathbf{r}, t) = E_0 \mathbf{u}(\mathbf{r})\exp\left[(-t/2\tau_c) + i\omega t\right] \qquad (4.2)$$

Here τ_c (the decay time of the square of the electric field amplitude) is called the cavity photon decay time. The property (ii) mentioned above means that, as we shall see later, the cavity resonant frequencies are closely spaced. Indeed, according to (2.14), the number of resonator modes N falling within the width $\Delta\nu_0$ of a laser line is given by $N = 8\pi\nu^2 V\Delta\nu_0/c^3$. As an example, if we take $\nu = 5 \times 10^{14}$ Hz (center of visible range), $V = 1\,\mathrm{cm}^3$, and $\Delta\nu_0 = 1.7 \times 10^9$ Hz [width of the 0.6328 μm Ne Doppler line, see (2.114)] we get $N \simeq 4 \times 10^8$ modes. If the resonator were closed, all these modes would have similar losses, and with this resonator used as a laser cavity, oscillation would occur on a very large number of modes. Such behavior would be undesirable since it would result in light from the laser being emitted in a wide spectral range and in all directions. This problem can be overcome to a large extent by the use of open resonators. In such resonators, only the very few modes corresponding to a superposition of waves traveling nearly parallel to the resonator axis will have low enough losses to allow laser oscillation. For all other modes the corresponding waves will be almost completely lost after a single pass through the resonator. This is the fundamental reason why open resonators are used in lasers.[†] Although the absence of lateral surfaces means that far fewer modes can oscillate, the number of modes oscillating may still be appreciably larger than unity, as we shall see later on.

    The most widely used laser resonators have either plane or spherical mirrors of rectangular (or more often circular) shape separated by some distance L. Typically, L may range from a few centimeters to a few tens of centimeters, while the mirror dimensions range from a fraction of a centimeter to a few centimeters. Of the various possible resonators we make particular mention of the following types:

    (*i*) *Plane-Parallel* (*or Fabry–Perot*) *Resonator* (*Fig.* 4.1). This consists of two plane mirrors set parallel to one another. To a first approximation the modes of this resonator can be thought of as the superposition of two plane e.m. waves propagating in opposite directions along the cavity axis, as shown schematically in Fig. 4.1. Within this approximation, the resonant

---

[†]The open resonator configuration is also useful for reasons of convenience: In the case of a flashlamp pumped laser, for example, a lateral surface would interfere with the pumping.

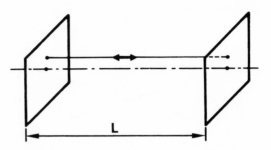

FIG. 4.1. Plane-parallel resonator.

frequencies can be readily obtained by imposing the condition that the cavity length L must be an integral number of half-wavelengths, i.e., $L = n(\lambda/2)$, where n is a positive integer. This is a necessary condition for the electric field of the e.m. standing wave to be zero on the two mirrors. It then follows that the resonant frequencies are given by

$$\nu = n(c/2L) \tag{4.3}$$

It is interesting to note that the same expression (4.3) can also be obtained by imposing the condition that the phase shift of a plane wave due to one round-trip through the cavity must equal an integral number times 2π, i.e., $2kL = 2n\pi$. This condition is readily obtained by a self-consistency argument. If the frequency of the plane wave is equal to that of a cavity mode, the phase shift after one round trip must be zero (apart from an integral number of 2π) since only in this case will the amplitudes at any arbitrary point, due to successive reflections, add up in phase so as to give an appreciable total field.

(*ii*) *Concentric (or Spherical) Resonator (Fig.* 4.2). This consists of two spherical mirrors having the same radius R and separated by a distance L such that the mirror centers of curvature C_1 and C_2 are coincident (i.e., $L = 2R$). The geometrical-optics picture of the modes of this resonator is also shown in the figure. In this case the modes are approximated by a

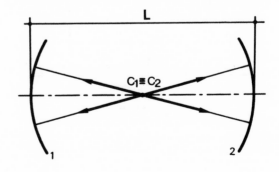

FIG. 4.2. Concentric (spherical) resonator.

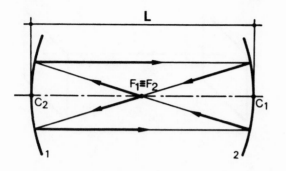

FIG. 4.3. Confocal resonator.

superposition of two oppositely traveling spherical waves originating from the point C. The application of the above self-consistency argument again leads to (4.3) as the expression for the resonant frequencies.

(*iii*) *Confocal Resonator* (*Fig.* 4.3). This consists of two spherical mirrors of the same radius of curvature R and separated by a distance L such that the mirror foci F_1 and F_2 are coincident. It then follows that the center of curvature C of one mirror lies on the surface of the second mirror (i.e., $L = R$). From a geometrical-optics point of view, we can draw a closed optical path as shown in Fig. 4.3. This path does not give any indication of what the mode configuration will be, however, and we shall see that in fact this configuration cannot be described either by a plane or by a spherical wave. For the same reason, the resonant frequencies cannot be readily obtained from geometrical-optics considerations.

(*iv*) *Resonators Using a Combination of Plane and Spherical Mirrors.* Examples of these are shown in Fig. 4.4 (hemiconfocal resonator) and Fig. 4.5 (hemispherical resonator).

Resonators formed by two spherical mirrors of the same radius of curvature R and separated by a distance L such that $R < L < 2R$ (i.e.,

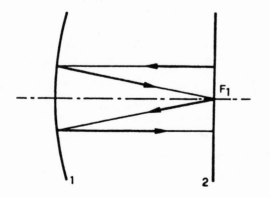

FIG. 4.4. Hemiconfocal resonator.

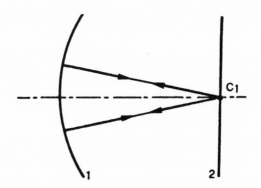

FIG. 4.5. Hemispherical resonator.

placed at an intermediate position between the confocal and concentric one) are also often used. In addition, we can have $L < R$. For these cases it is not generally possible to use a ray description in which a ray retraces itself after one or a few passes.

All of these resonators can be considered as particular examples of a general resonator consisting of two spherical mirrors of different radius of curvature (either positive or negative) spaced by some arbitrary distance L. These various resonators can be divided into two categories, namely, *stable* resonators and *unstable* resonators. A resonator will be described as unstable when an arbitrary ray, in bouncing back and forth between the two mirrors, will diverge indefinitely away from the resonator axis. An obvious example of an unstable resonator is shown in Fig. 4.6. Conversely, a resonator for which the ray remains bounded will be described as a stable resonator.

The purpose of the following sections is to calculate the mode configurations and the corresponding resonant frequencies and diffraction losses for the most commonly used resonators.

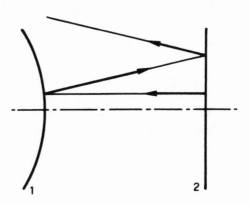

FIG. 4.6. Example of an unstable resonator.

4.2 PLANE-PARALLEL RESONATOR

4.2.1 Approximate Treatment of Schawlow and Townes[1]

The first study of a plane-parallel resonator appeared in the classic work of Schawlow and Townes,[1] in which they proposed an extension of the maser concept into the optical frequency range. Schawlow and Townes gave an approximate treatment of the problem in which they used an analogy with a closed rectangular cavity, whose solution is well known (see Section 2.1).

Before presenting the treatment of Schawlow and Townes, we should recall that the E-field components of the modes of a rectangular cavity such as shown in Fig. 2.1 can be written as

$$E_x = e_x \cos k_x x \sin k_y y \sin k_z z \sin \omega t$$

$$E_y = e_y \sin k_x x \cos k_y y \sin k_z z \sin \omega t \qquad (4.4)$$

$$E_z = e_z \sin k_x x \sin k_y y \cos k_z z \sin \omega t$$

where $k_x = l\pi/2a$, $k_y = m\pi/2a$, $k_z = n\pi/L$ (l, m, n being positive integers) and where the resonant frequencies are given by

$$\nu = \frac{c}{2} \left[\left(\frac{n}{L} \right)^2 + \left(\frac{m}{2a} \right)^2 + \left(\frac{l}{2a} \right)^2 \right]^{1/2} \qquad (4.5)$$

Note that (4.4) can be put in complex form by expressing the sine and cosine functions in terms of exponential functions. When this is done, each E-field component can be seen to be expressed as the sum of eight terms of the form $\exp[i(\pm k_x x \pm k_y y \pm k_z z - \omega t) + \text{c.c.}]$, i.e., as the sum of eight plane waves propagating along the directions of the eight wave vectors having components $\pm k_x$, $\pm k_y$, and $\pm k_z$. The direction cosines of these vectors are, therefore, $\pm(l\lambda/4a)$, $\pm(m\lambda/4a)$, and $\pm(n\lambda/2L)$, where λ is the wavelength of the given mode. The superposition of these eight plane waves gives the standing wave of (4.4).

Now, Schawlow and Townes assumed that, to a good approximation, the modes of the open cavity of Fig. 4.1 are described by those modes of the rectangular cavity of Fig. 2.1 having $(l, m) \ll n$ (the cavity of Fig. 4.1 being obtained from that of Fig. 2.1 by removing the lateral surface). The justification of this assumption can be seen when we note that, from what has been said above, the modes of this cavity can be expressed as the superposition of plane waves propagating at a very small angle to the z axis. Therefore, the removal of the lateral surface is not expected to drastically change these modes. On the other hand, those modes which correspond to values of l and m which are not small compared to n will be greatly affected by the removal of the resonator sides. Once the sides are removed, however, these modes have such high diffraction losses that they need not be considered further.

With the assumption that $(l, m) \ll n$, the resonant frequencies of the plane-parallel cavity can be obtained from (4.5) by a power series expansion of the expression within the square root, namely:

$$\nu \simeq \frac{c}{2} \left(\frac{n}{L} + \frac{1}{2} \frac{(l^2 + m^2)}{n} \frac{L}{4a^2} \right) \tag{4.6}$$

This expression can be compared with (4.3), which was derived using a simple one-dimensional argument. There is a well-defined cavity mode with a well-defined resonant frequency for each set of values of the three quantities l, m, and n.

The frequency difference between two modes having the same values of l and m and whose n values differ by 1 is

$$\Delta \nu_n = c/2L \tag{4.7}$$

as one can find immediately from (4.6). These two modes differ only in their field distribution along the z axis (i.e., longitudinally). For this reason $\Delta \nu_n$ is often referred to as the frequency difference between two consecutive *longitudinal modes*. The frequency difference between two modes which differ only by having a difference of unity in their m values (i.e., the frequency difference between two consecutive *transverse modes*)[†] is

$$\Delta \nu_m = \frac{cL}{8na^2} \left(m + \frac{1}{2} \right) \tag{4.8}$$

For typical values of L, $\Delta \nu_n$ is of the order of a few hundreds of megahertz while $\Delta \nu_m$ (or $\Delta \nu_l$) is of the order of a few megahertz. Figure 4.7 shows the frequency spectrum of a plane-parallel resonator. Note that modes having

---

[†] The usage of the terms "longitudinal mode" and "transverse mode" in the laser literature has sometimes been rather confusing, and can convey the (mistaken) impression that there are two distinct types of modes, *viz.* longitudinal modes (sometimes called axial modes) and transverse modes. In fact any mode is specified by three numbers, e.g., n, m, l of (4.5). The electric and magnetic fields of the modes are nearly perpendicular to the resonator axis. The variation of these fields in a transverse direction is specified by l, m while the field variation in a longitudinal (i.e., axial) direction is specified by n. When one refers, rather loosely, to a (given) transverse mode, it means that one is considering a mode with given values for the transverse indexes (l, m), regardless of the value of n. Accordingly a single transverse mode means a mode with a single value of the transverse indexes (l, m). A similar interpretation can be applied to the "longitudinal modes." Thus two consecutive longitudinal modes mean two modes with consecutive values of the longitudinal index n [i.e., n and $(n + 1)$ or $(n - 1)$].

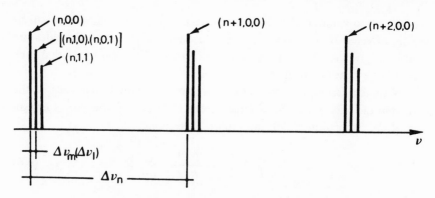

FIG. 4.7. Resonance frequencies of a plane-parallel resonator.

the same n but with different l and m values satisfying $l^2 + m^2 = \text{const}$ have the same frequency and are therefore said to be frequency degenerate.

So far we have not considered the cavity losses, and the cavity resonances have been assumed to be infinitely narrow. Actually, as already pointed out, optical resonators have unavoidable losses due to diffraction. A mode can, therefore, be represented as in (4.2), and this means that its resonance will have a linewidth (FWHM) given by

$$\Delta\omega_c = 1/\tau_c \qquad (4.9)$$

as can be shown by taking the Fourier transform of (4.2).

4.2.2 Fox and Li Treatment[2]

A more rigorous treatment of a plane-parallel resonator has been given by Fox and Li,[2] who studied the problem under the so-called scalar approximation, which is often used in optics. In this approximation, the e.m. field is assumed to be nearly transverse and uniformly (e.g., linearly or circularly) polarized. The field can then be described by a scalar quantity U representing, for instance, the magnitude of the electric field (or of the magnetic field). If we let U_1 be some arbitrary field distribution on mirror 1 (Fig. 4.8). This distribution will, due to diffraction, produce a field distribution on mirror 2 whose expression can be obtained by the Kirchhoff diffraction integral.[3] The field $U_2(P_2)$ at a general point P_2 of mirror 2 is then given by

$$U_2(P_2) = -\frac{i}{2\lambda} \int_1 \frac{U_1(P_1)\exp(ikr)(1 + \cos\theta)}{r}\, dS_1 \qquad (4.10)$$

where r is the distance between points P_1 and P_2, θ is the angle that $P_1 P_2$

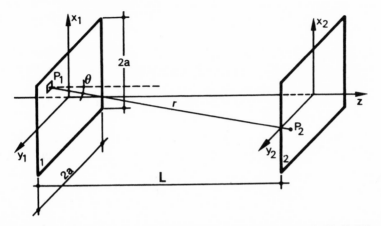

FIG. 4.8. Mode calculation for a plane-parallel resonator by use of the Kirchhoff diffraction integral.

makes with the normal to the surface at P_1, dS_1 is a surface element around P_1, and $k = 2\pi/\lambda$. The integral in (4.10) must be evaluated over the whole of surface 1.

Instead of considering a general distribution U_1, let us consider a distribution U corresponding to a cavity mode. In this case, if the two mirrors are identical, the field distribution on mirror 2, as calculated by (4.10), must again be equal to U apart from some constant factor. According to (4.10) we must, therefore, have

$$\sigma U(P_2) = -\frac{i}{2\lambda} \int_1 \frac{U(P_2)\exp(ikr)(1 + \cos\theta)}{r}\, dS_1 \qquad (4.11)$$

where σ is a constant. Equation (4.11) is a Fredholm homogeneous integral equation of the second kind. Its eigensolutions U give the cavity-mode field distributions over the mirrors. Since the integral operator of (4.11) is non-Hermitian, the eigenvalues σ are not real, and both the amplitude and phase have straightforward physical meanings. If we put $\sigma = |\sigma|\exp(i\phi)$, we can immediately see that $\gamma_d = 1 - |\sigma|^2$ gives the fractional power loss per pass due to diffraction. The quantity ϕ gives the phase delay of the wave in propagating from one mirror to the other, as can be more readily seen when it is realized that the time factor $\exp(i\omega t)$ has been omitted from both sides of (4.10) and (4.11). The quantity 2ϕ, therefore, gives the phase delay in one round-trip and it will be a function of k, i.e., of the wavelength. Upon equating 2ϕ to an integral number times 2π, we obtain the resonance frequencies (as already discussed for a simple case in Section 4.1). So we see that the eigensolutions and corresponding eigenvalues of (4.11) give all the quantities of interest, namely, field distribution on the mirrors, resonance frequencies, and diffraction losses. Once the field distribution U on the mirrors is known, it is possible through (4.10) to calculate the field distribution at any point inside (standing wave) or outside (traveling wave) the resonator.

When $L \gg a$, i.e., when the cavity length is much greater than its transverse dimensions, (4.11) can be considerably simplified. In fact we can put $\cos\theta \simeq 1$ and $r \simeq L$ in the amplitude factor appearing under the integral sign. To get a suitable approximate expression for the phase factor kr, we write r as

$$r = \left[L^2 + (x_1 - x_2)^2 + (y_1 - y_2)^2 \right]^{1/2}$$
$$= L + (1/2L)\left[(x_1 - x_2)^2 + (y_1 - y_2)^2 \right] + \epsilon \qquad (4.12)$$

where a power expansion of the expression appearing under the square root has been made. One can neglect ϵ, the remainder of the power series, provided that $k\epsilon \ll 2\pi$. Since ϵ consists of a converging series having terms of alternating sign, it follows that its value is smaller than the magnitude of the first term. It therefore follows that, for the condition $k\epsilon \ll 2\pi$ to be satisfied, it is sufficient that $ka^4/L^3 \ll 2\pi$ or, in terms of the Fresnel number[†] $N = a^2/L\lambda$, we require $N \ll L^2/a^2$. So, given the two assumptions $L \gg a$ and $N \ll L^2/a^2$, we can then write,

$$\exp(ikr) \simeq \exp\left\{ (ikL) + i(\pi N/a^2)\left[(x_1 - x_2)^2 + (y_1 - y_2)^2 \right] \right\} \qquad (4.13)$$

By using the dimensionless quantities

$$\xi = \left(\sqrt{N}/a \right)x$$
$$\eta = \left(\sqrt{N}/a \right)y \qquad (4.14)$$

and with the help of (4.13), we can now put (4.11) in the dimensionless form

$$\sigma^* U(\xi_2, \eta_2) = -i \int_1 U(\xi_1, \eta_1)\exp\left\{ i\pi\left[(\xi_1 - \xi_2)^2 + (\eta_1 - \eta_2)^2 \right] \right\} d\xi_1 d\eta_1$$
$$(4.15)$$

[†]The Fresnel number N is a dimensionless quantity often used in diffraction optics. A physical interpretation of this number can be obtained as follows. A plane e.m. wave of transverse dimension $2a$ has an angular spread due to diffraction $\theta_d \simeq \lambda/2a$ [see (1.11)]. On the other hand, for mirrors of transverse dimensions $2a$ and spaced by L, the geometrical angle θ_g subtended by one mirror at the center of the other is $\theta_g = a/L$. We then see that $N = \theta_g/2\theta_d$. High Fresnel numbers thus imply a diffraction spread small compared to the geometrical angle.

where we have defined

$$\sigma^* = \sigma \exp(-ikL) \tag{4.16}$$

For mirrors of square or rectangular shape, it is possible to separate the variables in (4.15). We in fact put

$$U(\xi, \eta) = U_\xi(\xi) U_\eta(\eta) \tag{4.17}$$

$$\sigma^* = \sigma_\xi^* \sigma_\eta^* \tag{4.18}$$

Then (4.15) gives the following two equations for $U_\xi(\xi)$ and $U_\eta(\eta)$:

$$\sigma_\xi^* U_\xi(\xi_2) = \exp\left[-i(\pi/4)\right] \int_{-\sqrt{N}}^{+\sqrt{N}} U_\xi(\xi_1) \exp\left[i\pi(\xi_1 - \xi_2)^2\right] d\xi_1 \tag{4.19a}$$

$$\sigma_\eta^* U_\eta(\eta_2) = \exp\left[-i(\pi/4)\right] \int_{-\sqrt{N}}^{+\sqrt{N}} U_\eta(\eta_1) \exp\left[i\pi(\eta_1 - \eta_2)^2\right] d\eta_1 \tag{4.19b}$$

It can be shown that the function U_ξ gives the field distribution for a resonator consisting of two plane mirrors with dimension $2a$ in the x direction and infinitely long in the y direction (strip mirrors). A similar interpretation holds for U_η. We will label the eigenfunctions and the eigenvalues of (4.19a) and (4.19b) by the corresponding m and l values, respectively. Therefore, according to (4.17) and (4.18), we will have

$$U_{ml}(\xi, \eta) = U_{\xi m}(\xi) U_{\xi l}(\xi) \tag{4.20}$$

$$\sigma_{ml}^* = \sigma_{\xi m}^* \sigma_{\eta l}^* \tag{4.21}$$

For circular mirrors, the treatment is somewhat similar. In this case, however, it is more convenient to express (4.11) as a function of cylindrical rather than rectangular coordinates, and the variables can again be separated in this coordinate system.

Although equations (4.19) are much simpler than the original equation (4.11), they are not amenable to an analytical solution. They have been solved by Fox and Li[2] with the help of a computer, for several values of the Fresnel number N. They used an iterative procedure based on the following physical argument. Let us consider a wave traveling back and forth in the cavity and assume that, at a given time, the field distribution $U_1(\xi_1)$ on mirror 1 is known. The field distribution $U_2(\xi_2)$ on mirror 2 which results from the field distribution U_1 can then be calculated through (4.19a). In fact, if we replace the function $U_\xi(\xi_1)$ in the right-hand side of (4.19a) by the function U_1 and then perform the integration, we will obtain the function $U_2 = U_\xi(\xi_2)$ which results from the first transit. Once U_2 is known, we can then calculate the new field distribution on mirror 1 due to the second transit, and so on. Fox and Li have shown that, after a sufficient number of passes, regardless of the initial field distribution on mirror 1, a

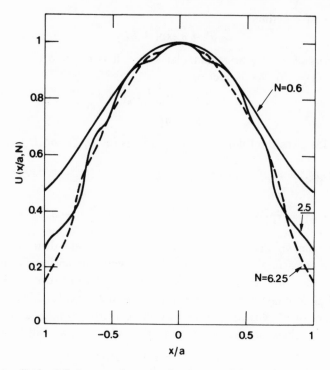

FIG. 4.9. Amplitude of the lowest-order mode of a plane-parallel resonator for three values of the Fresnel number (after Fox and Li[2]).

field distribution is reached which does not change any more from pass to pass. This distribution will then be an eigensolution of (4.19). This procedure also allows one to calculate the eigenvalue and hence, as explained above, the diffraction loss and resonant frequency of the given mode. If the initial field distribution is chosen to be an even function of ξ, one ends up with an even mode, while the odd modes are obtained by choosing the initial field distribution to be an odd function of ξ. As an example, Fig. 4.9 shows the results obtained for the amplitude of $U = U(x/a, N)$ when U_1 is initially chosen to be a uniform and symmetric field distribution (i.e., $U_1 = $ const). For the case $N = 6.25$, approximately 200 passes are needed to reach the stationary solution, as shown in Fig. 4.10. In a similar way, the lowest-order antisymmetric mode is obtained when one chooses a uniform and antisymmetric initial distribution (i.e., $U_1 = 1$ for $0 < x < a$ and $U_1 = -1$ for $-a < x < 0$). Figure 4.11 shows the field distributions $U(x/a, N)$ obtained in this way for two values of the Fresnel number. The mode which corresponds to the case where
According to (4.20), the overall field distribution $U_{ml}(x, y)$ is given by the product $U_m(x)U_l(y)$. The mode which corresponds to the case where

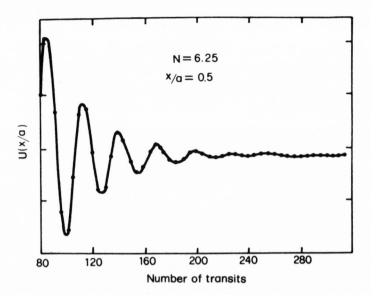

FIG. 4.10. Field amplitude U at the position $x/a = 0.5$ versus the number of transits (after Fox and Li[2]).

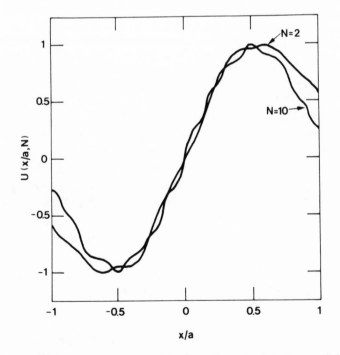

FIG. 4.11. Amplitude of the lowest-order antisymmetric mode of a plane-parallel resonator for two values of the Fresnel number (after Fox and Li[2]).

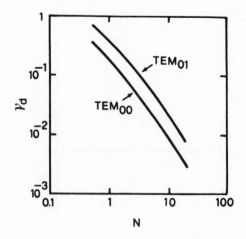

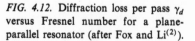

FIG. 4.12. Diffraction loss per pass γ_d versus Fresnel number for a plane-parallel resonator (after Fox and Li[2]).

both $U(x)$ and $U(y)$ are given by the lowest-order (i.e., $m = l = 0$) solution (Fig. 4.9) is called the TEM_{00} mode. The mode TEM_{01} is obtained when $U(x)$ is given by the lowest-order solution ($m = 0$, Fig. 4.9) and $U(y)$ by the next-higher-order solution (i.e., $l = 1$, Fig. 4.11) (and vice versa for the TEM_{10} mode). The letters TEM stand for transverse electric and magnetic field. For these modes, both the electric and magnetic fields of the e.m. wave are orthogonal to the resonator z axis.

It is readily seen from (4.19) and (4.21) that σ^* depends only on the Fresnel number N and on the mode indexes m and l. Accordingly the diffraction losses ($\gamma_d = 1 - |\sigma^*|^2$) will depend only on N, m, and l. Figure 4.12 shows the diffraction losses versus N for the lowest-order symmetric (TEM_{00}) and antisymmetric (TEM_{01}) modes. One can see from the figure that the losses rapidly decrease as N is increased. This can be easily understood when it is remembered that N is proportional to the ratio between geometrical (θ_g) and diffraction (θ_d) angles. This result can also be understood by noticing that, with increasing N, the field at the edge of the mirror ($x = \pm a$) decreases as shown in Figs. 4.9 and 4.11. In fact, it is this field which is mostly responsible for the diffraction losses. Note finally that, for a given Fresnel number, the loss of the TEM_{01} mode is always greater than that of the TEM_{00} mode.

The resonance frequencies are obtained by equating the phase of σ to an integral number times π. Thus, using (4.16), we get

$$kL + \phi_{m,l}^* = n\pi \tag{4.22}$$

where we have explicitly indicated that the phase ϕ^* of σ^* depends on the mode indexes m and l. Note that while k depends only on λ ($k = 2\pi/\lambda$), ϕ^*

depends both on λ (since it depends on the Fresnel number N) and on the mode indexes m and l. Equation (4.22) therefore allows a calculation of the resonance wavelengths λ (and hence the resonance frequencies ν) as a function of the mode indexes n, l, and m. The computer results of Fox and Li for σ^* confirm that, for sufficiently high values of the Fresnel number ($N > 10$), the resonant frequencies which are obtained in this way are in good agreement with the predictions of (4.6).

4.3 CONFOCAL RESONATOR [12]

The treatment of the confocal resonator using the scalar approximation was developed by Boyd and Gordon.[4] In this treatment, we again call the cavity length L and refer the points of the two mirror surfaces to coordinate systems (x_1, y_1) and (x_2, y_2), as shown in Fig. 4.13. For the sake of simplicity, the two mirrors will be taken to have square cross sections of dimension $2a$. In the scalar approximation, the eigensolutions are again given by (4.11). When $L \gg a$, we can again put $\cos\theta \simeq 1$ and $r \simeq L$ in the amplitude factor. To find a suitable approximation for the phase factor kr, we must first calculate the distance between P_1 and P_2 as a function of the coordinates of the two points. When this is done, the resulting expression

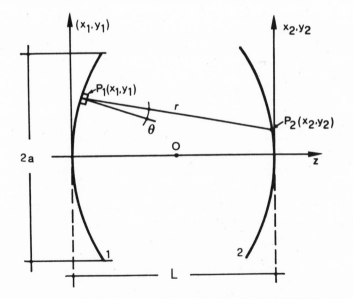

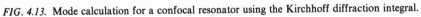

FIG. 4.13. Mode calculation for a confocal resonator using the Kirchhoff diffraction integral.

for r can be expanded in a power series to give

$$r \simeq L - (1/L)(x_1 x_2 + y_1 y_2) \qquad (4.23)$$

This expression provides a good approximation for kr provided that, as in the plane mirror case, the condition $N \ll L^2/a^2$ is satisfied. After introducing the dimensionless variables $\xi = \sqrt{N}(x/a)$ and $\eta = \sqrt{N}(y/a)$, (4.11) reduces to

$$\sigma^* U(\xi_2, \eta_2) = -i \int_1 U(\xi_1, \eta_1) \exp\left[-i2\pi(\xi_1 \xi_2 + \eta_1 \eta_2) \right] d\xi_1 d\eta_1 \qquad (4.24)$$

in which σ^* is again given by (4.16). We again look for a separable solution as in (4.17) and (4.18), which leads to

$$\sigma_\xi^* U_\xi(\xi_2) = \exp\left[-i(\pi/4) \right] \int_{-\sqrt{N}}^{+\sqrt{N}} U_\xi(\xi_1) \exp(-i2\pi\xi_1\xi_2) \, d\xi_1 \qquad (4.25)$$

$$\sigma_\eta^* U_\eta(\eta_2) = \exp\left[-i(\pi/4) \right] \int_{-\sqrt{N}}^{+\sqrt{N}} U_\eta(\eta_1) \exp(-i2\pi\eta_1\eta_2) \, d\eta_1 \qquad (4.26)$$

The physical meaning of the expressions in (4.25) and (4.26) is the same as for the Fabry–Perot resonator: They are the solutions for one-dimensional mirrors (strip mirrors). Equations (4.25) and (4.26) have a discrete set of eigensolutions which we will denote by the indexes m and l, i.e.,

$$U_{m,l}(\xi, \eta) = U_{\xi m}(\xi) U_{\eta l}(\eta) \qquad (4.27a)$$

$$\sigma_{ml}^* = \sigma_{\xi m}^* \sigma_{\eta l}^* \qquad (4.27b)$$

Unlike the plane mirror case, this integral equation can now be solved analytically. It can be shown in fact that $U_{\xi m}(\xi)$ and $U_{\eta l}(\eta)$ are proportional to the Flammer spheroidal angular functions, while the corresponding eigenvalues $\sigma_{\xi m}^*$ and $\sigma_{\eta l}^*$ are proportional to the Flammer spheroidal radial functions. These functions have been tabulated.[5]

As regards the eigenfunctions, a considerable simplification is possible when $N \gg 1$. In this case, the range of integration in (4.25) and (4.26) can be extended to cover the range from $-\infty$ to $+\infty$. In this case the right-hand sides of both (4.25) and (4.26), apart from a proportionality factor, are just the Fourier transforms of U_1 and U_2 respectively. Thus, according to (4.25) and (4.26), the required eigenfunctions must be invariant under a Fourier transform. The product of a Gaussian function with a Hermite polynomial is known to have this property. Returning to the original x and y coordinates, the eigenfunctions are then given by

$$U_{xm}(x) = H_m\left[x\left(\frac{2\pi}{L\lambda} \right)^{1/2} \right] \exp\left[-(\pi/L\lambda)x^2 \right] \qquad (4.28a)$$

$$U_{yl}(y) = H_l\left[y\left(\frac{2\pi}{L\lambda} \right)^{1/2} \right] \exp\left[-(\pi/L\lambda)y^2 \right] \qquad (4.28b)$$

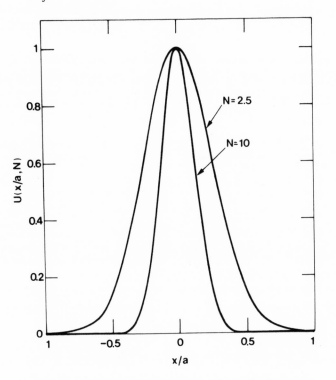

FIG. 4.14. Lowest-order symmetric mode of a confocal resonator.

where H_m and H_l are the Hermite polynomials of mth and lth order. The overall eigenfunction is then

$$U_{ml}(x, y) = H_m H_l \exp\left[-(\pi/L\lambda)(x^2 + y^2) \right] \qquad (4.29)$$

We will now consider a few examples. If $m = 0$, then $H_0 = 1$, and therefore from (4.28a) we have

$$U_{x0}(x) = \exp\left[-(\pi/L\lambda)x^2 \right] \qquad (4.30)$$

Figure 4.14 shows a plot of the behavior of U versus x/a for two values of the Fresnel number N. The electric field amplitude on the mirror is reduced to $1/e$ of its maximum value at a distance w_s from the center, where w_s is given by

$$w_s = (\lambda L/\pi)^{1/2} \qquad (4.31)$$

When $m = 1$, then $H_1 = (8\pi/L\lambda)^{1/2}x$, and Fig. 4.15 shows a normalized plot of U versus x/a for two values of the Fresnel number. Since the overall mode pattern is determined by (4.27a), the lowest-order modes will be as follows:

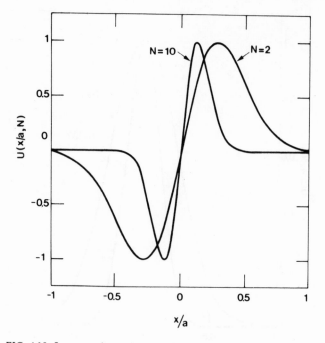

FIG. 4.15. Lowest-order antisymmetric mode of a confocal resonator.

(i) TEM_{00} *Mode* ($m = l = 0$). The eigensolution is $U_{00}(x, y) = \exp[-\pi(x^2 + y^2)/L\lambda]$, and the mode has a Gaussian radial profile both along the x and y directions. In this case the mode pattern corresponds to a circular luminous spot on the mirror (Fig. 4.16) with a dimension given by w_s. For this reason w_s is called the spot size at the mirror.[†] As an example, for $\lambda = 0.6\mu$m and $L = 0.5$ m we get $w_s \simeq 0.3$ mm.

(ii) TEM_{01} *Mode* ($m = 0, l = 1$). The eigensolution is $U_{01}(x, y) = H_1(y)\exp[-\pi(x^2 + y^2)/L\lambda]$, and the radial behavior of the field along the x direction is as in Fig. 4.14 while Fig. 4.15 shows the behavior along the y direction. The pattern of light formed on the mirror by this mode is shown in Fig. 4.16.

(iii) TEM_{11} *Mode* ($m = l = 1$). The eigenfunction is now $U_{11}(x, y) = H_1(x)H_1(y)\exp[-\pi(x^2 + y^2)/L\lambda]$, and the radial behavior is as in Fig. 4.15 along both the x and y directions. In a similar way we can find the

---

[†]We note here a new possible interpretation of the Fresnel number. With the help of (4.31) it is readily shown that $N = (1/\pi)(a^2/w_s^2)$. Apart from a proportionality constant N is seen to be given by the ratio of the mirror cross section (πa^2 for a circular mirror) and the mode cross section (πw_s^2 on the mirror).

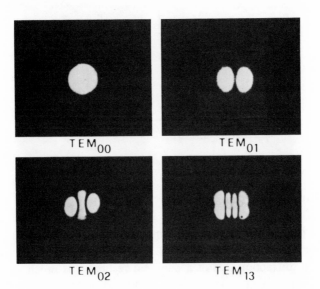

TEM$_{00}$ TEM$_{01}$

TEM$_{02}$ TEM$_{13}$

FIG. 4.16. Mode patterns of some low-order modes.

eigenfunctions and the mode patterns for the higher-order modes (see Fig. 4.16).

So far we have discussed only the eigenfunctions of (4.25) and (4.26). In discussing their corresponding eigenvalues, we will need to avoid the limitation posed above that $N \gg 1$ (the mirror cross section is much larger than the mode cross section). In fact, it can be shown that for $N \gg 1$, we have $|\sigma| \simeq 1$ and the diffraction losses vanish. So, for a meaningful discussion of the eigenvalues σ_{ml}^*, we will need to go back to the Flammer spheroidal radial functions. Fortunately, however, the expression for ϕ_{ml}^* turns out to be quite simple and so, using (4.22), the resonance frequencies turn out to be simply given by

$$\nu = \frac{c[2n + (1 + m + l)]}{4L} \tag{4.32}$$

The corresponding frequency spectrum is shown in Fig. 4.17. Note that modes having the same value of $2n + m + l$ have the same resonance frequency although they have different spatial configurations. These modes are said to be frequency degenerate. Note also that, unlike the plane wave case (Fig. 4.7), the frequency spacing is now $c/4L$. The frequency spacing between two modes with the same (l, m) values (e.g., TEM$_{00}$) and with n differing by 1 (i.e., the frequency spacing between two adjacent *longitudinal modes*) is, however, $c/2L$ as for the plane case. We now go on to consider

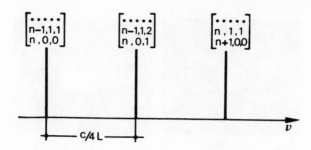

FIG. 4.17. Resonance frequencies of a confocal resonator.

the magnitude of σ, i.e., the diffraction losses. Figure 4.18 shows the behavior of the diffraction losses $\gamma_d = 1 - |\sigma|^2$ versus the Fresnel number as obtained from the value of the Flammer spheroidal radial functions. A comparison of Fig. 4.18 with Fig. 4.12 shows that, for a given Fresnel number, the diffraction loss of a confocal resonator is much smaller than that of a plane resonator. This can be easily understood by noting that, in a confocal resonator, as a result of the focusing properties of the spherical mirrors, the field tends to be much more concentrated along the resonator axis (compare, for instance, the curves of Figs. 4.9 and 4.14 or the curves of Figs. 4.11 and 4.15 at the same values of the Fresnel number).

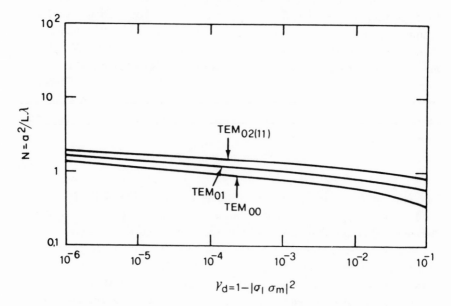

FIG. 4.18. Diffraction loss per pass γ_d versus Fresnel number for a confocal resonator (after Boyd and Gordon[4]).

Once the field distribution over the mirrors is known, the field distribution at any point inside as well as outside the cavity can be obtained by using the Kirchhoff integral. It can be shown[4] that this field distribution is given by

$$U(x, y, z) = \frac{w_0}{w(z)} H_m\left(\frac{\sqrt{2}\, x}{w(z)} \right) H_l\left(\frac{\sqrt{2}\, y}{w(z)} \right) \exp\left[-\frac{x^2 + y^2}{w^2(z)} \right]$$

$$\times \exp\left\{ -i\left[k\frac{(x^2 + y^2)}{2R(z)} + kz - (l + m + 1)\phi(z) \right] \right\} \quad (4.33)$$

If the resonator center is taken to be the origin (Fig. 4.19), the beam spot size $w(z)$ which appears in (4.33) is given by

$$w(z) = w_0\left[1 + (2z/L)^2 \right]^{1/2} \quad (4.34)$$

where w_0 is the spot size at the center of the resonator and is given by

$$w_0 = \left(\frac{L\lambda}{2\pi} \right)^{1/2} \quad (4.35)$$

In Fig. 4.19 the beam dimension (i.e., spot size) as a function of position along the resonator axis, as obtained from (4.34), is indicated by the solid curve. Note that the minimum spot size occurs at $z = 0$. The quantity w_0 is therefore usually referred to as the spot size at the *beam waist*. Note also that for $z = \pm L/2$ (i.e., on the mirrors), (4.34) gives $w = (L\lambda/\pi)^{1/2}$ in agreement with (4.31). The spot size at the mirrors is thus $\sqrt{2}$ larger than that at the resonator center. This result is readily understood when it is remembered that the mirrors tend to focus the beam at the resonator center.

We now consider the phase term appearing in the last exponential factor of (4.33). The functions $R(z)$ and $\phi(z)$ are given by[4]

$$R(z) = z\left[1 + \left(\frac{L}{2z} \right)^2 \right] \quad (4.36)$$

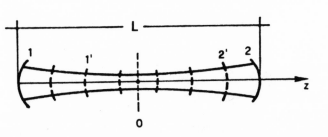

FIG. 4.19. Spot size and equiphase surfaces for a TEM$_{00}$ mode in a confocal resonator.

and

$$\phi(z) = \tan^{-1}\left(\frac{2z}{L}\right) \tag{4.37}$$

It can be shown from (4.33) that the equiphase surfaces are, to a good approximation, spherical with radius of curvature equal to $R(z)$. The sign of $R(z)$ is taken as positive when the center of curvature is to the left of the wavefront. In Fig. 4.19 the equiphase surfaces at a few points along the resonator axis are indicated by dashed curves. Note that for $z = 0$ (center of the resonator) we have $R = \infty$ and the wavefront is plane, as expected from symmetry considerations. Note also that for $z = \pm L/2$ (i.e., on the mirrors) we have $R = L$. This shows that, as expected, the two mirror surfaces are also equiphase surfaces. The expression for $\phi(z)$ in (4.37) allows one to calculate the mode frequencies. Thus, by substituting the phase term from (4.33) into (4.22) we find that $kL - (l + m + 1)[\phi(L/2) - \phi(-L/2)] = n\pi$. With the help of (4.37) we thus get (4.32).

4.4 GENERALIZED SPHERICAL RESONATOR

We will now consider the general case of a resonator consisting of two spherical mirrors with radii of curvature R_1 and R_2 separated by a distance L. The sign of the radius of curvature is taken to be positive for concave mirrors and negative for convex mirrors. Our aim is to calculate the mode amplitudes, diffraction losses, and resonance frequencies. Since R_1 and R_2 may take any values (either positive or negative), there will be some mirror combinations which constitute an unstable resonator configuration (see, for instance, Fig. 4.6). We are therefore also interested in finding the condition for the stability of a general spherical resonator. For the discussion that follows it is convenient to define two dimensionless quantities g_1 and g_2 as

$$g_1 = 1 - \frac{L}{R_1} \tag{4.38a}$$

$$g_2 = 1 - \frac{L}{R_2} \tag{4.38b}$$

4.4.1 Mode Amplitudes, Diffraction Losses, and Resonance Frequencies

To calculate the field distribution, let us first imagine the equiphase surfaces 1' and 2' of Fig. 4.19 to be replaced by two actual mirrors with the same curvatures as those of the equiphase surfaces. Let us also imagine the original mirrors 1 and 2 to have been removed. The resonator will now be formed by mirrors 1' and 2', and the field distribution inside the resonator will obviously not have changed. Accordingly, the spot size and equiphase surfaces, both inside and outside the resonator, will remain as in Fig. 4.19. On the other hand, we can see from (4.36) that the two equiphase surfaces 1' and 2' are no longer confocal. Therefore in order to find the modes of a resonator formed by the two mirrors 1' and 2' we can first calculate the position of the two corresponding confocal surfaces 1 and 2, thus reducing the problem to that of an *equivalent confocal resonator*. The location of this resonator can be obtained using (4.36) with L replaced by L_e, the length of the equivalent confocal resonator. Given the radii R_1 and R_2 of mirrors 1' and 2' and their spacing L, the quantities which can be determined are: (i) the distance of one of the two mirrors (say mirror 1') from the beam waist (i.e., the origin of the z axis); (ii) the length L_e of the equivalent confocal resonator. Having determined the above two quantities, the field distribution can be obtained from (4.33) with the help of (4.34), (4.35), (4.36), and (4.37) in which L has been replaced by L_e, namely,

$$w = w_0 \left[1 + \left(\frac{2z}{L_e} \right)^2 \right]^{1/2} \tag{4.39}$$

$$w_0 = \left(\frac{L_e \lambda}{2\pi} \right)^{1/2} \tag{4.40}$$

$$R(z) = z \left[1 + \left(\frac{L_e}{2z} \right)^2 \right] \tag{4.41}$$

$$\phi = \tan^{-1} \left(\frac{2z}{L_e} \right) \tag{4.42}$$

A particularly relevant case is that where $R_2 = -R_1 = R$ (symmetric resonator). In this case from (4.41) we find that

$$L_e^2 = (2R - L)L \tag{4.43}$$

The spot size at the mirror is obtained from (4.39), (4.40), and (4.43) as

$$w_s' = \left(\frac{\lambda L}{2\pi} \right)^{1/2} \left[\frac{4R^2}{(2R - L)L} \right]^{1/4} \tag{4.44}$$

The ratio of this spot size to that of a confocal resonator [see (4.31)] is

$$\frac{w_s'}{w_s} = \left[\frac{1}{(L/R)[2 - (L/R)]} \right]^{1/4} = \left[\frac{1}{1 - g^2} \right]^{1/4} \tag{4.45}$$

where (4.38a) and (4.38b) have also been used. The quantity w_s'/w_s is plotted versus L/R in Fig. 4.20. We see that: (i) The minimum spot size

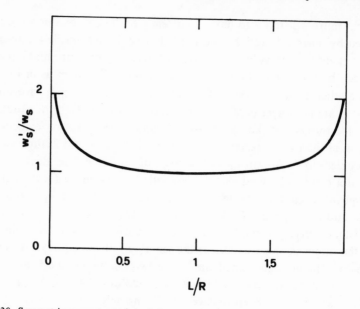

FIG. 4.20. Symmetric resonator: plot of the spot size w_s' on the mirror (normalized to the corresponding w_s for a confocal resonator of the same length) versus the ratio of resonator length L to mirror radius R.

occurs for $L/R = 1$ (confocal resonator). (ii) The spot size diverges for both $L/R = 0$ (plane resonator) and $L/R = 2$ (concentric resonator). Note, however, that, except when very near these two extremes, the spot size is not very different from that of a confocal resonator.

What has been said so far concerns only the calculation of the eigenfunctions, i.e., of the field distributions. To calculate the diffraction losses it is necessary to actually solve the Fredholm integral equation for the particular case under consideration. Figure 4.21 shows the calculated diffraction losses versus Fresnel number for a range of symmetric resonators (which are characterized by their corresponding g values). We note that, for a given Fresnel number, the confocal ($g = 0$) resonator has the lowest loss. To calculate the resonator frequencies, we consider a general resonator and let z_1 and z_2 be the z-coordinates of the two mirrors referred to the origin at the beam waist. From (4.22) and (4.33), one obtains the following expression, from which the resonance frequencies can be found:

$$kL - (l + m + 1)\big[\phi(z_2) - \phi(z_1)\big] = n\pi \tag{4.46}$$

where $\phi(z_1)$ and $\phi(z_2)$ are obtained from (4.42). Equation (4.46) gives

$$\nu = \frac{c}{2L}\left[n + (l + m + 1)\frac{\phi(z_2) - \phi(z_1)}{\pi}\right] \tag{4.47}$$

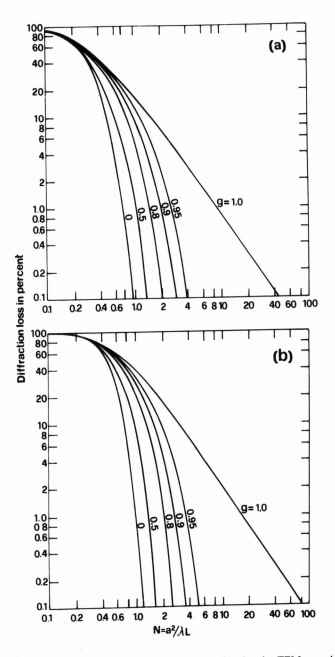

FIG. 4.21. Diffraction loss per transit versus Fresnel number for the TEM_{00} mode (a) and TEM_{10} mode (b) of several symmetric resonators (after Li[10]). Copyright 1965, American Telephone and Telegraph Company. Reprinted with permission.

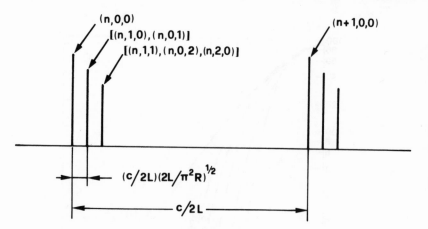

FIG. 4.22. Mode spectrum of a symmetric spherical mirror resonator when the radius of curvature R is much larger than the cavity length L.

After some lengthy algebra the following expression is obtained:

$$\nu = \frac{c}{2L}\left[n + (l + m + 1)\frac{\cos^{-1}(g_1 g_2)^{1/2}}{\pi} \right] \qquad (4.48)$$

where g_1 and g_2 are given by (4.38). Note that the frequency degeneracy which occurs for a confocal resonator (Fig. 4.17) is lifted in the case of a general spherical resonator. As an important example we consider a near-planar resonator with two identical and nearly flat mirrors, i.e., with $(L/R) \ll 1$. Then $\cos^{-1}(g_1 g_2)^{1/2} = \cos^{-1}[1 - (L/R)] \simeq (2L/R)^{1/2}$, and (4.48) becomes

$$\nu = \frac{c}{2L}\left[n + (l + m + 1)\frac{1}{\pi}\left(\frac{2L}{R}\right)^{1/2} \right] \qquad (4.49)$$

The resulting frequency spectrum is indicated in Fig. 4.22 (compare with Fig. 4.7).

4.4.2 Stability Condition

The stability condition can be obtained by an argument based on geometrical optics.[6] With reference to Fig. 4.23, let us consider a ray leaving point P_0 of some general plane β inside the resonator. This ray, after reflection from mirrors 1 and 2, will intersect the plane β at P_1. If we let x_0 and x_1 be the coordinates of P_0 and P_1 with respect to the resonator axis, and θ_0 and θ_1 the angles that the corresponding rays make with the

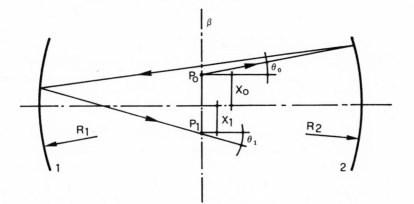

FIG. 4.23. Matrix method for finding the stability condition for a general spherical resonator.

axis, then for small values of x and θ, the quantities x_1 and θ_1 are obtained from the corresponding quantities x_0 and θ_0 by a linear transformation. Thus, in matrix form

$$\begin{vmatrix} x_1 \\ \theta_1 \end{vmatrix} = \begin{vmatrix} A & B \\ C & D \end{vmatrix} \begin{vmatrix} x_0 \\ \theta_0 \end{vmatrix} \tag{4.50}$$

where the matrix elements A, B, C, and D will depend only on the resonator geometry. The ray leaving point $P_1(x_1, \theta_1)$ will, after two reflections, intersect the plane β at point $P_2(x_2, \theta_2)$ given by

$$\begin{vmatrix} x_2 \\ \theta_2 \end{vmatrix} = \begin{vmatrix} A & B \\ C & D \end{vmatrix} \begin{vmatrix} x_1 \\ \theta_1 \end{vmatrix} = \begin{vmatrix} A & B \\ C & D \end{vmatrix}^2 \begin{vmatrix} x_0 \\ \theta_0 \end{vmatrix} \tag{4.51}$$

Therefore, after n round trips, the point $P_n(x_n, \theta_n)$ is given by

$$\begin{vmatrix} x_n \\ \theta_n \end{vmatrix} = \begin{vmatrix} A & B \\ C & D \end{vmatrix}^n \begin{vmatrix} x_0 \\ \theta_0 \end{vmatrix} \tag{4.52}$$

For the resonator to be stable, we require that, for any initial point (x_0, θ_0), the point (x_n, θ_n) should not diverge as n increases. This means that the matrix

$$\begin{vmatrix} A & B \\ C & D \end{vmatrix}^n$$

must not diverge as n increases. Since the determinant of the matrix, $AD - BC$, can be shown to be unity, one then has from matrix calculus[11] that

$$\begin{vmatrix} A & B \\ C & D \end{vmatrix}^n = \frac{1}{\sin\theta} \begin{vmatrix} A\sin n\theta - \sin(n-1)\theta & B\sin n\theta \\ C\sin n\theta & D\sin n\theta - \sin(n-1)\theta \end{vmatrix} \tag{4.53}$$

where

$$\cos\theta = \tfrac{1}{2}(A + D) \tag{4.54}$$

From (4.54) we see that, for the matrix (4.53) not to diverge, we require that

$$-1 < \tfrac{1}{2}(A + D) < 1 \tag{4.55}$$

In fact, if (4.55) is not satisfied, θ will be a complex number and $\sin(n\theta)$ will diverge as n increases.

By calculating the coefficients A and D for a generalized resonator and then using (4.55) we finally arrive at a very simple expression for the stability, namely,

$$0 < g_1 g_2 < 1 \tag{4.56}$$

This stability condition is depicted in Fig. 4.24. In this figure, the stable regions correspond to the shaded area. A particularly interesting class of spherical resonators is that corresponding to points on the straight

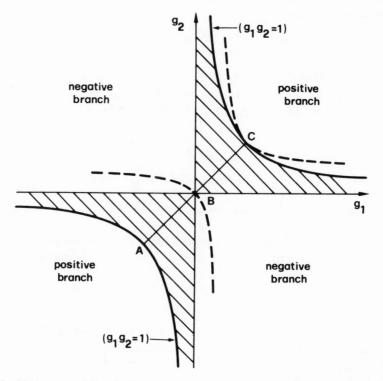

FIG. 4.24. g_1, g_2 stability diagram for a general spherical resonator. The stable region corresponds to the shaded parts of the figure. The dashed curves correspond to the possible confocal resonators.

line AC, making an angle of $45°$ with the g_1 and g_2 axes. This line corresponds to resonators having mirrors of the same radius of curvature (symmetric resonators). As particular examples of these symmetric resonators, we notice that those corresponding to points A, B, and C of the figure are the concentric, confocal, and plane resonators, respectively. Therefore all three of these resonators lie on the boundary between the stable and unstable regions. The disadvantages of a concentric resonator are that: (i) It produces a very small spot size at the resonator center (Fig. 4.2), which can be a problem in high-power lasers. (ii) It is rather sensitive to mirror misalignment. Concentric resonators are therefore seldom used. Confocal resonators, on the other hand, typically give a spot size [see (4.35)] that is too small for effective use of all the available cross section of the laser medium. For this reason confocal resonators are not often used. Plane-parallel resonators can make good use of the cross section (see Fig. 4.9). Like concentric resonators, however, they are rather sensitive to mirror misalignment. For the various reasons discussed above, the most commonly used laser resonators make use of either two concave mirrors of large radius of curvature (say from two to ten times the resonator length) or a plane mirror and a concave mirror of large radius. These resonators give a spot size somewhat larger than that of confocal resonators (see Fig. 4.20), and a reasonable stability against misalignment. Such resonators lie in the stable region near point C of Fig. 4.24.

4.5 UNSTABLE RESONATORS [7,8]

The stability condition for a generalized spherical resonator was discussed in the previous section [see, in particular, (4.56)], and the unstable regions were shown to correspond to the unshaded regions of the g_1-g_2 plane in Fig. 4.24. Unstable resonators can be separated into two classes: (i) positive branch resonators, which correspond to the case $g_1 g_2 > 1$, and (ii) negative branch resonators, which correspond to the case $g_1 g_2 < 0$.

Before going on to a quantitative discussion of unstable resonators, it is worth pointing out here the reasons why these resonators are of interest in the laser field. First, we note that, for a stable resonator, the spot size w is typically of the order of that given for the case of a confocal resonator (see Fig. 4.20). This implies that for a resonator length of the order of a meter and for a wavelength in the visible range, the spot size will be of the order of or smaller than 1 mm. With such a small cross section the output power (or energy) available in a single transverse mode is necessarily rather small.

For unstable resonators, on the contrary, the field does not tend to be confined to the axis (see, for example, Fig. 4.6), and a large mode volume in a single transverse mode is possible. With unstable resonators, however, there is the problem that rays tend to walk off out of the cavity. The corresponding modes, therefore, have substantially greater (geometrical) losses than those of a stable cavity (where the losses are due to diffraction). This fact can, however, be used to advantage if these walk-off losses are turned into useful output coupling.

To find the modes of an unstable resonator, we can start by using a geometrical-optics approximation, as first done by Siegman.[8] To do this, we begin by recalling the two main results which were obtained for the eigensolutions of a stable resonator (see Fig. 4.19): (i) The amplitude is given by the product of a Hermite polynomial with a Gaussian function. (ii) The phase distribution is such as to give a spherical wavefront. The presence of the Gaussian function limits the beam spot size and essentially arises from the focusing properties of a stable spherical resonator. The fact that the wavefront is spherical is, on the other hand, connected with the boundary conditions set by a spherical mirror. In the unstable case no Hermite–Gaussian solution is possible (Problem 4.16) since the beam is no longer focused toward the resonator axis. It is, therefore, natural to assume, as a first approximation, that the solution in this case has an amplitude corresponding to uniform illumination, while the wavefront is still spherical.

After this preliminary discussion, let us consider a general unstable resonator such as that of Fig. 4.25a. As explained above, we will assume the mode to be made up of a superposition of two spherical waves of uniform intensity.[7] The centers P_1 and P_2 of the two waves are *not* the centers of curvature of mirrors M_1 and M_2, and their positions are easily calculated by a self-consistent argument: The spherical wave originating from P_1, upon reflection at mirror M_2, must give a spherical wave originating from P_2 and *vice versa*. The positions of points P_1 and P_2 are then obtained by a straightforward calculation based on geometrical optics. The results for the quantities r_1 and r_2, indicated in Fig. 4.25a, are

$$r_1 = g_2 \left\{ \left[g_1 g_2 (g_1 g_2 - 1) \right]^{1/2} + g_1 g_2 - g_2 \right\}^{-1} \quad (4.57a)$$

$$r_2 = g_1 \left\{ \left[g_1 g_2 (g_1 g_2 - 1) \right]^{1/2} + g_1 g_2 - g_1 \right\}^{-1} \quad (4.57b)$$

where g_1 and g_2 are given by (4.38).

So far only the mode configuration has been considered. To calculate the loss of this mode, we will limit ourselves to a consideration of the

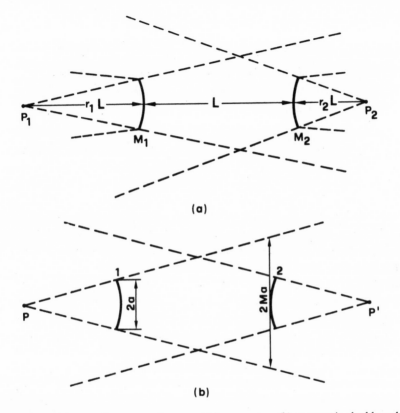

FIG. 4.25. (a) A general convex mirror unstable resonator; (b) symmetric double-ended unstable resonator.

symmetric (i.e., $R_1 = R_2 = R$, where R is the radius of the mirrors), double-ended (i.e., $a_1 = a_2 = a$, where $2a$ is the mirror aperture), unstable resonator (Fig. 4.25b). In this case, it can be readily shown that, on passing from one mirror to the other, the spot of each spherical wave becomes magnified by a factor M given by

$$M = g + \left(g^2 - 1 \right)^{1/2} \tag{4.58}$$

where we have set $g = g_1 = g_2$. The quantity M is, therefore, called the *one-way* (*symmetric*) *magnification factor*. Since we have assumed uniform illumination, the loss per pass is then seen to be

$$\gamma = \frac{S_2 - S_1}{S_2} = \frac{M^2 - 1}{M^2} \tag{4.59}$$

where S_1 and S_2 are the cross sections at the mirrors 1 and 2, respectively, of the beam originating from point P. As already mentioned, this loss per pass also gives the fractional output coupling from each end. Note that both M and γ are independent of mirror diameter $2a$.

So far, only one mode (which turns out to be the lowest-loss mode) has been considered. To find the higher-order modes, still working within the geometrical-optics approximation, we will again restrict ourselves to considering the symmetric double-ended case. In this case, the field at position x of mirror 2 is due to the field at position x/M of mirror 1. If we let U_2 and U_1 be the corresponding field distributions, we can write

$$U_2(x) = \frac{1}{M^{1/2}} U_1\left(\frac{x}{M} \right) \tag{4.60}$$

where the amplitude factor $1/M^{1/2}$ on the right-hand side of (4.60) accounts for the fact that the beam dimension is increased by a factor M on passing from mirror 1 to mirror 2. For $U(x)$ to be a cavity mode, we require (since the cavity is symmetrical) that $U_2(x) = \sigma_x U_1(x)$. So, from (4.60) we get

$$\sigma_x U(x) = \frac{1}{M^{1/2}} U\left(\frac{x}{M} \right) \tag{4.61}$$

which is an eigenvalue equation. A similar equation applies to the y coordinate. The overall eigensolution is then $U(x, y) = U(x)U(y)$, and the corresponding eigenvalue is $\sigma = \sigma_x \sigma_y$. One can immediately verify that the zeroth-order solution of (4.61) is $U_0 = $ const and $\sigma_x = 1/M^{1/2}$. Combining these solutions for both x and y coordinates, we get $U(x, y) = $ const and $\sigma = 1/M$. This is just the mode which was previously considered and whose losses are given by (4.59). It is, however, easy to show that the higher-order solutions of (4.61) are of the form

$$U_n(x) = x^n \tag{4.62a}$$

where $n > 0$ and the corresponding eigenvalues are

$$\sigma_{xn} = 1/M^{n+1/2} \tag{4.62b}$$

Note that the case $n = 0$ (zeroth-order solution) corresponds to the lowest-loss solution.

What has been said so far can be readily generalized to an asymmetric unstable resonator. We will limit the discussion to a consideration of a particularly important class of asymmetric resonators, namely, the confocal resonator. This class can be further subdivided into: (i) negative branch (Fig. 4.26a) and (ii) positive branch (Fig. 4.26b) confocal resonators. These two branches are represented in the g_1-g_2 plane by the two branches of the hyperbola shown as dotted curves in Fig. 4.24 [the equation of the hyper-

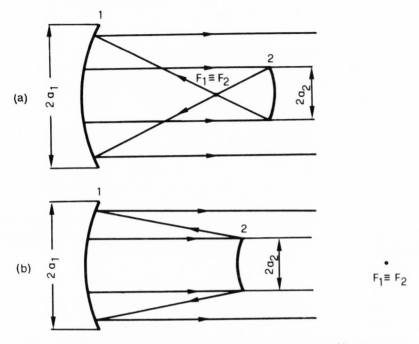

FIG. 4.26. (a) Negative-branch and (b) positive-branch confocal unstable resonators.

bola is $(2g_1 - 1)(2g_2 - 1) = 1$]. Of these various resonators, only the (symmetric) confocal one ($g_1 = g_2 = 0$) and the plane-parallel one ($g_1 = g_2 = 1$) lie on the boundary between the stable and unstable regions. All other confocal resonators are unstable. The mode of an unstable confocal resonator is made up of the superposition of a spherical wave (originating from the common focus) with a plane wave. In this case we can define a round-trip magnification factor M given by $M = |R_1 / R_2|$, where R_1 and R_2 are the two mirror radii. The quantity M gives the increase in diameter of the plane wave after one round trip. If the diameter $2a_1$ of mirror 1 is made sufficiently large ($2a_1 > 2Ma_2$), only the plane beam will escape out of the cavity. The round-trip loss (or fractional output coupling) of this single-ended resonator is then given by (4.59).

The discussion so far has been based on a geometrical-optics approximation. To get a more realistic picture of the modes of an unstable resonator one must use a wave approach (e.g., use the Kirchhoff diffraction integral again). This will not be discussed at any length here. We will just present and discuss a few relevant results. As far as the eigensolutions are concerned, the wave approach shows the following: (i) The phase of the solution corresponds to a wavefront that is close to spherical, with radius almost equal (though always a little larger) than that predicted by geometrical optics. (ii) The amplitude of the solution shows a radial variation which differs considerably from the geometrical-optics result [i.e., equation (4.62a)]. The radial variation shows a characteristic ring pattern which arises from diffraction effects. As an example, one such pattern is shown in Fig. 4.27. The wave theory does show, however, that different modes, i.e., different self-reproducing spatial patterns, do exist. These modes differ from each other in the number of rings they display and also in their location and strength. A clear-cut distinction between the lowest-order and higher-order modes is no longer possible. A distinction is still possible, however, when the eigenvalues of the equation, which give the diffraction losses, are considered. In fact, a new characteristic feature appears: At each half-integer value of a suitably defined equivalent Fresnel number (N_{eq}) a different and distinct mode becomes the "lowest-order" (i.e., the lowest-loss) mode. This is shown in Fig. 4.28 where the magnitude of the eigenvalue σ is plotted versus N_{eq} for three consecutive modes (the corresponding loss is then given by $1 - |\sigma|^2$). Note that, for each half-integer value of N_{eq}, there is a large difference between the losses of the lowest-order mode and those of other modes. This shows that a large transverse-mode discrimination can be obtained under these conditions. For a symmetric double-ended resonator N_{eq} is given by $N_{eq} = [(M^2 - 1)/2M]N$, where N is the usually defined Fresnel number $N = a^2/L\lambda$. Note that, when $M \simeq 1$ (i.e., for a low-loss resonator), we have $N_{eq} \ll N$. For a positive branch single-ended confocal resonator, N_{eq} is given by $N_{eq} = [(M - 1)/2] \cdot (a_2^2/L\lambda)$, while for a negative branch it is given by $N_{eq} = [(M + 1)/2] \cdot (a_2^2/L\lambda)$. In Fig. 4.28, the geometrical-optics value of $|\sigma|$ for the zeroth-order solution is also indicated [according to (4.59), this value is $|\sigma| = 1/M$, independent of mirror dimension and hence of N_{eq}]. Note that, at each half-integer value of N_{eq}, the lowest-order mode (i.e., the

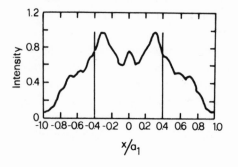

FIG. 4.27. Typical example of the radial behavior of mode intensity distribution in an unstable cavity obtained by use of the Kirchhoff integral. The calculation refers to a positive branch confocal resonator with $M = 2.5$ and $N_{eq} = 0.6$. The vertical lines mark the edge of the output mirrors [after Rensch and Chester[(9)]].

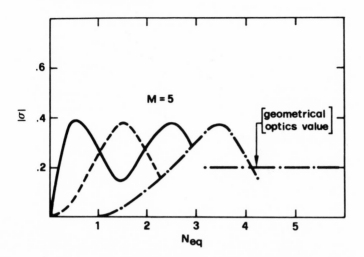

FIG. 4.28. Typical example of the oscillatory behavior of the magnitude of the eigenvalue σ versus the equivalent Fresnel number for three consecutive modes.

one whose curve displays a maximum for that value of N_{eq}), has an appreciably smaller loss $\gamma = (1 - |\sigma|^2)$ than that predicted by geometrical optics. This is also apparent in Fig. 4.29, where the loss γ is plotted versus the magnification factor M. In this figure, the solid curves (which apply to successive half-integer values of N_{eq}) are obtained by diffraction theory, while the dashed curve corresponds to the geometrical-optics result. The fact that the true losses are smaller than those predicted by geometrical optics is again an effect arising from diffraction: Diffraction effects produce a field amplitude with such a ring structure that the losses are minimized.

As a conclusion to this section we list the main advantages and disadvantages of unstable as compared to stable resonators. The main useful properties of an unstable resonator can be summarized as follows: (i) large, controllable mode volume (ii), good transverse-mode discrimination, and (iii) all reflective optics (which is particularly attractive in the infrared, where metallic mirrors can be used). The main disadvantages are as follows: (i) The output beam cross section is in the form of a ring (i.e., it has a dark hole in its center). For example, in a confocal resonator (Fig. 4.26), the inner diameter of the ring is $2a_2$ while its outer diameter is $2Ma_2$. Although this hole disappears in the focal plane of a lens used to focus the beam (far-field pattern), the peak intensity in this focal plane turns out to decrease with decreasing ring thickness. In fact, for a given total power, the peak intensity for an annular beam is reduced by $(M^2 - 1)/M^2$ from that of a uniform-intensity beam with a diameter equal to the large diameter of

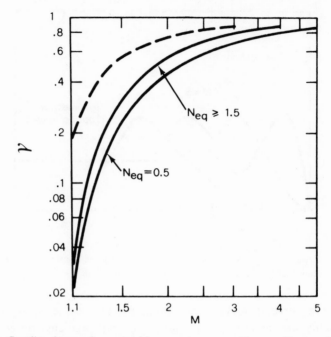

FIG. 4.29. Coupling losses of an unstable resonator versus the magnification factor M; dashed curve: geometrical-optics result; solid lines: wave theory results (after Siegman[8]).

the annular beam. (ii) The intensity distribution in the beam does not follow a smooth curve, but exhibits diffraction rings. (iii) An unstable resonator has greater sensitivity to cavity perturbations compared to a stable resonator. The above advantages and disadvantages mean that unstable resonators find their applications in high-gain lasers (so that M can be relatively large), especially in the infrared, and when high-power (or high-energy) diffraction-limited beams are required.

PROBLEMS

4.1 Consider a confocal resonator of length $L = 1$ m used for a He–Ne laser at a wavelength $\lambda = 0.6328$ μm. Calculate the spot size at the resonator center and on the mirrors.

4.2 For the above resonator calculate the frequency difference between two adjacent longitudinal modes.

4.3 For the resonator of Problem 4.1 calculate how many different mode frequencies fall within the width (FWHM) of the Ne line [see equation (2.114)].

4.4 Consider a hemiconfocal resonator of length $L = 2$ m used for a CO_2 laser at wavelength $\lambda = 10.6$ μm. Calculate the spot size on both mirrors.

4.5 For the above resonator calculate the frequency difference between two adjacent TEM_{00} modes. Given that the width (FWHM) of the CO_2 laser line is 50 MHz, find how many TEM_{00} modes fall within this width.

4.6 A laser operating at $\lambda = 0.6$ μm has a power gain of 2×10^{-2} per pass and is provided with a symmetric resonator consisting of two mirrors each of radius $R = 10$ m and spaced by $L = 1$ m. Choose an appropriate size of mirror aperture in order to suppress TEM_{01} mode operation while allowing TEM_{00} mode operation.

4.7 Consider a resonator consisting of two concave spherical mirrors both with radius of curvature 4 m and separated by a distance of 1 m. Calculate the spot size of the TEM_{00} mode at the resonator center and on the mirrors when the cavity is oscillating at the wavelength $\lambda = 514.5$ nm [one of the Ar^+ laser wavelengths].

4.8 How are the spot sizes at the two mirrors modified if one of the mirrors of the above problem is replaced by a plane mirror?

4.9 One of the mirrors in the resonator of Problem 4.7 is replaced by a concave mirror of 1.5 m radius of curvature. Calculate: (i) the position of the beam waist; (ii) the spot size at the beam waist and on the two mirrors.

4.10 A resonator is formed by a convex mirror of radius $R_1 = -1$ m and a concave mirror of radius $R_2 = 1.5$ m. What is the maximum possible mirror separation if this is to remain a stable resonator?

4.11 A confocal unstable resonator is to be used for a CO_2 laser at a wavelength of $\lambda = 10.6$ μm. The resonator length is chosen to be $L = 1$ m. Which branch would you choose for this resonator if the mode volume is to be maximized? Calculate the mirror apertures $2a_1$ and $2a_2$ so that: (i) $N_{eq} = 7.5$, (ii) single-ended output is achieved, and (iii) a 20% round-trip output coupling is obtained. Then find the two mirror radii R_1 and R_2.

4.12 Using a geometrical-optics approach (and assuming lowest-order mode oscillation), calculate the round-trip loss of the resonator designed in the above problem. What are the shape and dimensions of the output beam?

4.13 What is the radial dependence of the energy density within the resonator (or of the intensity of the output beam) for a TEM_{00} mode? What is the value of the intensity spot size w_I?

4.14 Show that the total power in a Gaussian beam is $P = I_0(\pi w_I^2)$, where I_0 is the peak (on-axis) beam intensity.

4.15 By direct substitution, show that (4.25) has the eigensolution $U = \exp(-\pi\xi^2)$ when $N = \infty$. Find the corresponding eigenvalue σ_ξ^*.

4.16 Show that an equivalent confocal resonator can only be found when the g_1, g_2 parameters of a generalized spherical resonator satisfy (4.56).

4.17 Calculate the $(ABCD)$ ray transfer matrix for free-space propagation of a ray between two planes β and β' separated by a distance L. Calculate the determinant of the matrix.

4.18 Calculate the $(ABCD)$ matrix for a ray which is reflected by a spherical mirror when planes β and β' are coincident and immediately in front of the mirror. Calculate the determinant of the matrix.

4.19 Show that, when the planes in the above problem are coincident and at a distance L from a spherical mirror, the corresponding matrix can be obtained as the product of the matrices calculated in Problems 4.17 and 4.18. Calculate the determinant of the matrix.

4.20 Using the results of Problems 4.17, 4.18, and 4.19, derive the stability condition (4.38).

4.21 Using the geometrical-optics relationship between the conjugate points of a spherical mirror, prove equation (4.57).

REFERENCES

1. A. L. Schawlow and C. H. Townes, *Phys. Rev.* **112**, 1940 (1958).
2. A. G. Fox and T. Li, *Bell Syst. Tech. J.* **40**, 453 (1961).
3. Max Born and Emil Wolf, *Principles of Optics*, 4th ed. (Pergamon Press, London, 1970). pp. 370–382.
4. G. D. Boyd and J. P. Gordon, *Bell Syst. Tech. J.* **40**, 489 (1961).
5. D. Slepian and H. O. Pollak, *Bell Syst. Tech. J.* **40**, 43 (1961).
6. H. Kogelnik and T. Li, *Proc. IEEE* **54**, 1312 (1966).
7. W. H. Steier, in *Laser Handbook*, ed. by M. L. Stitch (North-Holland Publishing Company, Amsterdam, 1979), pp. 3–39.
8. A. E. Siegman, *Laser Focus* **7**, 42 (May 1971).
9. D. B. Rensch and A. N. Chester, *Appl. Opt.* **12**, 997 (1973).
10. T. Li, *Bell Syst. Tech. J.* **44**, 917 (1965).
11. Max Born and Emil Wolf, *Principles of Optics*, 4th ed. (Pergamon Press, London, 1970), Section 1.6.5.
12. Herwig Kogelnik, in *Lasers*, Vol. 1, ed. by A. K. Levine (Marcel Dekker, New York, 1968), Chapter 5.

5

Continuous Wave and Transient Laser Behavior

5.1 INTRODUCTION

In the previous chapters we have discussed several features of the components which make up a laser. These are the laser medium itself (whose interaction with an e.m. wave was considered in Chapter 2), the pumping system (Chapter 3), and the passive optical resonator (Chapter 4). In this chapter we will make use of the results from earlier chapters in developing the theoretical background necessary for a description of the behavior of a laser both for continuous wave (cw) and transient operation. The theory developed here uses the so-called rate-equation approximation. In this approximation the laser equations are derived on the basis of a simple notion that there should be a balance between the rate of change of total population and total number of laser photons. This theory has the advantage of providing a very simple and intuitive picture of laser behavior. Furthermore, it gives sufficiently accurate results for most practical purposes. For a more refined treatment one should use either the semiclassical approach (in which the matter is quantized and the e.m. waves are treated classically, i.e., by Maxwell's equations) or the fully quantum approach (in which both matter and fields are quantized). We refer the reader elsewhere for these more advanced treatments.[1]

5.2 RATE EQUATIONS[(2,3)]

5.2.1 Four-Level Laser

We will consider first a laser operating on a four-level scheme, in which we assume, for the sake of simplicity, that there is only one pump band (band 3 of Fig. 5.1). The following analysis remains unchanged, however, even if more than one pump band (or level) is involved provided that the decay from these bands to the upper laser level 2 is very fast. Let the populations of the four levels 0, 1, 2, and 3 be N_g, N_1, N_2, and N_3, respectively. We will assume the laser to be oscillating on only one cavity mode and we let q be the corresponding total number of photons in the cavity. With the further assumption of a very fast decay between levels 3 and 2 and levels 1 and 0, we can put $N_1 \simeq N_3 \simeq 0$, and so write the following rate equations:

$$N_g + N_2 = N_t \tag{5.1a}$$

$$\dot{N}_2 = W_p N_g - B q N_2 - (N_2/\tau) \tag{5.1b}$$

$$\dot{q} = V_a B q N_2 - (q/\tau_c) \tag{5.1c}$$

In (5.1a), N_t is the total population of active atoms (or molecules). In (5.1b), the term $W_p N_g$ accounts for pumping [see equation (1.10)]. Explicit expressions for the pumping rate W_p have already been derived in Chapter 3, both for optical and electrical pumping. The term $B q N_2$ in (5.1b) accounts for stimulated emission. It was shown in Chapter 2 that the stimulated emission rate W is in fact proportional to the square of the electric field of the e.m. wave, and it follows that W can also be taken to be proportional to q. The coefficient B will therefore be referred to as the

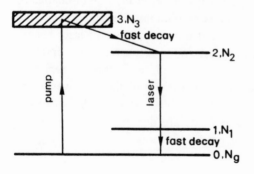

FIG. 5.1. Four-level laser scheme.

stimulated transition rate per photon and per mode. The quantity τ is the lifetime of the upper laser level, and it is, in general, given by (2.93). In (5.1c), V_a is the mode volume within the active material, and a general expression for this will be given in Appendix A. In fact, as discussed in Section 4.4 of Chapter 4, it is often the case that one has a symmetric laser resonator consisting of two spherical mirrors whose curvature is much larger than the cavity length. The mode spot size w will then be approximately constant along the resonator length and hence equal to the value w_0 at the resonator center. For a TEM$_{00}$ mode, the volume V_a is then

$$V_a = \pi w_0^2 l / 4 \tag{5.2}$$

where l is the length of the active material. The appearance of a factor of 4 in the denominator of (5.2) is the result of the following two circumstances: (i) w_0 is the spot size defined for the field amplitude U, whereas the spot size for the field intensity (i.e., for U^2) is, obviously, $\sqrt{2}$ smaller, which contributes a factor $\frac{1}{2}$ in (5.2). (ii) Another factor $\frac{1}{2}$ is due to the standing-wave character of the mode (so that $\langle \sin^2 kz \rangle = \frac{1}{2}$). The term $V_a B q N_2$ in (5.1c) is opposite in sign to the analogous term appearing in (5.1b) and has been included on the basis of a simple consideration of balance: Each stimulated-emission process creates a photon, and each absorption process removes a photon. Finally, the term q/τ_c accounts for the removal of photons through the cavity losses.

Before proceeding it is worth pointing out that, in (5.1a), the term corresponding to spontaneous emission has been omitted. Since, as already mentioned in Chapter 1, laser action is actually initiated by spontaneous emission, we would expect that equation (5.1) is not able to account correctly for the onset of laser oscillation. Indeed, if we put $q = 0$ at time $t = 0$ in (5.1c), we get $\dot{q} = 0$, and laser action therefore cannot start. To account for spontaneous emission, one could again try to apply simple considerations of balance, starting with the term N_2/τ_{sp} which is included in the term N_2/τ of (5.1b). It might then be thought that the appropriate term in (5.1c) to account for spontaneous emission would be $V_a(N_2/\tau_{sp})$. This is wrong, however. In fact, as seen in Section 2.3 [see (2.91), in particular], the spontaneously emitted light is distributed over the entire frequency range described by the line shape function $g(\Delta\omega)$. A spontaneous-emission term in (5.1c) should, however, only include that fraction of the spontaneously emitted light which contributes to the given mode. The correct expression for this term can only be obtained by a quantized treatment of the e.m. field of the cavity mode. The result is particularly simple and instructive.[4] When spontaneous emission is taken into account, the term $V_a BqN_2$ in (5.1c) becomes instead $V_a B(q + 1)N_2$. Everything behaves as if there were an "extra photon" in the term describing stimulated emission. For the sake of simplicity, however, in the following analysis we will not introduce this extra term arising from spontaneous emission, and instead we will assume that an arbitrary small number of photons q_i is initially present in the cavity. In fact, the treatment that follows is not changed by this small number of photons, which is only needed to let laser action start.

We are now interested in deriving explicit expressions for the quantities B and τ_c to be used in (5.1b) and (5.1c). These expressions can be obtained by means of a simple argument. To this purpose we will consider a resonator of length L in which an active material of length l and refractive index n is inserted. We can think of the cavity mode as being due to the superposition of two waves traveling in opposite directions. Let I be the intensity of one of these waves. The change in intensity, dI, when the wave travels a distance dz in the active material is, according to (1.7), given by $dI = \sigma(N_2 - N_1)I\,dz$. Here, σ is the transition cross section at the frequency of the given cavity mode. We now introduce the following notation: (i) T_1 and T_2 are the power transmission of the two cavity mirrors, (ii) a_1 and a_2 are the corresponding fractional mirror losses, and (iii) T_i is the fractional internal loss per pass. The change in intensity ΔI for a cavity round-trip will then be

$$\Delta I = \left\{ (1 - a_1 - T_1)(1 - a_2 - T_2)(1 - T_i)^2 \right.$$
$$\left. \times \exp\left[2\sigma(N_2 - N_1)l \right] - 1 \right\}I \qquad (5.3)$$

We will now assume that the mirror losses are equal ($a_1 = a_2 = a$) and so small that we can put $(1 - a - T_1) \simeq (1 - a)(1 - T_1)$ and $(1 - a - T_2) \simeq (1 - a)(1 - T_2)$. The analysis that follows is then simplified by introducing some new quantities, γ, which can be described as the logarithmic losses per pass, namely,

$$\gamma_1 = -\ln(1 - T_1) \qquad (5.4a)$$

$$\gamma_2 = -\ln(1 - T_2) \qquad (5.4b)$$

$$\gamma_i = -\left[\ln(1 - a) + \ln(1 - T_i) \right] \qquad (5.4c)$$

γ_1 and γ_2 are the logarithmic losses per pass due to mirror transmission and γ_i is the logarithmic internal loss. For brevity, however, we will simply call γ_1 and γ_2 the mirror losses and γ_i the internal loss. We can also define a total loss per pass γ as

$$\gamma = \gamma_i + \frac{\gamma_1 + \gamma_2}{2} \tag{5.5}$$

If (5.5) and (5.4) are substituted into (5.3) and the additional assumption is made that

$$\left[\sigma(N_2 - N_1)l - \gamma\right] \ll 1 \tag{5.6}$$

then the exponential function in (5.3) can be expanded as a power series to yield

$$\Delta I = 2\left[\sigma(N_2 - N_1)l - \gamma\right]I \tag{5.7}$$

We divide both sides of (5.7) by the time Δt taken for the light to make one cavity round trip, i.e., $\Delta t = 2L'/c_0$, where L' is given by

$$L' = L + (n - 1)l \tag{5.7a}$$

If the approximation $\Delta I/\Delta t \simeq dI/dt$ is used, we get

$$\frac{dI}{dt} = \left[\frac{\sigma l c_0}{L'}(N_2 - N_1) - \frac{\gamma c_0}{L'}\right]I \tag{5.8}$$

Since the number q of photons in the cavity is proportional to I, a comparison of (5.8) with (5.1c) gives

$$B = \frac{\sigma l c_0}{V_a L'} = \frac{\sigma c_0}{V} \tag{5.9a}$$

$$\tau_c = \frac{L'}{\gamma c_0} \tag{5.9b}$$

where V is the effective cavity mode volume. For a cavity of the type referred to earlier [see the discussion preceding (5.2)], V is given by

$$V \simeq \pi w_0^2 L'/4 \tag{5.10}$$

Thus far our discussion has provided a justification of equation (5.1c) and derived explicit expressions for B and τ_c in terms of measurable parameters of the laser. Note, however, that we had to make use of the approximation (5.6), which implies that the difference between gain and loss is small (i.e., that the laser is not too far above threshold). If this is not true, the dynamic behavior of the laser must be analyzed using (5.3) on a pass-by-pass basis. Note, finally, that according to (5.5), (5.9b) can also be written in the following form:

$$\frac{1}{\tau_c} = \frac{\gamma_i c_0}{L'} + \frac{\gamma_1 c_0}{2L'} + \frac{\gamma_2 c_0}{2L'} \tag{5.11}$$

Equations (5.1) together with the explicit expressions for B and τ_c [equations (5.9)] describe the static and dynamic behavior of a four-level laser. Note that, instead of expressing the equation in terms of the upper-level population N_2, the population inversion

$$N = N_2 - N_1 \tag{5.12}$$

is more commonly used. Given our assumption of rapid decay from level 1, it follows that $N \simeq N_2$, and equations (5.1) can readily be reduced to just two equations in the variables $N(t)$ and $q(t)$:

$$\dot{N} = W_p(N_t - N) - BqN - (N/\tau) \tag{5.13a}$$

$$\dot{q} = \left[V_a BN - (1/\tau_c) \right] q \tag{5.13b}$$

A quantitative description of the laser's behavior then requires that these equations be solved with the appropriate initial conditions. For instance, if pumping is initiated at $t = 0$, the initial conditions will be $N(0) = 0$ and $q(0) = q_i$, where q_i is a very small number of initial photons (e.g., $q_i = 1$) simulating the effect of spontaneous emission. Once $q(t)$ is known, we can readily calculate the output power through one of the two cavity mirrors (e.g., mirror 1). In fact, if we substitute (5.11) into (5.13b) we realize that the term $(\gamma_1 c_0/2L')q$ is the rate of photon loss due to the transmission of the output mirror. The output power is therefore

$$P_1 = \left(\frac{\gamma_1 c_0}{2L'} \right) \hbar \omega q \tag{5.14}$$

Before ending this section we wish to point out again that the results derived so far apply only when the laser is oscillating on a single mode. For a laser oscillating on more than one mode the calculation is, in principle, much more complicated. If we consider, for instance, a laser oscillating on two modes, we would need to have separate rate equations for the photon numbers q_1 and q_2 in the two modes. Actually, a description in terms of the corresponding electric fields would be more appropriate since one could then account properly for effects due to beating between the two modes (see Section 5.4.3 on mode locking). However, when many modes are oscillating the picture can again be simplified by considering just the total number of photons q summed over all modes. In this case the equations given thus far can still be applied in an approximate fashion, with the volume V_a now given by

$$V_a = Al \tag{5.2a}$$

where A is the cross-sectional area of the laser medium, occupied by the oscillating modes.

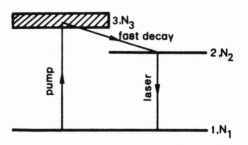

FIG. 5.2. Three-level laser scheme.

5.2.2 Three-Level Laser

The analysis for a three-level laser proceeds in a similar way to that for the four-level case. Referring now to Fig. 5.2, we again assume only one pump band and, if the $3 \rightarrow 2$ transition is fast enough, we can again put $N_3 \simeq 0$. The rate equations can then be written in much the same way as for the four-level case, *viz.*:

$$N_1 + N_2 = N_t \qquad (5.15a)$$

$$\dot{N}_2 = W_p N_1 - Bq(N_2 - N_1) - (N_2/\tau) \qquad (5.15b)$$

$$\dot{q} = V_a Bq(N_2 - N_1) - q/\tau_c \qquad (5.15c)$$

With the help of (5.12) these equations can also be reduced to just two equations in the variables $N(t)$ and $q(t)$:

$$\dot{N} = W_p(N_t - N) - 2BqN - (N_t + N)/\tau \qquad (5.16a)$$

$$\dot{q} = \left[V_a BN - (1/\tau_c) \right] q \qquad (5.16b)$$

These equations, together with the explicit expressions for B and τ_c [see (5.9)], describe the static and dynamic behavior of a three-level laser. Note that the photon rate equations for four-level (5.13b) and three-level (5.16b) lasers are the same. The rate equations for the population inversion are somewhat different, however. Note, in particular, that the stimulated-emission term is given by $-2BqN$ for a three-level laser, whereas it is given by $-BqN$ for a four-level laser. The factor of 2 difference arises from the fact that the emission of a photon implies a change of 2 in the population inversion of a three-level laser (N_2 decreases by 1 and N_1 increases by 1), while it only implies a population change of 1 in a four-level laser. In the latter case, in fact, while N_2 again decreases by 1, N_1 remains approximately unchanged (i.e., zero) on account of the fast $1 \rightarrow 0$ decay.

5.3 CW LASER BEHAVIOR

In this section, we will investigate the laser behavior for a steady-state pump (i.e., W_p independent of time). Since, as we shall see later on, a steady-state pump gives a steady behavior, this case will be referred to as cw laser behavior.

5.3.1 Four-Level Laser

We begin by considering the threshold condition for laser action. Suppose that, at time $t = 0$, an arbitrarily small number q_i of photons is present in the cavity due to spontaneous emission. From (5.13b) we then see that, to have $\dot{q} > 0$, the condition $V_a BN > 1/\tau_c$ must be satisfied. Laser action is, therefore, produced when the population inversion N reaches a critical value N_c given by

$$N_c = \frac{1}{V_a B\tau_c} = \frac{\gamma}{\sigma l} \qquad (5.17)$$

where use has been made of (5.9). The critical pump rate W_p is then obtained by putting $\dot{N} = 0$, $N = N_c$, and $q = 0$ in (5.13a). We thus see that the critical pump rate corresponds to the situation where the total rate of pump transitions, $W_{cp}(N_t - N_c)$, equals the spontaneous transition rate from level 2, N_c/τ, i.e.,

$$W_{cp} = N_c/(N_t - N_c)\tau \qquad (5.18)$$

The physical significance of equation (5.17) can also be appreciated when we notice that, with the help of (5.5) and (5.4), it can be rearranged in the form

$$(1 - T_1)(1 - T_2)(1 - a)^2(1 - T_i)^2 \exp 2\sigma N_c l = 1 \qquad (5.19)$$

Equation (5.19) [and hence also (5.17)] implies that N_c must be large enough for the gain to compensate the total laser losses [see also Eq. (1.9), in which, for simplicity, the losses a and T_i were neglected].

If $W_p > W_{cp}$, the photon number q will grow from the initial value determined by spontaneous emission, and, if W_p is independent of time, it will eventually reach some constant value q_0. This steady-state value and the corresponding steady-state value N_0 for the inversion are obtained from (5.13) by setting $\dot{N} = \dot{q} = 0$. This gives

$$N_0 = 1/V_a B\tau_c = N_c \qquad (5.20a)$$

$$q_0 = V_a \tau_c \left[W_p(N_t - N_0) - \frac{N_0}{\tau} \right] \qquad (5.20b)$$

Equations (5.20) describe the cw behavior of a four-level laser. We will now examine these equations in some detail. First, it should be noted that equation (5.20a) shows that, even when $W_p > W_{cp}$, the relation $N_0 = N_c$ still holds. The steady-state inversion N_0 is always equal to the critical inversion N_c. To get a better understanding of the physical implication of this result, let us suppose the pump rate W_p to be increased from the critical value W_{cp}. When $W_p = W_{cp}$, we obviously have $N = N_c$ and $q_0 = 0$, whereas, if we now make $W_p > W_{cp}$, we see from (5.20) that, while N_0 remains fixed at the critical inversion, we have $q_0 > 0$. In other words, the increase in pump rate above the critical value produces an increase in the number of photons in the laser cavity (i.e., its e.m. energy) rather than an increase in the inversion (i.e., of the energy stored in the material). This situation is illustrated in Fig. 5.3 which shows a plot of the behavior of both N and q versus the pump rate W_p. Secondly it should be noted that equation (5.20b), with the help of (5.18) and (5.20a), can be recast in a somewhat more informative form:

$$q_0 = (V_a N_0) \frac{\tau_c}{\tau} (x - 1) \tag{5.21}$$

where

$$x = W_p / W_{cp} \tag{5.22}$$

is the amount by which threshold is exceeded. From (5.14) and (5.21), and making use also of (5.17) and (5.9b), we then find that the output power through one of the two cavity mirrors is given by

$$P_1 = \left(\frac{V_a \hbar \omega}{\sigma l \tau} \right) \left(\frac{\gamma_1}{2} \right) (x - 1) \tag{5.23}$$

This expression agrees with that originally given by Rigrod[5] for the case

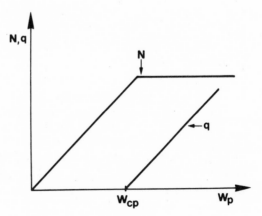

FIG. 5.3. Qualitative behavior of the population inversion N and total number of cavity photons q as a function of the pump rate W_p.

where mirror 2 is 100% reflecting. Equation (5.23) can be further simplified if we write $V_a = A_e l$, where A_e is the equivalent cross-sectional area of laser medium occupied by the oscillating mode (or modes). According to (5.2) and (5.2a) we have $A_e = \pi w_0^2 / 4$ or $A_e = A$ depending on whether the laser is oscillating on one or on many modes. Furthermore, for both optical and electrical pumping, we can write $x = P_{in}/P_{th}$, where P_{in} is the input power (to the lamp or to the discharge) and P_{th} is its threshold value. Thus, (5.23) can be written as

$$P_1 = (A_e I_s) \frac{\gamma_1}{2} \left[\frac{P_{in}}{P_{th}} - 1 \right] \tag{5.23a}$$

where $I_s = \hbar\omega/\sigma\tau$ is the gain saturation intensity for a four-level system [see (2.134)]. A plot of P_1 versus P_{in} should therefore yield a straight line which intercepts the P_{in} axis at $P_{in} = P_{th}$. We can therefore define the slope efficiency η_s of the laser as

$$\eta_s = \frac{dP_1}{dP_{in}} \tag{5.24}$$

and η_s turns out to be a constant for a given laser configuration. Before ending this section we stress again that the results we have derived are valid only when level 1 can be considered to be empty. This applies when $\tau_1 \ll \tau$, where τ_1 is the lifetime of level 1. When τ_1 is comparable with τ the previous equations need to be modified. A particularly simple case occurs when the lifetime τ_{21} (radiative plus nonradiative) for the $2 \to 1$ transition is equal to the overall lifetime of level 2 (i.e., $\tau_{2g} \to \infty$). In this case it can be shown by a somewhat lengthy but straightforward calculation that (5.17), (5.20a), (5.21), (5.22), and (5.23) still apply while (5.18), given the assumption that $N_c \ll N_t$, becomes

$$W_{cp} = \frac{N_c}{N_t(\tau - \tau_1)} \tag{5.18a}$$

With the help of the previous equations we can get two useful and informative expressions for η_s both for optical and electrical pumping respectively. For the optical pumping case, using (5.18) and (5.17) we get $W_{cp} = \gamma/\sigma l N_t \tau$, and with the help of (3.15) we obtain

$$P_{th} = \frac{\gamma}{\eta_p} A I_s \tag{5.25}$$

From (5.23a), (5.24), and (5.25) we see that η_s can be expressed in an instructive form in which the various sources of inefficiency can be separately identified (compare with that given in Reference 12)

$$\eta_s = \eta_p \eta_c \eta_A \tag{5.24a}$$

The symbols have the following meaning: (i) η_p is the pump efficiency as given by (3.15). (ii) $\eta_c = \gamma_1/2\gamma$ may be called the output coupling efficiency (it is indeed $\leqslant 1$, having the value 1 when $\gamma_2 = \gamma_i = 0$). (iii) $\eta_A = A_e/A$ may be called the mode cross-section efficiency. For the case of electrical pumping, from (5.18), (5.17), and (3.25) we get

$$P_{th} = \frac{\gamma}{\eta_p} \frac{A\hbar\omega_p}{(\tau - \tau_1)} \qquad (5.25a)$$

With the help of (5.23a) and (5.25a), equation (5.24) gives the following expression for the slope efficiency η_s in which, again, the various sources of inefficiency can be separately identified:

$$\eta_s = \eta_p \eta_c \eta_A \eta_d \eta_q \qquad (5.24b)$$

The symbols have the following meaning: (i) η_p is the pump efficiency as given by (3.25). (ii) η_c, the coupling efficiency, and η_A, the cross-section efficiency, are defined as above. (iii) $\eta_d = (\tau - \tau_1)/\tau$ may be termed the decay efficiency of the lower laser level. (iv) $\eta_q = \hbar\omega_0/\hbar\omega_p$ may be termed the laser quantum efficiency. Note that η_q is not present in the corresponding expression for optical pumping as a result of the slight difference in definition of pump efficiency η_p for the two cases [compare (3.15) with (3.25)].

We now conclude this section by deriving a necessary condition for cw oscillation to be achieved in a four-level laser. Thus, we observe that, in the absence of oscillation, the cw population of level 1 must be given by the following equation, which simply balances the populations entering and leaving level 1: $(N_1/\tau_1) = (N_2/\tau_{21})$. To get laser oscillation we require $N_2 > N_1$, which, from the previous expression, implies

$$\tau_1 < \tau_{21} \qquad (5.26)$$

If this inequality is not satisfied, then laser action can only be possible on a pulsed basis provided the pumping pulse is shorter than or comparable to the lifetime of the upper level. Laser action will then begin and last until the number of atoms accumulated in the lower level is sufficient to wipe out the population inversion. For this reason these lasers are called self-terminating.

5.3.2 Three-Level Laser

The calculation for a three-level laser proceeds in a similar way to the four-level case, starting now from (5.16).

The threshold inversion is obtained by putting $\dot{q} = 0$ in (5.16b), thus giving

$$N_c = \frac{1}{BV_a\tau_c} = \frac{\gamma}{\sigma l} \qquad (5.27)$$

which is the same as for the four-level laser. The critical pump rate is then obtained from (5.16a) [setting $\dot{N} = 0$, $q = 0$, and $N = N_c$] as

$$W_{cp} = (N_t + N_c)/(N_t - N_c)\tau \qquad (5.28)$$

Note that, in practice, one has (both for three- and four-level lasers) $N_c \ll N_t$. Equation (5.28) then reduces to

$$W_{cp} \simeq 1/\tau \qquad (5.29)$$

A comparison of (5.29) with (5.18) shows that, for the same value of τ, the critical pump rate for a four-level laser is smaller by a factor (N_c/N_t) than for a three-level laser. This is the basis of the superior performance obtained with the four-level scheme.

The cw inversion N_0 and cw photon number q_0 when above threshold are then obtained from (5.16) by putting $\dot{N} = \dot{q} = 0$. Just as for the three-level laser, N_0 is again seen to be equal to N_c while q_0, obtained using (5.29) and (5.22/), is given by

$$q_0 = \frac{V_a(N_t + N_0)\tau_c}{2\tau}(x - 1) \qquad (5.30)$$

The output power through one mirror is then obtained, with the help of (5.14), as

$$P_1 = \frac{V_a(N_t + N_0)\hbar\omega}{2\tau}\left(\frac{\gamma_1}{2\gamma}\right)(x - 1) \qquad (5.31)$$

5.3.3 Optimum Output Coupling[6]

For a fixed pump rate, there is some value of the transmission T_1 of the output mirror which maximizes the output power. Physically the reason for this optimum stems from the fact that, as T_1 is increased, we have the following two contrasting circumstances: (i) The output power would tend to increase due to the increased transmission. (ii) The output power would tend to decrease since the increased cavity losses would cause the number of cavity photons q_0 to decrease.

To find the optimum transmission, we can use either (5.23) (four-level laser) or (5.31) (three-level laser) and impose the condition $dP_1/d\gamma_1 = 0$. We must obviously take into account the fact that x, N_0, and γ are also functions of γ_1. The problem is particularly simple for a four-level laser, and we will therefore limit ourselves to this case only. If we assume for simplicity $W_{cp} = N_c/N_t\tau$, then (5.23a), with the help of (5.22) and (5.17), can be written as

$$P_1 = \left[A_e I_s \left(\gamma_i + \frac{\gamma_2}{2} \right) \right] S \left(\frac{x_{min}}{S+1} - 1 \right) \tag{5.32}$$

where

$$S = \frac{\gamma_1}{\gamma_2 + 2\gamma_i} \tag{5.33a}$$

and

$$x_{min} = \frac{2 W_p \sigma l N_t \tau}{\gamma_2 + 2\gamma_i} \tag{5.33b}$$

The quantity x_{min} is the ratio between the actual pump rate (W_p) and the minimum pump rate (i.e., the pump rate to reach threshold for zero output coupling, $\gamma_1 = 0$). Since the first term in the square brackets of (5.32) is independent of γ_1, we can readily see, by imposing the condition $dP_1/dS = 0$, that the optimum value of S is

$$S_{op} = (x_{min})^{1/2} - 1 \tag{5.34}$$

and the corresponding output power is

$$P_{op} = \left[A_e I_s \left(\gamma_i + \frac{\gamma_2}{2} \right) \right] \left[(x_{min})^{1/2} - 1 \right]^2 \tag{5.35}$$

The reduction in power as a result of nonoptimum operating conditions becomes particularly important when working very close to threshold (i.e., when $x_{min} \simeq 1$). Well above threshold, however, the output power becomes rather insensitive to a change of output coupling around the optimum value. In the examples to be considered in the next section, we will in fact see that changes of the output coupling by as much as 50% only result in $\sim 10\%$ reduction of the output power.

5.3.4 Reasons for Multimode Oscillation

A number of the results obtained in the previous sections are strictly valid only when the laser is oscillating on a single mode. It is therefore appropriate, at this point, to consider the conditions under which single-mode or multimode oscillation is achieved. This forms the subject of both this and the next section.

Lasers generally tend to oscillate on many modes. The reason for this behavior arises essentially from the fact that the frequency separation of the modes is usually smaller (and often much smaller) than the width of the gain profile. This seemingly straightforward statement needs to be looked at more carefully, however. In fact, in the early days of laser development, it

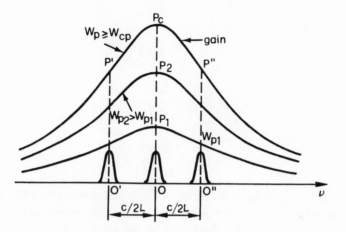

FIG. 5.4. Frequency dependence of laser gain versus pump rate W_p (homogeneous line).

was argued that lasers should always tend to oscillate on a single mode, provided the gain line was homogeneously broadened. The argument can be followed with the help of Fig. 5.4, in which it has been assumed that one of the cavity modes is coincident with the peak of the gain curve. For the sake of simplicity we will consider a plane-parallel resonator so that the modes are separated by $c/2L$ (only the lowest-order modes are considered, see Fig. 4.7). The laser gain coefficient is given by (2.65a). Oscillation will start on the central mode when the inversion $N = N_2 - N_1$ reaches a critical value N_c giving a gain equal to the cavity losses. Equation (5.17) is the quantitative statement of this condition. However, even when W_p is increased above the threshold value, in the steady state the inversion N remains fixed at the critical value N_c. The peak gain, represented by the length OP in Fig. 5.4, will therefore remain fixed at the value OP_c when $W_p \geqslant W_{cp}$. If the line is homogeneously broadened, its shape cannot change and the whole gain curve will remain the same for $W_p \geqslant W_{cp}$, as indicated in Fig. 5.4. The gain for other modes, represented by the lengths $O'P'$, $O''P''$, etc., will always remain smaller than the value OP_c, for the central mode. If all modes have the same losses, then, in the steady state, only the central mode should oscillate. The situation is quite different for an inhomogeneous line (Fig. 5.5). In this case, in fact, it is possible to "burn holes" in the gain curve (see Section 2.6.3 and, in particular, Fig. 2.20). Therefore, when W_p is increased above W_{cp}, the gain on the central mode remains fixed to the critical value OP_c, while the gain for other modes $O'P'$, O'', P'', etc. can keep on increasing up to the corresponding threshold value. In this case, if the laser is operating somewhat above threshold, then more than one mode can be expected to oscillate.

Shortly after the discovery of the laser, what was actually observed

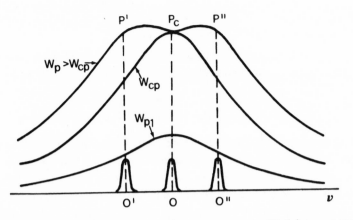

FIG. 5.5. Frequency dependence of laser gain versus pump rate W_p (inhomogeneous line): frequency hole burning effect.

experimentally was that multimode oscillation occurred both for inhomogeneous (e.g., gas laser) and homogeneous (e.g., ruby laser) lines. This last result appears to be in conflict with the argument given above. This inconsistency was later removed[7] by taking into account the fact that each mode has a well-defined standing-wave pattern in the active material. For the sake of simplicity, we will consider two modes whose standing-wave patterns are shifted by $\lambda/4$ in the active material (Fig. 5.6). We will assume that mode 1 in Fig. 5.6 is the center mode of Fig. 5.4, so that it is the first to reach threshold. However, when oscillation on mode 1 sets in, the inversion at those points where the electric field is zero (points A, B, etc.) will be left undepleted. At these points the inversion can continue growing beyond the critical value N_c. Mode 2, which initially had a lower gain, can now reach a gain equal to or even larger than that of mode 1 since it uses inversion from those regions which have not been depleted by mode 1. Mode 2, therefore, can oscillate as well as mode 1. The fact that the laser oscillates on many modes for a homogeneous line is, therefore, not due to holes burned in the gain curve (frequency hole burning) but to holes burned in the spatial distribution of inversion within the active material (spatial hole burning).

The conclusion then is that a laser always tends to oscillate on many modes. For a homogeneous line this is due to spatial hole burning, while for

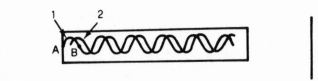

FIG. 5.6. Spatial hole burning effect in a laser material.

an inhomogeneous line this is due to both spatial hole burning (Fig. 5.6) and frequency hole burning (Fig. 5.5). There are, however, several methods for constraining a laser to oscillate on a single mode, and they will be discussed briefly in the next section.

5.3.5 Single-Line and Single-Mode Oscillation

Lasers often exhibit gain on more than one transition, the strongest of which usually results in laser oscillation. To make the laser oscillate on one of the other transitions one can use a dispersive prism (Fig. 5.7a) or a diffraction grating (Fig. 5.7b) in the so-called Littrow configuration. For a given angular setting of either the prism or grating there is only one wavelength (labeled λ_1 in each figure) which is reflected back in the resonator. Tuning is thus achieved by rotation of the grating, in the configuration of Fig. 5.7b, and by rotation of either the prism or mirror 2 in the configuration of Fig. 5.7a. Assuming the laser to be oscillating on a single line, we can now consider the conditions under which single-mode oscillation can be achieved.

It is usually fairly easy to make a laser oscillate on some particular transverse mode, i.e., one with prescribed values of the transverse mode indexes m and l (see Chapter 4). For example, to produce TEM_{00} mode oscillation, a diaphragm having an aperture of suitable size is usually inserted at some point on the axis of the resonator. If the radius a of this aperture is sufficiently small, the Fresnel number of the cavity $N = a^2/L\lambda$ will be determined by this aperture. As a decreases, the difference in loss between the TEM_{00} mode and higher-order modes increases (see Fig. 4.18 and 4.21). So, by an appropriate choice of aperture, we can obtain oscillation on the TEM_{00} mode alone. Note that this mode selecting scheme inevitably introduces some loss of the TEM_{00} mode itself. Another way of

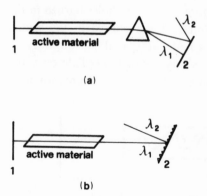

FIG. 5.7. Single-line operation of a laser using the wavelength dispersive behavior of: (a) a prism, (b) a diffraction grating.

producing single transverse-mode oscillation is to use an unstable resonator and choose the resonator parameters so that the equivalent Fresnel number has a half-integer value. As discussed in Section 4.5 (see, in particular, Fig. 4.28) there is a large loss discrimination between the lowest-order and the higher-order modes for these half-integer values of N_{eq}. In this case, however, the output beam is in the form of a ring, and this is not always convenient.

Even when a laser is oscillating on a single transverse mode (i.e., with m and l fixed), it can still oscillate on several longitudinal modes (i.e., modes differing in their value of the longitudinal mode index n). These modes are separated in frequency by $\Delta\nu_n = c/2L$ (Fig. 4.22). To isolate a single longitudinal mode, it is sometimes possible to use such a short cavity length that $\Delta\nu_n > \Delta\nu_0$, where $\Delta\nu_0$ is the width of the gain curve. In this case, if a mode is tuned to coincide with the center of the gain curve, the next longitudinal mode is far enough away from line center that (for a laser not too far above threshold) it cannot oscillate. This method can be used effectively with a gas laser where the laser linewidths are relatively small (a few gigahertz or smaller). However since L must be small, the volume of active material is also small, and this results in a low output power. The laser linewidths for solids or liquids are usually much larger (100 GHz or more), and the above method cannot be applied. In this case, and also for high-power single-mode gas lasers, two other longitudinal mode selection techniques are used[8] (Fig. 5.8). The first method makes use of a so-called Fabry–Perot transmission etalon inserted in the laser cavity (Fig. 5.8a). It consists of two plane-parallel reflectors (indicated by R in the figure) spaced by a distance d' and inclined at an angle θ to the resonator axis. Often the etalon consists of a solid block of transparent material (e.g., glass or quartz) with high-reflectivity coatings (e.g., $R = 80\%$) on its two parallel faces. The lowest-loss modes will be those for which the amplitude of the reflected beam U is zero. This beam is produced by interference of the beam OAU with the beam OBU (plus all the multiple reflections, such as $OBA'B'U$, etc.). The beam OAU undergoes a phase shift of π upon reflection,[†] while the phase shift of the beam OBU is $2kd'\cos\theta$. The difference between the phase shifts of the two beams is thus $(2kd'\cos\theta) - \pi$. For minimum loss the two beams must have opposite phases so as to interfere destructively. This condition implies that $2kd'\cos\theta - \pi = (2m - 1)\pi$,

---

[†]The fact that there is a phase shift π on reflection at A while there is no phase shift on reflection at B is the result of the well-known law of reflection at the boundary between two media. A π phase shift only occurs when the reflected ray lies in the medium of lower refractive index. In this case OA is assumed to be in air ($n \simeq 1$), while the etalon is made of some solid material ($n > 1$).

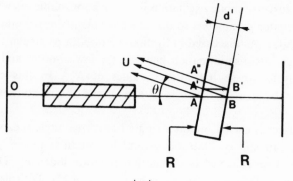

(a)

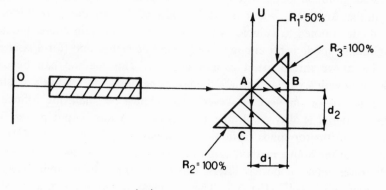

(b)

FIG. 5.8. Longitudinal mode selection: (a) use of a transmission Fabry–Perot etalon, (b) use of the Fox–Smith type reflection interferometer.

where m is a positive integer.[†] Therefore, since $k = 2\pi n\nu / c_0$ (where n is the refractive index of the etalon material), the frequencies corresponding to minimum loss are given by the expression $\nu = mc_0/2nd' \cos\theta$, and the frequency separation between two consecutive low-loss modes is $\Delta\nu = c_0/2nd' \cos\theta$. Since d' can be made very small, $\Delta\nu$ can be made very large, and the angle θ can be adjusted so as to make a low-loss mode coincident with the center of the gain line while the next one is outside this line. The second method makes use of a so-called Fox–Smith reflection interferometer, and it is illustrated in Fig. 5.8b. It is made by adding two extra mirrors R_1 and R_2 as indicated in the figure. For our purposes, we will

[†]By taking account of all multiple reflections, it can actually be shown that the overall reflectivity is in this case zero irrespective of the reflectivity of the etalon surfaces (provided that the reflectivity is the same for the two surfaces).

consider the interferometer to be made of a solid block of transparent material (the shaded block in Fig. 5.8b) coated on its three faces so as to provide the three mirrors R_1, R_2, and R_3. In this case again the lowest-loss modes are those for which the amplitude of the reflected beam U is zero. This beam is produced by the interference of the beam OAU with the beam $OBACU$ (plus all multiple reflections, e.g., $OBACABACU$, etc.). Upon reflection, the beam OAU undergoes a phase shift π, while the phase shift of the beam $OBACU$ is $2k(d_1 + d_2)$. The difference in phase shifts is $2k(d_1 + d_2) - \pi$, and this must be equal to an odd number of π, i.e., $2k(d_1 + d_2) - \pi = (2m - 1)\pi$. The frequency difference between two consecutive low-loss modes is now $\Delta\nu = c_0/2n(d_1 + d_2)$, where n is the refractive index of the block material. Here again $d_1 + d_2$, just like $d'\cos\theta$ in the previous case, can be made small enough to provide mode selection without any need to affect the length of the active material. Actually, these two longitudinal mode selection techniques really require a more detailed discussion than that given above. In fact, one should take into account both the frequency behavior of the Fabry–Perot etalon (or Fox–Smith interferometer) and the frequency behavior of the cavity modes (which are separated by $c/2L$). One should also take into account the fact that both of these frequency filters (i.e., the Fabry–Perot transmission filter and the Fox–Smith reflection filter) are not infinitely narrow in frequency. These more detailed points will not be discussed any further here. We refer the reader elsewhere for further details.[8]

5.3.6 Two Numerical Examples

As a first example, we will consider the case of a cw Nd:YAG laser. The active material is the Nd^{3+} ion in a crystal of $Y_3Al_5O_{12}$ (the crystal is known as YAG, an acronym from yttrium aluminum garnet).[10,11] The Nd^{3+} ions substitute for some of the Y^{3+} ions. A more detailed description of this laser material will be given in Chapter 6, and it is enough for our discussion here to note that this laser works on a four-level scheme and has an emission wavelength of $\lambda = 1.06$ μm (near infrared). We assume a 1% Nd^{3+} concentration (1% of Y^{3+} substituted by Nd^{3+}), which corresponds to a population in the ground-state (i.e., the lowest level of the $^4I_{9/2}$ state) of $N_g = 6 \times 10^{19}$ Nd^{3+} ions/cm^3. At this concentration, the lifetime of the upper laser level (which depends upon concentration due to a concentration dependent nonradiative channel) is $\tau = 0.23 \times 10^{-3}$ s. The lifetime of the lower laser level is much shorter than this ($\sim$30 ns). To calculate the effective cross section, we note that the upper laser level is actually made up of two strongly coupled levels separated by $\Delta E = 88\,\text{cm}^{-1}$ (see Fig. 6.2). Laser action takes place between the R_2 sublevel of the upper level to a sublevel of the lower ($^4I_{11/2}$) laser level. The cross section for this transition is $\sigma = 8.8 \times 10^{-19}\,\text{cm}^2$. Since, however, the two sublevels of the upper state are strongly coupled, then, according to (2.142m), the effective cross section to be used is equal to

$$\sigma_{21} = z_{2l}\sigma = 3.5 \times 10^{-19}\,cm^2 \tag{5.36}$$

where $z_{2l} = \exp(-\Delta E/kT)/[1 + \exp(-\Delta E/kT)] = 0.4$ is the partition function for the R_2 sublevel.[†]

We now consider a laser system as shown in Fig. 5.9 and assume the rod to be pumped by a high-pressure Kr lamp in an elliptical pump cavity. A typical curve of power output P_1 (for multimode oscillation) versus input power P_{in} to the Kr lamp would be as shown in Fig. 5.10.[(9)] Except for input powers just above threshold, the experimental points of Fig. 5.10 indeed show a linear dependence of the output power versus input power as predicted by (5.23a). The nonlinear portion of the curve close to threshold is very likely due to the focusing action of the elliptical pump cavity (see Section 3.2.2 of Chapter 3) since this would mean that at first lasing would occur only at the center of the rod. The linear portion of the curve gives an extrapolated threshold of $P_{th} = 2.2\ \text{kW}$, and it can be fitted by the equation (P_1 expressed in watts)

$$P_1 = 53\left(\frac{P_{in}}{P_{th}} - 1 \right) \tag{5.37}$$

The theoretical prediction can be readily obtained from (5.23a) once we note that the whole cross section of the rod is lasing so that we can take

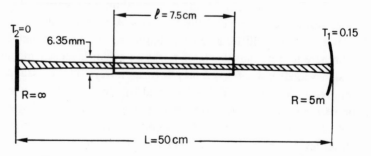

FIG. 5.9. Possible cavity configuration for a cw Nd:YAG laser.

[†]The author is indebted to Dr. Hanna for pointing out this interesting feature to him.

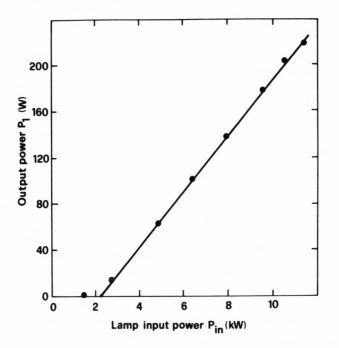

FIG. 5.10. Continuous output versus lamp input of a powerful Nd:YAG laser [after Koechner[9]].

$A_e = A = 0.31$ cm^2. Using the previous values for τ and σ_{21}, we get $I_s = \hbar\omega/\sigma_{21}\tau = 2.27$ kW/cm^2 and from (5.23a) $P_1 = 57[(P_{in}/P_{th}) - 1]$, in close agreement with the experimental results.

To compare the measured extrapolated threshold ($P_{th} = 2.2$ kW) and slope efficiency ($\eta_s = 2.4\%$) with the values predicted by calculation, we need to know γ, i.e., γ_i. Now, since $\gamma_2 = 0$, equation (5.25) can be rearranged as

$$\frac{-\ln R_1}{2} + \gamma_i = \eta_p \frac{P_{th}}{AI_s} \qquad (5.38)$$

where $R_1 = (1 - a_1 - T_1) \simeq (1 - T_1)$ is the reflectivity of the output mirror. We have neglected the mirror absorption a_1 since, for a good multilayer coating, it is certainly smaller than 0.5%. If several measurements are made of the threshold input power at different mirror reflectivities R_1, a plot of P_{th} versus $-\ln R_1$ should yield a straight line. In fact, this is what is found experimentally, as shown in Fig. 5.11. The extrapolation of the straight line in Fig. 5.11 to $P_{th} = 0$ gives, according to (5.38), the value of the internal

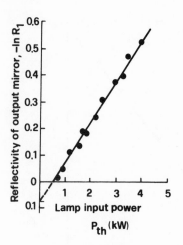

FIG. 5.11. Threshold power input as a function of mirror reflectivity [after Koechner[9]].

losses. In this way we get $\gamma_i \simeq 0.038$, which gives a total loss $\gamma = (\gamma_1/2) + \gamma_i = 0.1192$.

Once the internal losses are known, equation (5.24a) can be used to calculate the pumping efficiency η_p. From the slope efficiency of the curve of Fig. 5.10 (and taking $\eta_A = 1$) we get $\eta_p = 3.5\%$, which is a reasonable number for this type of pumping configuration (see also Table 3.1 of Chapter 3). A knowledge of the total losses also allows one to calculate the threshold inversion. From (5.17) one finds

$$N_c = 4.5 \times 10^{16} \text{Nd}^{3+} \text{ ions/cm}^3 \qquad (5.39)$$

Thus $N_c/N_g = 7 \times 10^{-4}$, which confirms that the population inversion is a very small fraction of the total population. We now calculate the optimum transmission of the output mirror when the laser is pumped three times above threshold $(x = 3)$, i.e., at a lamp input power of 6.6 kW. From (5.33b) we can see that $x_{\min} = x(\gamma/\gamma_i) = 9.4$. From (5.34) we thus get $(\gamma_1)_{\mathrm{op}} = 0.157$, which corresponds to an optimum transmission of $(T_1)_{\mathrm{op}} \simeq 14.5\%$, i.e., very close to the value used in the example.

As a final problem we calculate the expected output power in TEM_{00} operation at a lamp input power $P_{\mathrm{in}} = 10$ kW. First we find that, according to (4.41), the length L_e of the equivalent confocal resonator is 316 cm and the spot size at the plane mirror of Fig. 5.9 is $w_0 = (L_e\lambda/2\pi)^{1/2} = 0.73$ mm. To obtain TEM_{00} mode operation, we assume a circular aperture to be inserted near the spherical mirror, the diameter $2a$ of the aperture being just small enough to prevent oscillation of the TEM_{10} mode. The total losses for this mode must therefore be at least as much as $\gamma' = \gamma(P_{\mathrm{in}}/P_{\mathrm{th}}) = 0.54$, and the diffraction loss introduced by the diaphragm must be

$\gamma_d = \gamma' - \gamma = 0.42$. In a round trip the diffraction loss is therefore $2\gamma \simeq 0.84$, which corresponds, according to (5.4c), to a loss $T_i = 57\%$ in a double pass. To calculate the required aperture size we note that the round-trip loss of the system shown in Fig. 5.9 is the same as the single-pass loss of a symmetric resonator consisting of two mirrors both of radius $R = 5$ m and aperture $2a$, spaced by a length $L_s = 2L = 1$ m. From Fig. 4.21b we then see that, since $g = 0.8$ and since the loss needs to be 57%, we require $N = a^2/\lambda L_s = 0.5$, which gives $a = 0.73$ mm. From Fig. 4.21a we see that this aperture introduces a 28% loss for the TEM_{00} mode of the equivalent symmetric resonator. This is also therefore the round-trip diffraction loss of our resonator, which means that, according to (5.4c), the single-pass loss is $\gamma_d \simeq 0.164$. Total losses for the TEM_{00} mode now rise to $\gamma' = \gamma + \gamma_d = 0.283$, and the expected threshold power is $P'_{th} = 5.2$ kW. According to (5.37) the expected output power at $P_{in} = 10$ kW is $P_1 = 53 \ (A'_e/A_e)$ $[(P_{in}/P'_{th}) - 1] = 1.3$ W, where $A'_e = \pi w_0^2/2 = 0.84$ mm^2.

As a second example we will take the case of a high-power CO_2 laser. We will consider a laser system such as that indicated in Fig. 5.12, having a positive-branch unstable confocal resonator. The length of the resonator is $L = 175$ cm, while the length of the laser medium is $l = 140$ cm. Excitation of the CO_2 gas is provided by an electric discharge between two plane electrodes as indicated in the figure (see also Fig. 6.15). Typical performance data for output power P_1 versus input power P_{in} to the electrical discharge are shown in Fig. 5.13.[13] The data points can be fitted by the equation

$$P_1 = 6.66\left[\frac{P_{in}}{P_{th}} - 1 \right] \tag{5.40}$$

where P_1 is given in kilowatts and P_{th} is the extrapolated threshold input power ($P_{th} \simeq 44$ kW).

Since the CO_2 laser operates as a four-level laser, equation (5.40) can be compared with (5.23a). To do this, we need to know the transmission T_1

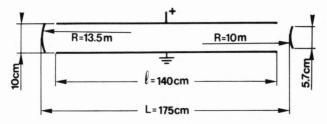

FIG. 5.12. Possible cavity configuration for a high-power CO_2 TE laser.

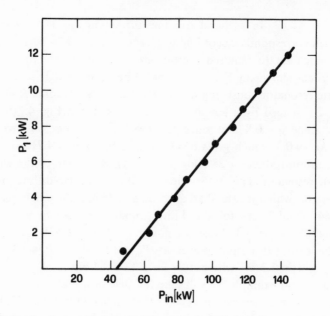

FIG. 5.13. Continuous output power P_1 versus electrical discharge power P_{in} for a powerful CO_2 TE laser.

of the output mirror. In the geometrical-optics approximation we have [see (4.59)]

$$T_1 = \frac{M^2 - 1}{M^2} = 0.45 \qquad (5.41)$$

In this expression M is the round-trip magnification factor and it is given by $M = R_1/R_2 = 1.35$, where R_1 and R_2 are the radii of the two mirrors. Using a wave-theory calculation (see Fig. 4.29) would yield $T_1 = 0.2$ for the lowest-order mode. We will take the value given by geometrical optics as the more realistic value in our case for the following two reasons: (i) The equivalent Fresnel number is rather large ($N_{eq} = 7.4$) and a few transverse modes are expected to have comparable losses (see Fig. 4.28). (ii) The laser is excited sufficiently far above threshold (by a factor 2.8 at 12 kW output, see Fig. 5.13) that most of the above modes may actually be oscillating. We will in fact see, in the calculation that follows, that the geometrical-optics value of T_1 leads to a much better agreement with experiment than that obtained using the wave-theory value for T_1. A comparison of (5.40) with (5.23a) using $T_1 = 0.45$ then yields $A_e I_s = 22.3$ kW. The beam diameter in the laser cavity is (see also Fig. 4.2.6b) $D = 2Ma_2 = 7.6$ cm, thus giving $A_e = \pi D^2/4 \simeq 45$ cm^2 and hence $I_s \simeq 500$ W/cm^2. This value is in agreement with theoretical estimates.[14]

From the data of Fig. 5.13 we can now go on to evaluate the (unsaturated) gain g_0 expected for the laser medium at an input power $P_{in} \simeq 140$ kW. In fact, we have

$$g_0 = N_2\sigma = \frac{P_{in}}{P_{th}} N_{20}\sigma = \frac{P_{in}}{P_{th}} \frac{\gamma}{l} \tag{5.42}$$

where N_2 and N_{20} are the population of level 2 at $P_{in} = 140$ kW and $P_{in} = P_{th}$ respectively. To calculate γ we assume a mirror loss (absorption plus scattering) of 2%, while we neglect internal losses. From (5.4) and (5.5) we then get $\gamma_1 = 0.598$, $\gamma_2 = 0$, $\gamma_i = 0.02$, and $\gamma = 0.319$. Substituting this last value into (5.42) gives $g_0 = 6.3 \times 10^{-3}$ cm^{-1}, which is in fair agreement with the measured values for this type of laser.[15]

We now compare the experimental value of the slope efficiency in Fig. 5.13 with the theoretical prediction. Since $\eta_p \simeq 0.7$ (see Section 3.3.3) and since $\eta_q = 0.4$, we obtain from (5.24*b*)

$$\eta_s = 0.22\eta_A\eta_d \tag{5.43}$$

which is to be compared with the value $\eta_s \simeq 0.12$ obtained from Fig. 5.13. We are therefore led to a value for $\eta_A\eta_d$ of ~ 0.55. It is, however, quite possible that $\eta_A\eta_d$ is appreciably higher than this value since the actual value of pumping efficiency may be somewhat smaller than 0.7. The data of Fig. 5.13 refer in fact to a partially closed-cycle system and, in this case, the products of the discharge are likely to accumulate in the gas mixture, thus reducing the pumping efficiency.

We can finally go on to calculate the optimum output coupling at $P_{in} = 140$ kW, i.e., at $x = 2.8$ above the threshold condition of Fig. 5.13. Since $x_{min} = x(\gamma/\gamma_i) = 44.6$, from (5.34) we get $(\gamma_1)_{op} = 0.23$, which corresponds to $(T_1)_{op} = 20\%$. This implies that the laser is appreciably overcoupled. This may well be done deliberately since, while it slightly reduces (by $\sim 10\%$) the power of the output beam, it improves its focusing behavior. In fact, an increased value for T_1 is achieved by increasing M and hence also increasing the annular width of the output beam [$\simeq (M - 1)a_2$, see Fig. 4.26]. This results in an improved focusing behavior for the beam.

5.3.7 Frequency Pulling and Limit to Monochromaticity

We now look at two phenomena which cannot be described within the rate-equation approximation used so far but which are, nevertheless, very important and should be considered here. For this discussion we refer to Fig. 5.14 which shows the resonance curves of both the laser line (centered at ω_0 and of width $\Delta\omega_0$) and of a cavity mode (centered at ω_c and of width

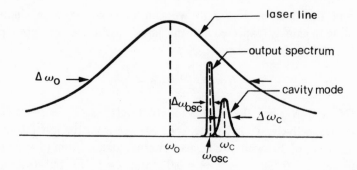

FIG. 5.14. Frequency pulling and spectral output in a single-mode laser.

$\Delta\omega_c$). We assume oscillation to occur on this mode, and we address ourselves to the question of finding the oscillation frequency ω_{osc} and the width of the output spectrum $\Delta\omega_{osc}$.

The calculation of ω_{osc} can be carried out within the semiclassical approximation. It can be shown[1] that ω_{osc} will be in some intermediate position between ω_0 and ω_c. Thus, ω_{osc} is not coincident with ω_c but it is pulled toward the laser line center frequency ω_0. To first order, for an inhomogeneous line (and rigorously for a homogeneous line), the oscillation frequency is given by a weighted average of the two frequencies ω_0 and ω_c. The weighting factors are proportional to the inverse of the corresponding linewidths. Thus we have

$$\omega_{osc} = \frac{(\omega_0/\Delta\omega_0) + (\omega_c/\Delta\omega_c)}{(1/\Delta\omega_0) + (1/\Delta\omega_c)} \tag{5.44}$$

The value of $(\Delta\omega_0/2\pi)$ may range from ~ 1 GHz [for Doppler-broadened transitions in the visible, see (2.114)] to as much as 300 GHz for solid-state lasers (see Fig. 2.14). On the other hand, for a 1-m-long cavity, $(\Delta\omega_c/2\pi)$ $= 1/2\pi\tau_c = \gamma c_0/2\pi L$ [see (4.9) and (5.9b)] may range from ~ 1 MHz to a few tens of MHz (for γ ranging from $\sim 10^{-2}$, typical of a low-gain laser medium such as He–Ne, to values of the order of 5×10^{-1} for higher-gain materials). Since, therefore, $\Delta\omega_c \ll \Delta\omega_0$, the effect of frequency pulling is generally very small.

We now turn our attention to the calculation of the width $\Delta\omega_{osc}$ of the laser output spectrum when oscillating in this single mode. Its ultimate limit is established by spontaneous emission noise, or equivalently by the zero-point fluctuations of the laser mode field. Since these fluctuations can only be accounted for by a full quantum mechanical approach to the problem (see Section 2.3.2), this limit cannot be derived within our present treatment. It can be shown that zero-point fluctuations produce a spectral broadening of the output with a Lorentzian line shape mostly arising from frequency fluctuations of the output beam.[16] If, for simplicity, the internal losses γ_i are neglected, then the spectral width (FWHM) of the laser output is given by

$$\Delta\omega_{osc} = \frac{4\hbar\omega_{osc}(\Delta\omega_c)^2}{P} \qquad (5.45)$$

where P is the output power. Even for modest output powers (e.g., $P \simeq 1$ mW) the value of $\Delta\omega_{osc}$ predicted by (5.45) is so small that in practice other spectral broadening mechanisms dominate in establishing the actual linewidth $\Delta\omega_{osc}$. From (5.45) we have in fact $(\Delta\omega_{osc}/\omega_{osc}) = 4\hbar(\Delta\omega_c)^2/P$ which, for $(\Delta\omega_c/2\pi) \simeq 10^7$ Hz, gives $(\Delta\omega_{osc}/\omega_{osc}) \simeq 10^{-15}$. To appreciate the significance of such a spectral purity, let us examine the requirements on the stability of the cavity length in order to keep the resonator frequency ω_c stable to within the same limits. From (4.3), for $n = $ const, we find $(\Delta L/L) = -(\Delta\omega_c/\omega_c) \simeq 10^{-15}$. Hence for $L = 1$ m $= 10^{10}$ Å we have $|\Delta L| \simeq 10^{-5}$ Å, i.e., much smaller than a typical atomic dimension (~ 1Å). This indicates that, in practice, the limit to monochromaticity is likely to be set by changes of cavity length induced by vibrations or thermal effects.[17] If the two cavity mirrors are supported by massive Invar bars as spacers, acoustic vibrations can result in values of $(\Delta\omega_{osc}/2\pi)$ of the order of a few to a few tens of kilohertz $(\Delta\omega_{osc}/\omega_{osc} = 10^{-10}$–$10^{-11})$. A temperature change ΔT of the cavity contributes an amount $(\Delta\omega_{osc}/\omega_{osc}) = \alpha\Delta T$, where α is the expansion coefficient of the spacer material. For Invar $\alpha \simeq 10^{-7}/°$K so that $(\Delta\omega_{osc}/\omega_{osc}) = 10^{-7}\Delta T$. Temperature changes of even $10^{-3}°$K thus result in a drift of the mode frequency (and hence of the laser output frequency), which is larger than the width established by acoustic vibrations. The effects of acoustic vibrations (short-term frequency stability) and of temperature changes (long-term frequency stability) can, however, be drastically reduced using techniques for active stabilization of the cavity frequency. These are described in the next section.

5.3.8 Lamb Dip and Active Stabilization of Laser Frequency

Another interesting effect which cannot be explained within the present treatment based on the rate-equation approximation is the *Lamb dip* phenomenon, which takes its name from the physicist W. E. Lamb, who predicted it theoretically.[18] It occurs in any gas laser oscillating in a single mode when inhomogeneous broadening, due to the Doppler effect, domi-

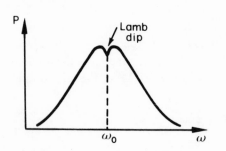

FIG. 5.15. Lamb dip phenomenon.

nates over homogeneous broadening. The phenomenon is illustrated in Fig. 5.15, which shows the variation of output power with oscillation frequency for a fixed pumping rate. An output-power curve of this form can be observed experimentally with a single-mode laser in which the output frequency is changed by fine tuning of the cavity length (over a range of one half-wavelength). As indicated in Fig. 5.15, the output power shows a dip at the center frequency ω_0 of the transition, a behavior which, at first sight, seems paradoxical.

   To understand this behavior, we first consider the experimental situation of Fig. 5.16, in which the saturation induced by the laser field in the active material is monitored by a low-intensity (i.e., nonsaturating) probe beam propagating at a small angle to the cavity axis (compare with Fig. 2.16). Let us begin by considering the case in which the laser is oscillating at a frequency $\omega \neq \omega_0$ (for example, $\omega < \omega_0$). The laser radiation will interact only with those atoms having a velocity v in the opposite direction to that of the radiation, the value of v being such that $\omega[1 + (v/c)] = \omega_0$ (Doppler effect). Now, in a laser cavity, the beam propagates forwards and backwards between the two mirrors. Therefore, for the beam traveling to the right the interaction will involve atoms traveling to the left, whereas for the left-traveling beam the interaction involves right-traveling atoms. Thus the mode we are considering will saturate the populations of two groups of atoms: those with velocity $+v$ and those with velocity $-v$. From what was said in Section 2.6.3 (see Fig. 2.19) it is then clear that, due to the saturation arising from the intense laser beam, the probe beam will see two "holes" in the gain line at the frequencies corresponding to the atoms with velocity

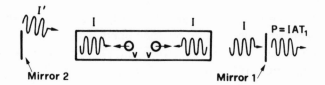

FIG. 5.16. Gain saturation in a gas laser with a Doppler-broadened transition.

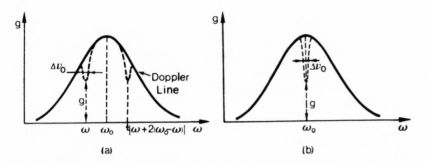

FIG. 5.17. Holes burned in the gain profile of a gas laser oscillating at a frequency (a) $\omega \neq \omega_0$ and (b) $\omega = \omega_0$.

$\pm v$. The two frequencies will therefore be ω and the mirror image of ω with respect to ω_0 (Fig. 5.17a). The width of each hole is of the order of the homogeneous linewidth. At each of the two frequencies, the saturated gain $g(\omega)$ will be [see (2.135)]

$$g = \frac{g_0(\omega)}{1 + \dfrac{I(\omega)}{I_s(\omega)}} \qquad (5.46)$$

where g_0 is the unsaturated gain and I is the intensity of each of the two oppositely traveling waves (which, for simplicity, we assume to have the same intensity). If we now consider the case where $\omega = \omega_0$, the laser beam will interact with those atoms for which $v = 0$. The two holes of Fig. 5.17a then coalesce into a single hole at the line center (Fig. 5.17b). In this case the saturated gain $g(\omega_0)$ will be

$$g(\omega_0) = \frac{g_0(\omega_0)}{1 + \dfrac{2I(\omega_0)}{I_s(\omega_0)}} \qquad (5.47)$$

The factor 2 in the denominator of (5.47) accounts for the fact that both beams are now saturating the same set of atoms.

The output power expected in the two cases can be obtained from the condition that the saturated gain must equal the cavity losses (which we denote by α_c). Since $P = IAT_1$ (see Fig. 5.16), from (5.46) and (5.47) we then find, for the two cases, respectively [compare with (5.23a)]

$$P(\omega) = AT_1 I_s(\omega) \left[\frac{g_0(\omega)}{\alpha_c} - 1 \right] \qquad (5.48)$$

$$P(\omega_0) = AT_1 \frac{I_s(\omega_0)}{2} \left[\frac{g_0(\omega_0)}{\alpha_c} - 1 \right] \qquad (5.49)$$

where A is the cross-sectional area of the beam and T_1 is the transmission of the output mirror. Under the appropriate experimental conditions (which are easily met in practice), the factor 2 which is present in the denominator of (5.49) but absent from (5.48) results in $P(\omega_0) < P(\omega)$. A dip in the laser output is therefore expected at $\omega = \omega_0$, as indicated in Fig. 5.15. Note that, from the discussion relating to Fig. 5.17, it follows that the width of this dip is of the order of the homogeneous linewidth $\Delta\nu_0$.

The Lamb dip phenomenon can be exploited as a very effective way of stabilizing the laser frequency.[19] Since, in fact, the width of the Lamb dip is usually much smaller than the laser transition linewidth [e.g., compare (2.106) with (2.114)], the location of the bottom of the Lamb dip is very well defined. Suppose therefore that one of the cavity mirrors is mounted on a piezoelectric transducer so that the cavity length can be finely tuned by changing the voltage applied to the transducer. With a suitable electronic feedback circuit the laser frequency can now be stabilized with respect to the minimum of the Lamb dip. He–Ne lasers that have been stabilized in this way have shown a frequency stability and reproducibility of one part in 10^9. This figure for stability is limited by the fact that the center frequency of the transition is itself not perfectly stable since it depends (although only to a small degree) on the gas pressure and the value of the excitation current. An even better method of stabilization is to make use of a phenomenon similar to the Lamb dip which occurs when a gas (which is not pumped) having an absorption line which coincides exactly with the center frequency of the gain line is put in a separate cell inside the laser cavity. According to the discussion above, such a gas would show an absorption which, under saturation conditions (i.e., during laser oscillation), has a minimum at $\omega = \omega_0$. For $\omega = \omega_0$ one would then have a dip in the gain curve for the active medium and also a dip in the absorption curve of the absorber gas. By choosing the operating parameters so that the second effect predominates, one obtains a peak in the output power for $\omega = \omega_0$, called the inverted Lamb dip (Fig. 5.18). By locking the output frequency

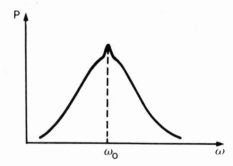

FIG. 5.18. Output power P as a function of frequency for a gas laser having a Doppler-broadened absorber (at frequency ω_0) placed inside the cavity (inverted Lamb dip).

to this peak, a better stability and reproducibility is achieved (10^{-12}–10^{-13}).[19] The center frequency of the absorber is much more stable than that of the active material because no current is passed through the absorber and also because the absorber can be kept at a lower pressure. For a He–Ne laser oscillating at $\lambda = 3.39$ μm, methane gas is used as the absorber, while for oscillation at $\lambda = 0.633$ μm, $^{129}I_2$ is used.

5.4 TRANSIENT LASER BEHAVIOR

The study of the transient behavior of a laser implies solving equations (5.13) or (5.16) for a four-level or three-level laser, respectively. For a given time-dependent pump rate $W_p(t)$ we thus find the time behavior of $q(t)$ and $N(t)$ once the initial conditions are given. In the following discussion we will look into a few interesting examples of transient behavior of lasers. Since the equations describing this behavior are nonlinear in the variables $q(t)$ and $N(t)$ (they in fact involve products of the form qN) a general analytical solution is not possible, and we will limit ourselves to discussing a few important results.

5.4.1 Spiking Behavior of Single-Mode and Multimode Lasers

The first case we consider is that of a step-function pump rate. We thus assume that $W_p = 0$ for $t < 0$ and $W_p(t) = W_p$ (independent of time) for $t > 0$. We will first assume the laser to be oscillating in a single mode since, strictly speaking, this is necessary if equations (5.13) and (5.16) are to be valid.

As a representative example, Fig. 5.19 shows the computed time behavior of $N(t)$ and $q(t)$ for a three-level laser such as a ruby laser. The initial conditions are $N(0) = -N_t$ and $q(0) = q_i$, where q_i is some small integer which is needed only to allow laser action to start. Several features of this figure are worth pointing out: (i) The cavity photon number $q(t)$ displays a regular sequence of peaks (spikes) of decreasing amplitude with consecutive spikes separated by a few microseconds. The output power would therefore show a similar time behavior. A regular oscillation of this sort is usually referred to as regular spiking. (ii) The population inversion $N(t)$ oscillates about the steady-state value N_0. (iii) Both $N(t)$ and $q(t)$ eventually reach the steady-state values predicted by (5.27) and (5.30) respectively. The oscillatory behavior of both $N(t)$ and $q(t)$ is due to the delay which it takes for the photons to follow a given change of population inversion. Thus, when $N(t)$ first passes through the value N_0 (at $t \simeq 4$ μs in

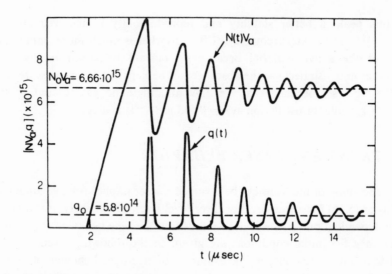

FIG. 5.19. Example of the temporal behavior of total inversion $V_a N(t)$ and photon number $q(t)$ for a three-level laser (from Reference 3).

the figure), the threshold condition is reached, and the laser can start oscillating. However, it takes some time for the cavity photon number to build up from its initial value produced by spontaneous emission, and during this time $N(t)$ can continue to grow above N_0 due to the continuing process of pumping. When, however, $q(t)$ has grown to a sufficiently high value, $N(t)$ begins to decrease due to the rapid rate of stimulated emission. At the time corresponding to the maximum of $q(t)$, $N(t)$ has dropped back to N_0. This can readily be shown from (5.16b) since, when $(dq/dt) = 0$, we have $N = 1/V_a B\tau_c = N_0$. The value of $N(t)$ is then driven below N_0 by the still rapid rate of stimulated emission. Thus the laser goes below threshold and laser action tends to cease. The population $N(t)$ is then driven upwards again by the pumping process until threshold is again reached ($t \simeq 6 \ \mu s$) and the cavity photons can grow again, and so on. It should be noted that, since the steady-state solution predicted by (5.27) and (5.30) is eventually reached, the computer calculation confirms that these solutions correspond to a stable operating condition.

The discussion so far applies only to single-mode oscillation and, in this case, it has been found that the experimental results are in good agreement with the above theoretical predictions. Actually, it is not always easy to achieve single-mode oscillation, particularly when the linewidth of the laser transition is much greater than the mode separation (which applies, for instance, to solid-state and liquid lasers). For multimode

FIG. 5.20. Typical time behavior of a multi-mode solid-state laser. The output in this case is from a ruby laser, and the time scale is 10 μs per division.

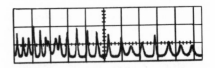

oscillation the theoretical treatment becomes much more involved. It is not enough to specify simply the total number of photons summed over all oscillating modes. In fact, to take account of the temporal and spatial interference of the modes, it is necessary to write as many equations for the electric field of the e.m. wave (both amplitude and phase) as there are oscillating modes. In this case the time behavior of the output is generally not as simple as in Fig. 5.19, and a typical example of the time behavior observed with solid-state lasers is shown in Fig. 5.20. It can be seen that the output consists of a train of pulses irregularly spaced in time and of random amplitude (*irregular spiking*). Furthermore, the oscillation does not tend to a steady-state value as in Fig. 5.19. This behavior is due to the fact that the oscillating modes usually change from one spike to the next or from one set of spikes to the next set, a phenomenon known as mode hopping. In this case, the laser does not behave in a regular and reproducible way.

The output from a multimode laser can still behave in a regular fashion, as in Fig. 5.19, under certain conditions. This applies when the number of oscillating modes is very large and, at the same time, the phases of the corresponding electric fields are random. In this case, in fact, the total light intensity is the sum of the individual mode intensities, and we can, therefore, continue to talk in terms of a total number of photons, q, in the cavity. This can occur when (i) the frequency separation between the modes is very small compared to the laser linewidth (long resonator); (ii) the loss for each mode is large, and mode linewidths are then comparable to or larger than the frequency spacing of the modes; (iii) the loss is approximately the same for all modes. In this case, actually, the concept of a cavity mode has little physical significance, and one should, instead, treat the cavity as a nonresonant feedback system.[26]

5.4.2 Q Switching[23]

The technique of Q switching[20] allows the generation of laser pulses of short duration (from a few nanoseconds to a few tens of nanoseconds) and high peak power (from a few megawatts to a few tens of megawatts). The principle of the technique is as follows. Suppose a shutter is introduced into the laser cavity. If the shutter is closed, laser action cannot occur and the population inversion can reach a very high value. If the shutter is opened suddenly, the laser will have a gain far in excess of the losses, and the stored energy will be released in the form of a short and intense light pulse. Since this technique involves switching the cavity Q factor from a low to a high value, it is known as Q switching. Provided the opening of the shutter only takes a short time compared to the buildup time of the laser pulse (fast switching), the output does indeed consist of a single giant pulse. In the case of slow switching, however, multiple pulses may occur. In fact, the energy stored in the medium prior to switching becomes depleted in a series of steps, each step corresponding to the emission of a pulse. Each pulse drives the gain below the instantaneous threshold, thus inhibiting further oscillation until the switch again decreases the loss in the laser cavity and hence decreases the threshold.

5.4.2.1 Methods of Q switching

The following switching systems are the most widely used.

(i) *Electro-optical Shutters.* These exploit a suitable electro-optical effect such as the Pockels effect. Referring the reader elsewhere for details,[21] we mention here that a cell based on the Pockels effect (Pockels cell) is a device which, when subjected to a dc applied voltage, becomes birefringent. This induced birefringence is proportional to the applied voltage. Figure 5.21 shows a Q-switched laser using a combination of a polarizer and Pockels cell. The Pockels cell is oriented and biased in such a way that the axes X' and Y' of the induced birefringence are lying in the plane orthogonal to the axis of the laser resonator. The polarizer axis makes an angle of 45° to the birefringence axes. Consider now a light wave

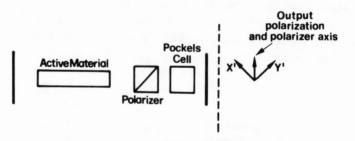

FIG. 5.21. Polarizer–Pockels-cell combination for Q switching. The right-hand side of the figure (after the dashed line) is a view along the resonator axis of the output polarization, of the polarizer axis, and of the birefringence axes of the Pockels cell (X', Y').

propagating from the active material toward the polarizer–Pockels-cell combination. If the voltage applied to the Pockels cell has a suitable value (of the order of 1–5 kV), the induced birefringence will be such that the linearly polarized light passed by the polarizer will be converted to circularly polarized light after passing through the Pockels cell. Upon reflection at the mirror, this light is further transformed by the Pockels cell to linearly polarized light whose polarization is orthogonal to its original direction. This light is therefore blocked by the polarizer. Thus, in this condition, the Q switch is closed. The switch is opened by removing the bias voltage, since the birefringence then vanishes and the incoming light is transmitted without change of polarization.

(ii) Mechanical Shutters. One mechanical means of Q switching consists of rotating one of the end mirrors of the laser resonator about an axis perpendicular to the resonator axis (Fig. 5.22). To avoid multiple pulsing it is necessary to use a high speed of rotation. For a resonator length $L = 50$ cm, the speed required is of the order of 30,000 rpm.

(iii) Shutter Using Saturable Absorbers. This provides the simplest Q-switching method. The shutter in this case consists of a cell containing some suitable saturable absorber which absorbs at the laser wavelength. It is usually in the form of a solution of a saturable dye (e.g., the dye known as BDN for the case of Nd:YAG). Such an absorber can be treated as a two-level system with a very large peak absorption cross section (10^{-16} cm^2 is typical for a saturable dye). It then follows from (2.128) that the corresponding saturation intensity I_s is comparatively small, and the absorber becomes almost transparent (due to saturation) for a comparatively low incident-light intensity. Now suppose that a cell containing a solution of this absorber, having a peak absorption wavelength coincident with the

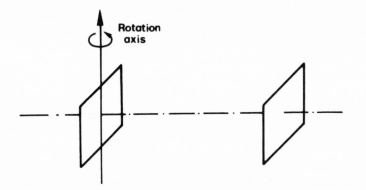

FIG. 5.22. Rotating mirror system for Q switching.

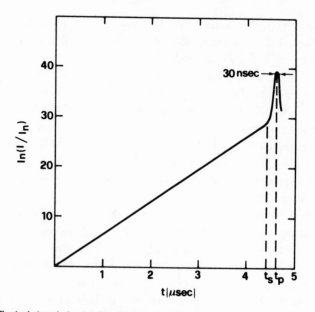

FIG. 5.23. Typical time behavior for the laser beam intensity I in a 60-cm-long cavity which is passively Q switched by a saturable absorber. The quantity I_n is the noise intensity in the given mode due to spontaneous emission. The pulsewidth (FWHM, ~ 30 ns) is also shown.

laser wavelength, is introduced into the laser cavity. Suppose also, for the sake of argument, that the initial (i.e., unsaturated) absorption of the cell is 50%. Laser action can start only when the gain of the active material compensates the loss of the cell plus the unsaturable cavity losses. Owing to the large absorption of the cell, the critical population inversion is very high. When laser action eventually starts, the laser intensity will build up from the starting noise due to spontaneous emission (Fig. 5.23). When the intensity becomes comparable to I_s (which occurs at time $t = t_s$ in Fig. 5.23), the absorber begins to bleach due to saturation. The rate of growth of laser intensity is thus increased, and this in turn results in an increased rate of absorber bleaching, and so on. Since I_s is comparatively small, the inversion still left in the laser medium after the absorber is bleached is essentially the same as the initial inversion (i.e., very large). After bleaching the laser will therefore have a gain well in excess of the losses, and a giant pulse will develop (Fig. 5.23).

(iv) Acousto-optic Q Switches. An acousto-optic modulator consists of a block of transparent optical material (e.g., fused silica is used in the visible, or germanium in the infrared) in which an ultrasonic wave is launched by a piezoelectric transducer. Due to the presence of the ultrasonic wave, the

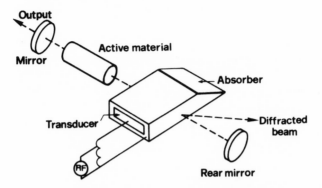

FIG. 5.24. A laser Q switched by an acousto-optic modulator.

material acts like a phase grating. In fact the strain induced by the ultrasonic wave results in local changes of the material refractive index (photoelastic effect). The grating has a period equal to the acoustic wavelength and an amplitude proportional to the sound amplitude. If an acousto-optic cell is inserted in a laser cavity (Fig. 5.24), an additional cavity loss will be present while the driving voltage to the transducer is applied. A fraction of the laser beam is in fact diffracted out of the cavity by the induced phase grating. If the driving voltage is high enough, this additional loss will be sufficient to prevent the laser from oscillating. The laser is then returned to its high-Q condition by switching off the transducer voltage.

5.4.2.2 Operating Regimes

Q-switched lasers can operate in either of the following two ways: (i) Pulsed operation (Fig. 5.25). In this case the pump rate $W_p(t)$ is in the form of a pulse of suitable duration. The population inversion $N(t)$ prior to Q switching grows to a maximum value and then tends to decrease. The cavity Q is switched at the time when the maximum of $N(t)$ occurs ($t = 0$ in the figure). From $t > 0$, the number of photons begins to grow, leading to a pulse with the peak occurring at some time t_d after switching. Due to the growth of photon number, the population inversion $N(t)$ will decrease from its initial value N_i (at $t = 0$) to a final value N_f which is left after the pulse has finished. (ii) Continuously pumped, repetitively Q-switched operation (Fig. 5.26). In this case a cw pump (W_p) is applied to the laser and the cavity losses are periodically switched to a low value. The laser output will in this case consist of a continuous train of light pulses, while the inversion

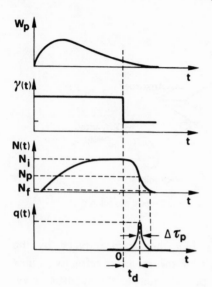

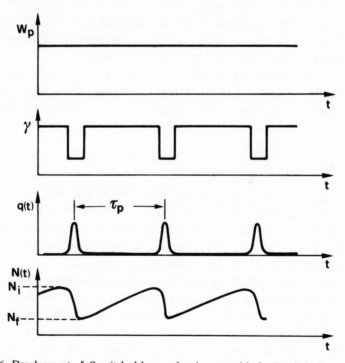

FIG. 5.25. Development of a Q switched laser pulse in a pulsed operation. The figure shows the time behavior of the pump rate W_p, resonator losses γ, population inversion N, and number of photons q.

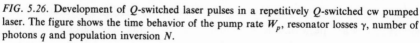

FIG. 5.26. Development of Q-switched laser pulses in a repetitively Q-switched cw pumped laser. The figure shows the time behavior of the pump rate W_p, resonator losses γ, number of photons q and population inversion N.

will periodically oscillate from the initial value N_i (before Q switching) to a final value N_f (after the Q switching pulse).

Electro-optical and mechanical shutters as well as saturable absorbers are commonly used for pulsed operation. For repetitive Q switching of continuously pumped lasers (which have lower gain than pulsed lasers) mechanical shutters or, more commonly, acousto-optic Q switches are used.

5.4.2.3 Theory of Q Switching

Provided the laser operates in a single mode, its dynamic behavior during Q switching can be obtained from (5.13) or (5.16) for four- and three-level lasers, respectively. For the sake of simplicity we will only consider the so-called fast-switching case, in which the cavity losses are switched in a time much shorter than the buildup time for laser radiation.[22]

We will first consider a four-level laser operated in a pulsed regime (Fig. 5.25) and assume that for $t < 0$ the losses are so high that the laser is below threshold (i.e., $q = 0$ for $t < 0$). If Q switching is carried out when $N(t)$ has attained its maximum value, the corresponding initial inversion N_i can be obtained from (5.13a) by setting $q = 0$ in the right-hand side of this equation. If we assume $N \ll N_t$, we can readily see from (5.13a) that, for a given time behavior of the pump rate W_p, if the amplitude of W_p is doubled, $N(t)$ will also double while retaining the same time behavior. Thus, if we let E_p be the pump energy corresponding to the given pump rate ($E_p \propto \int W_p \, dt$), we can write $N_i \propto E_p$. If we also let N_c and E_{cp} be the critical inversion and pump energy, respectively, when the laser is operated just at threshold, then we can write

$$(N_i/N_c) = (E_p/E_{cp}) \tag{5.50}$$

For $t > 0$, the time evolution of the system will again be described by the two equations of (5.13) with the initial conditions $N(0) = N_i$ and $q(0) = q_i$. Here again q_i is just some small number of photons needed to let laser action start. The equations can, however, be considerably simplified since we expect the time evolution of both $N(t)$ and $q(t)$ to occur in a time so short that the pump term $W_p(N_t - N)$ and the decay term N/τ in (5.13a) can be neglected. Equations (5.13) then reduce to

$$\dot{N} = -BqN \tag{5.51a}$$

$$\dot{q} = \left(V_a BN - \frac{1}{\tau_c}\right)q \tag{5.51b}$$

It is worth noting here that, according to (5.51b), the population N_p corresponding to the peak of the photon pulse (i.e., when $\dot{q} = 0$) is

$$N_p = 1/V_a B\tau_c = \gamma/\sigma l \tag{5.52}$$

which is just the same as the laser critical inversion. This result allows us to put (5.50) in a form which is more convenient for the analysis that follows, namely

$$(N_i/N_p) = x \tag{5.53}$$

where $x = (E_p/E_{cp})$.

To calculate the number of photons q_p in the cavity at the peak of the pulse, we take the ratio of (5.51a) and (5.51b). Using (5.52) also, we get

$$\frac{dq}{dN} = -V_a\left(1 - \frac{N_p}{N}\right) \tag{5.54}$$

which can be readily integrated to give

$$q = V_a\left[N_i - N - N_p \ln \frac{N_i}{N}\right] \tag{5.55}$$

where the small number q_i has, for simplicity, been neglected. At the peak of the pulse we then get

$$q_p = V_a N_p\left[\frac{N_i}{N_p} - \ln \frac{N_i}{N_p} - 1\right] \tag{5.56}$$

which readily gives q_p once N_p [through (5.52)] and (N_i/N_p) [through (5.53)] are known. The peak output power P_{1p} is then obtained through (5.14) as

$$P_{1p} = \frac{\gamma_1}{2}\left(\frac{V_a}{\sigma l}\right)\left(\frac{\hbar\omega}{\tau_c}\right)\left[\frac{N_i}{N_p} - \ln \frac{N_i}{N_p} - 1\right] \tag{5.57}$$

The output energy is then

$$E = \int_0^\infty P_1\,dt = \left(\frac{\gamma_1 c_0}{2L'}\right)\hbar\omega\int_0^\infty q\,dt \tag{5.58}$$

The integration in (5.58) can be carried out easily by integrating both sides of (5.51b) and by using both (5.51a) and the condition $q(0) = q(\infty) = 0$. We find $\int_0^\infty q\,dt = V_a\tau_c(N_i - N_f)$ so that (5.58) becomes

$$E = \left(\frac{\gamma_1}{2\gamma}\right)(N_i - N_f)(V_a\hbar\omega) \tag{5.59}$$

where N_f is the final inversion (see Fig. 5.25). Note that equation (5.59) could have been written down at once by noting that $(N_i - N_f)$ is the available inversion and this inversion can produce a number of photons

$(N_i - N_f)V_a$. Out of this number of photons emitted by the medium, only the fraction $[\gamma_1/2\gamma]$ is available as output energy. To calculate E from (5.59) it is necessary to know N_f. This can be obtained from (5.55) by putting $t = \infty$. Since $q(\infty) = 0$ we get

$$\frac{N_i - N_f}{N_i} = \frac{N_p}{N_i} \ln \frac{N_i}{N_f} \tag{5.60}$$

which gives N_f/N_i as a function of N_p/N_i. The quantity $(N_i - N_f)/N_i$ in (5.60) is called the inversion (or energy) utilization factor. In fact, although the initial inversion is N_i, the inversion actually used is $(N_i - N_f)$. Figure 5.27 shows a plot of the energy utilization factor versus N_i/N_p. Note that, for large values of N_i/N_p, this factor tends to unity.

Once the output energy and peak power are known, we can get an approximate value, $\Delta\tau_p$, for the width of the pulse by defining it as $\Delta\tau_p = E/P_{1p}$. From (5.57) and (5.59) we get

$$\Delta\tau_p = \tau_c \frac{N_i - N_f}{N_p[(N_i/N_p) - \ln(N_i/N_p) - 1]} \tag{5.61}$$

The time delay τ_d between the peak of the pulse and the time of Q switching (see Fig. 5.25) can be considered to be approximately equal to the time required for the pulse to reach, say, $(q_p/10)$. Since no appreciable saturation of the inversion occurs up to this point, we can put $N(t) = N_i$ in (5.51b). With the help of (5.52) and (5.53), equation (5.51b) then gives

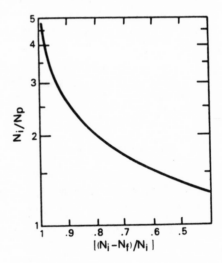

FIG. 5.27. Energy utilization factor $(N_i - N_f)/N_i$ versus N_i/N_p (from Reference 22).

$\dot{q} = (x - 1)q/\tau_c$, which upon integration gives

$$q = q_i \exp\left[\frac{(x - 1)t}{\tau_c} \right] \tag{5.62}$$

The time delay τ_d is obtained from (5.62) by putting $q \simeq q_p/10$. Setting $q_i \simeq 1$ we get

$$\tau_d = \frac{\tau_c}{x - 1} \ln\left(\frac{q_p}{10} \right) \tag{5.63}$$

The calculation for a continuously pumped repetitively Q-switched laser (Fig. 5.26) proceeds in a similar way. First we need to calculate the quantities N_i and N_f. One relation between N_i and N_f has already been established in (5.60). A second relation is obtained from the condition that in the time τ_p between consecutive pulses, the pump rate must re-establish the initial inversion N_i starting from N_f. From (5.13a) [putting $W_p(N_t - N) \simeq W_p N_t$ and $q = 0$] we get

$$N_i = (W_p N_t \tau) - (W_p N_t \tau - N_f)\exp(-\tau_p/\tau) \tag{5.64}$$

From (5.18) ($N_c \ll N_t$) and (5.22) we have $W_p N_t \tau = x N_c = x N_p$, and (5.64) then becomes

$$1 = \frac{x N_p}{N_i} - \left(x \frac{N_p}{N_i} - \frac{N_f}{N_i} \right)\exp(-\tau_p/\tau) \tag{5.65}$$

where x is the amount by which the pump rate exceeds the threshold pump rate. Equations (5.60) and (5.65) give N_i/N_p and N_i/N_f once x and τ_p/τ are known. Since N_p can be calculated [using (5.52)], the three quantities N_i, N_p and N_f are found in this way. The peak power, output energy, and pulse duration are then obtained through (5.57), (5.59), and (5.61) respectively.

The calculations for a three-level laser proceed in a similar way starting from (5.16). To save space these calculations are not presented here.

5.4.2.4 A Numerical Example

Figure 5.28 shows a typical plot of laser output energy, E, versus input energy, E_p, to the flashlamp for a Nd:YAG Q-switched laser. The laser is operated in a pulsed regime and is Q-switched using a KD*P (deuterated potassium dihydrogen phosphate, KD_2PO_4) Pockels cell.[24] The rod and cavity dimensions are also indicated in the figure. From the figure we note that the laser has a threshold energy $E_{cp} \simeq 3.4$ J and gives an output energy

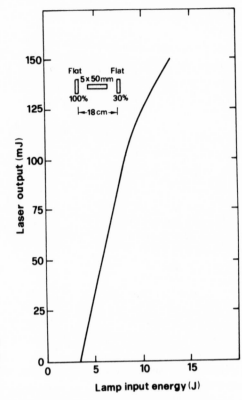

FIG. 5.28. Laser output energy versus input energy to the flashlamp for a Nd: YAG Q-switched laser [after Koechner[24]].

$E \simeq 0.12$ J for $E_p \simeq 10$ J (i.e., at $x = E_p/E_{cp} = 2.9$). At this value of pump energy the pulsewidth is experimentally found to be ~6 ns.

We can now carry out a comparison of these experimental results with those predicted from the equations of the previous section. We will neglect mirror absorption and so put $\gamma_1 \simeq -\ln R_1 = 1.2$ and $\gamma_2 \simeq 0$. Internal losses of the polarizer–Pockels cell combination are estimated to be $T_i \simeq 15\%$, while the internal losses of the rod can be neglected. We thus get $\gamma_i = -\ln(1 - T_i) = 0.162$ and $\gamma = [(\gamma_1 + \gamma_2)/2] + \gamma_i = 0.762$. The laser energy can be obtained from (5.59) with the help of (5.53) and (5.52) as

$$E = \left(\frac{\gamma_1}{2} x\eta_E \right)\left(\frac{A}{\sigma} \right)\hbar\omega \qquad (5.66)$$

where η_E is the energy utilization factor and A is the laser rod cross section. For $x = N_i/N_p = 2.9$ we find $\eta_E = 0.94$ from Fig. 5.27. Since $A \simeq 0.19$ cm^2, we find from (5.66) with the help of (5.36) that $E \simeq 160$ mJ, which is in reasonable agreement with the experimental value ($E = 120$ mJ). The time delay τ_d can now be obtained from (5.63). According to (5.7a) the effective length of the resonator is $L' = L + (n - 1)l \simeq 22$ cm, where $n \simeq 1.83$ for Nd:YAG, so that $\tau_c = L'/c_0\gamma \simeq 1$ ns. The calculation of q_p is then made using (5.56) with $V_a \simeq Al \simeq 1$ cm^3 and $N_p = \gamma/\sigma l = 4.35 \times 10^{17}$ cm^{-3}. We get $q_p \simeq 3.5 \times 10^{17}$ photons and, from (5.63), $\tau_d \simeq 20$ ns. The laser pulse-width is obtained through (5.61) as $\Delta\tau_p = \tau_c \eta_E x/(x - \ln x - 1) \simeq 3.3$ ns. The discrepancy between this value and the experimental value (~ 6 ns) is attributed to two factors: (i) Multimode oscillation. The time delay τ_d is expected to be different for the different modes, and this should appreciably broaden the pulse duration. (ii) The conditions for fast switching are not satisfied. The calculated time delay (~ 20 ns) assuming fast switching is in fact comparable to the typical switching time of a conventional Pockels cell, and the pulsewidth is expected to be somewhat increased by slow switching.

5.4.3 Mode Locking[25]

The technique of mode locking allows the generation of laser pulses of ultrashort duration (from a fraction of a picosecond to a few tens of picoseconds) and very high peak power (a few gigawatts). Mode locking refers to the situation where the cavity modes are made to oscillate with comparable amplitudes and with locked phases.

As a first example we will consider the case of $2n + 1$ longitudinal modes oscillating with the same amplitude E_0 (Fig. 5.29a). We will assume the phases ϕ_l of the modes to be locked according to the relation

$$\phi_l - \phi_{l-1} = \phi \tag{5.67}$$

where ϕ is a constant. The total electric field $E(t)$ of the e.m. wave (at any

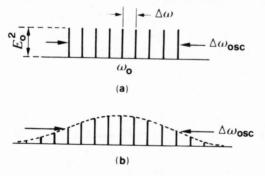

(a)

(b)

FIG. 5.29. Mode amplitude (represented by vertical lines) versus frequency for a mode-locked laser. (a) Uniform amplitude; (b) Gaussian amplitude distribution over a bandwidth (FWHM) $\Delta\omega_{osc}$.

point inside or outside the resonator) can then be written as

$$E(t) = \sum_{-n}^{n} {}_lE_0\exp\left\{i\left[(\omega_0 + l\Delta\omega)t + l\phi\right]\right\} \tag{5.68}$$

where ω_0 is the frequency of the central mode and $\Delta\omega$ is the frequency difference between two consecutive modes. For simplicity, we shall consider the field at that point where the phase of the center mode is zero. We recall (see Chapter 4) that the frequency separation $\Delta\omega$ between two consecutive longitudinal modes is

$$\Delta\omega = \pi c / L \tag{5.69}$$

where L is the cavity length. If the summation in (5.68) is carried out, we find that

$$E(t) = A(t)\exp(i\omega_0 t) \tag{5.70}$$

where

$$A(t) = E_0 \frac{\sin\left[(2n + 1)(\Delta\omega t + \phi)/2\right]}{\sin\left[(\Delta\omega t + \phi)/2\right]} \tag{5.71}$$

$E(t)$ therefore behaves like a sinusoidal carrier wave at the center-mode frequency ω_0 with an amplitude $A(t)$ which changes with time according to (5.71). The corresponding output power is proportional to $A^2(t)$. An example is shown in Fig. 5.30 for $2n + 1 = 7$ oscillating modes.

As a result of the phase-locking condition (5.67), the oscillating modes interfere to produce short light pulses. The pulse maxima occur at those

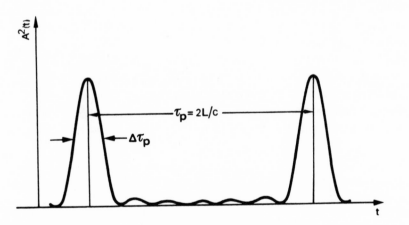

FIG. 5.30. Time behavior of the squared amplitude of the electric field for the case of seven oscillating modes with locked phases and equal amplitude.

times for which the denominator of (5.71) vanishes. Two successive pulses are, therefore, separated by a time

$$\tau_p = 2\pi/\Delta\omega = 2L/c \tag{5.72}$$

This is the time taken for a round-trip of the cavity. The oscillation behavior can, therefore, also be visualized as a pulse which propagates back and forth in the cavity. From (5.71) one finds that the width $\Delta\tau_p$ (FWHM) of $A^2(t)$ (i.e., of each laser pulse) is approximately given by

$$\Delta\tau_p \simeq 1/\Delta\nu_{osc} \tag{5.72a}$$

where $\Delta\nu_{osc} = (2n + 1)\Delta\omega/2\pi$ is the total oscillating bandwidth (see Fig. 5.29a). Thus it can be seen that a large oscillating bandwidth is required for very short pulses. Obviously this bandwidth cannot exceed the gain bandwidth of the laser, and this means that with typical gas lasers one cannot get pulses shorter than about 0.1 ns. With solid state and dye lasers, however, one can obtain pulses of 1 ps or even shorter. In addition, very large peak powers can be obtained in this way. The peak power is in fact proportional to $(2n + 1)^2 A^2$, whereas, for random phases, the power is the sum of powers in the modes and hence is proportional to $(2n + 1)A^2$. The peak power enhancement due to mode locking is therefore equal to the number of locked modes, which, for a solid-state laser, may range typically between 10^3 and 10^4. At the same time, the average power is essentially unaffected by mode locking.

The oscillation behavior of Fig. 5.30 can be readily understood if we consider the various modes to be represented by vectors in the complex plane. The lth mode would thus correspond to a complex vector of amplitude E_0 and rotating at the angular velocity $(\omega_0 + l\Delta\omega)$. If we now refer to axes rotating at angular velocity ω_0, the central mode will appear fixed relative to these axes and the lth mode rotating at velocity $l\Delta\omega$. If, at time $t = 0$, all vectors lie in the same direction, the situation of these vectors at some general time t will be as indicated in Fig. 5.31 (five modes). If the time t is such that mode 1 has made a 2π rotation (i.e., $\Delta\omega t = 2\pi$), mode $1'$ will also have rotated (anticlockwise) by 2π, while modes 2 and $2'$ will have

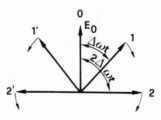

FIG. 5.31. Representation of cavity modes in the complex plane.

rotated by 4π. All these vectors will therefore be aligned again with that at frequency ω_0, and the total electric field will again be $(2n + 1)E_0$. Thus the time interval τ_p between two consecutive pulses will be such that $\Delta\omega\tau_p = 2\pi$, as indeed shown by (5.72).

As a second example of mode locking we will consider a Gaussian distribution of mode amplitudes with a bandwidth (FWHM) $\Delta\nu_{\mathrm{osc}}$ (Fig. 5.29b), i.e.,

$$E_l^2 = E_0^2 \exp\left[-\ln 2\left(\frac{2l\Delta\nu}{\Delta\nu_{\mathrm{osc}}} \right)^2 \right] \tag{5.73}$$

while the phases are still assumed to be locked according to (5.67). If we take, for simplicity, $\phi = 0$, the total electric field $E(t)$ can be written as

$$E(t) = \exp(i\omega_0 t) \sum_{-\infty}^{+\infty} {}_l E_l \exp i(l\Delta\omega t)$$

$$= A(t) \exp(i\omega_0 t) \tag{5.74}$$

If the sum is approximated by an integral [i.e., $A(t) \simeq \int E_l \exp i(l\Delta\omega t)\, dl$], the field amplitude $A(t)$ is seen to be proportional to the Fourier transform of the spectral amplitude E_l. We then get

$$A^2(t) \propto \exp\left[-\ln 2\left(\frac{2t}{\Delta\tau_p} \right)^2 \right] \tag{5.75}$$

where the pulsewidth $\Delta\tau_p$ (FWHM) is

$$\Delta\tau_p = 2\ln 2 / \pi\Delta\nu_{\mathrm{osc}} = 0.441 / \Delta\nu_{\mathrm{osc}} \tag{5.76}$$

As a conclusion to the two examples given above, we can say that, when the mode-locking condition (5.67) holds, the field amplitude is proportional to the Fourier transform of the magnitude of the spectral amplitude. The pulsewidth $\Delta\tau_p$ is related to the width of the spectral intensity $\Delta\nu_{\mathrm{osc}}$ by the relation $\Delta\tau_p = k/\Delta\nu_{\mathrm{osc}}$, where k is a numerical factor (of the order of unity) which depends on the particular shape of the spectral intensity distribution. A pulse of this sort is said to be *transform limited*.

Under locking conditions different from (5.67) the output pulse may be far from being transform limited. If for instance we take $\phi_l = l\phi + l^2\phi_2$ [note that (5.67) can be written as $\phi_l = l\phi$] and assume again a Gaussian amplitude distribution [(5.73)], we get

$$E(t) = A(t) \exp i\left[\omega_0 t + \beta t^2 \right] \tag{5.77}$$

For this expression $A^2(t)$ can again be expressed as in (5.75) (i.e., it is still a

Gaussian function), where now

$$\Delta\tau_p = \left(\frac{2\ln 2}{\pi\Delta\nu_{osc}} \right)\left[1 + \frac{\left(\beta\Delta\tau_p^2 \right)^2}{2\ln 2} \right]^{1/2} \qquad (5.77a)$$

In this case, therefore $\Delta\tau_p\Delta\nu_{osc}$ is larger (and sometimes much larger) than 0.441. The physical reason for this can be traced back to the term βt^2 in (5.77), which corresponds to a linear sweep of the carrier frequency (linear frequency chirp). In this case the spectral amplitude [which is the Fourier transform of (5.77)] arises both from the amplitude and the frequency modulation of $E(t)$, and $\Delta\nu_{osc}$ will be larger than $0.441/\Delta\tau_p$.

5.4.3.1 Methods of Mode Locking

The most commonly used mode-locking methods belong to one of the following two categories: (i) mode locking by an active modulator driven by an external signal (active mode locking) and (ii) mode locking by means of a suitable nonlinear optical material (passive mode locking).

To illustrate the first method, suppose we insert in the cavity a modulator driven by an external signal, thus producing a sinusoidal time-varying loss at frequency $\Delta\omega'$. If $\Delta\omega' \neq \Delta\omega$, this loss will simply amplitude-modulate the energy of each cavity mode. If $\Delta\omega' = \Delta\omega$, however, each mode will have amplitude-modulation side bands which coincide with adjacent mode frequencies. The field equation of a given cavity mode will thus contain terms arising from the modulation of the two adjacent modes. The cavity modes therefore become coupled and the phases tend to lock. This type of locking, often referred to as amplitude-modulation (AM) mode locking, can be shown to lead to the phase relation given in (5.67) if the modulator is placed very close to one of the end mirrors. There is an alternative way of mode locking with an active modulator where one makes use of a modulator whose optical pathlength (rather than optical loss) is modulated at frequency $\Delta\omega$. In this case it can be shown that the phases are again locked but in a different relationship from that given in (5.67). Nevertheless, one again obtains short pulses whose duration is of the order of the inverse of the oscillating bandwidth. Since this type of modulator produces a cavity-length modulation and hence a modulation of its resonance frequencies, this method of mode locking is often referred to as frequency-modulation (FM) mode locking.

The operation of both AM and FM mode locking can perhaps be more readily understood in the time domain rather than in the frequency domain. In Fig. 5.32a, for the AM case, we show the time behavior of the cavity losses γ which are being modulated at frequency $\Delta\omega'$. We will assume the modulator to be placed at one end of the cavity. If $\Delta\omega' = \Delta\omega$, the modulation period T' will be equal to the cavity round-trip time $2L/c$. In this case light pulses will develop in the cavity, as indicated in Fig. 5.32a, since a pulse which passes through the modulator at a time t_m for minimum loss will return to the modulator after a time $(2L/c)$ when the loss is again at a minimum. It can also be shown that, if the pulse maximum initially occurs at a time slightly different from t_m, the pulse will be reshaped by the time-varying loss γ in such a way that the time of its maximum tends to t_m. A similar argument can be used for the FM mode locking (Fig. 5.32b). In this case the modulator refractive index n rather than the modulator loss is sinusoidally changed, and the light pulses tend to occur either at a minimum of $n(t)$ (solid lines) or at a maximum of $n(t)$ (dotted lines).

To explain how passive mode locking works, let us consider what happens when the laser cavity contains a saturable absorber. It suffices here to consider an idealized absorber having just two levels whose transition frequency coincides with the laser frequency. To understand how a satura-

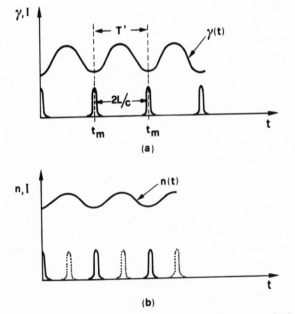

FIG. 5.32. (a) AM-type of locking. Time behavior of cavity losses γ and of output intensity I. (b) FM-type of locking. Time behavior of modulator refractive index n and of output intensity I.

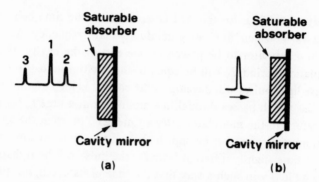

FIG. 5.33. Time-domain description of passive mode locking.

ble absorber can lead to mode locking, let us consider two adjacent axial laser modes. If both of these modes are oscillating, the interaction of their respective fields with the saturable absorber leads to a population difference between lower and upper levels containing a term oscillating at the difference frequency between the two modes.[†] This term effectively represents a time-varying loss within the cavity and it will therefore couple each mode to its two neighbors. Evidently one can only produce a time-varying population difference in the absorber if the absorber decay time τ is much faster than the inverse of the mode-separation frequency. An alternative way of visualizing the passive mode-locking process is provided by a description in the time domain rather than, as above, in the frequency domain. Let us suppose then that the saturable absorber is contained in a thin cell placed in contact with one of the cavity mirrors (Fig. 5.33a). If the modes are initially unlocked, the intensity of each of the two traveling waves in the cavity will be made up of a random sequence of light bursts (indicated as 1, 2, and 3 in Fig. 5.33a). As a result of absorber saturation, pulse 1 in the figure, being the most intense, will suffer the least attenuation in the absorber. This pulse will grow faster than the others, and after many round trips the situation depicted in Fig. 5.33b will eventually be established where a single intense mode-locked pulse remains.

We have so far considered mode locking as produced by a modulation of the cavity losses. It is also possible, however, to achieve mode locking by modulating the laser gain rather than the losses. In the case of lasers pumped by another laser this is commonly achieved by pumping with a

---

[†]In fact, according to (2.126), we have (for $I \ll I_s$) $\Delta N \simeq N_1 \, [1 - (I/I_s)]$. Since the total electric field $E(t)$ is the sum of the fields of the two modes, the intensity $I \propto E^2(t)$ and hence N will have a term oscillating at the difference frequency between the two modes.

mode-locked laser and adjusting the length L of the second laser cavity so that the pulse repetition time of this second laser, $(2L/c)$, is equal to that of the pump laser. The mode-locked pulses of the second laser are then in synchronism with those of the pump laser, and this method is usually referred to as mode locking by synchronous pumping. Note that, for this scheme to work, the decay time of the inversion in the second laser must be fast enough (i.e., of the order of the cavity transit time), so that the corresponding gain can actually be significantly modulated. The method is therefore often used for dye and color-center lasers which have short lifetimes (a few nanoseconds) for the upper state.

5.4.3.2 Operating Regimes

Mode-locked lasers can be operated either with a pulsed pump or cw pump (Fig. 5.34). In a pulsed situation, the overall duration $\Delta\tau_p'$ of the train of mode-locked pulses is in some cases determined by the duration of the pump pulse. This is, for instance, true for pulsed dye lasers where $\Delta\tau_p'$ may be a few microseconds. In some cases, however (e.g., solid-state lasers where a saturable absorber is used), the presence of the saturable absorber will result in both Q-switching and mode-locking operation. In this case the duration $\Delta\tau_p'$ of the mode-locked train will be given by the duration $\Delta\tau_p$ of the Q-switched pulse as calculated in Section 5.4.2.3 (a few nanoseconds). In the pulsed case, a Pockels cell electro-optical modulator (e.g., the configuration of Fig. 5.21 in which the voltage to the Pockels cell is sinusoidally modulated) or a saturable absorber cell are the most commonly used mode-locking elements.

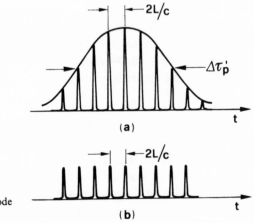

FIG. 5.34. (a) Pulsed and (b) cw mode locking.

TABLE 5.1. Mode-Locking Systems

Active material	Mode-locking element	Type of operation	$\Delta\tau_p$
Gas			
He–Ne	Acoustic modulator (quartz)	cw	1 ns
He–Ne	Saturable absorber	cw	
	Neon cell		0.35 ns
	Cresyl violet meth.		0.22 ns
Ar^+	Quartz acoustic modulator	cw	0.15 ns
CO_2 (low pressure)	Germanium acoustic modulator	cw	10–20 ns
	Saturable absorber (SF_6)	cw	10–20 ns
CO_2 (TEA)	Germanium acoustic modulator	Pulsed	1 ns
	Saturable absorber (SF_6)	Pulsed	1 ns
Solid			
Nd:glass	Saturable absorber (Kodak 9860, 9840 dyes)	Pulsed	5 ps
Nd:YAG	Electro-optic modulator ($LiNbO_3$)	cw, pulsed	40 ps
Ruby	Saturable absorber (DDI dye)	Pulsed	10 ps
Semiconductor	Saturable absorber	cw	5 ps
Color center	Synchronous pumping	cw	5 ps
Liquid			
Rhodamine 6G	Saturable absorber (DODCI dye)	$\begin{cases} \text{cw, } Ar^+ \text{ pumped} \\ \text{flash pumped} \end{cases}$	0.03 ps / 1 ps
	Synchronous pumping	cw, Ar^+ pumped	1 ps

For cw mode locking (Fig 5.34b), the laser is pumped cw and mode locking is usually achieved either by a saturable absorber or by an acousto-optic modulator (i.e., the configuration of Fig. 5.24 in which the transducer is continuously driven at the frequency $\Delta\omega$ corresponding to the frequency difference between two consecutive longitudinal modes).

Table 5.1 summarizes the operating conditions for some of the most commonly used mode-locked lasers. For a detailed description of each of these lasers, the reader is referred to the next chapter.

5.5 LIMITS TO THE RATE EQUATIONS

In this chapter, the cw and transient laser behavior has been discussed within the lowest-order approximation: the (spatially averaged) rate equations. In order of increasing accuracy (and complexity), the following treatments should be used: (i) Rate equations in which the spatial variation of both the inversion and e.m. energy density is taken into account. This will be discussed in Appendix A. (ii) Fully semiclassical treatment, in which the matter is quantized while the e.m. fields of the cavity are treated classically, i.e., by Maxwell's equations. It can be shown[1] that the corresponding equations reduce to the rate equations for the cw case. This is also true for the transient case provided that the duration of any transients is much longer than the inverse of the linewidth of the laser transition. Hence all transient cases considered in this chapter (except perhaps mode locking) can be adequately described by a rate-equation treatment. (iii) Fully quantum treatment in which both matter and fields are quantized. This would of course be the most complete treatment of all. It is needed to correctly describe the start of laser oscillation as well as laser noise. It can be shown, however, that, when the number of photons of the given cavity mode is much larger than 1, the (average) results of the quantum treatment coincide with those of a semiclassical treatment. Thus, except when dealing with aspects such as laser noise, one can avoid the difficulties of a fully quantum treatment. It should finally be noted that the rate equations, in their simplest form as given here, only apply to relatively few cases. In most cases there are more than just three or four levels involved, and the rate-equation treatment is correspondingly more complicated. In fact it can be said in general that each laser has its own particular set of rate equations. However, the equations considered in this chapter provide a model which can be readily extended to handle more complicated situations.

PROBLEMS

5.1. Which expression would you use for the mode volume V_a in the laser medium if many longitudinal modes all with the same transverse field distribution (TEM$_{00}$) are oscillating?

5.2. Calculate the logarithmic loss γ corresponding to a mirror transmission of $T = 80\%$.

5.3. Prove equation (5.18a).

5.4. A He–Ne laser oscillating on its red ($\lambda = 632.8$ nm) transition has a 2% gain per pass. The resonator consists of two concave spherical mirrors both of radius $R = 5$ m and spaced by $L = 1$m. Identical apertures are inserted at both ends of the resonator to obtain TEM$_{00}$ mode operation. Calculate the required aperture diameter.

5.5. The linewidth, $\Delta\nu_0^* = 50$ MHz, of a low-pressure CO_2 laser is predominantly due to Doppler broadening. The laser is operated with an input power twice the threshold value. Calculate the maximum mirror spacing which would still allow single longitudinal mode operation.

5.6. For the Nd:YAG laser considered in Fig. 5.9, calculate the threshold input power and the output power at $P_{in} = 10$ kW when the output coupling is reduced to 10%. Calculate also the corresponding slope efficiency.

5.7. For the CO_2 laser considered in Fig. 5.12, calculate the threshold input power and the output power at $P_{in} = 140$ kW for the case of optimum output coupling.

5.8. A He–Ne laser is oscillating on two consecutive longitudinal modes, one of them being coincident with the center of the laser transition ω_0. The cavity length is 1 m and the output coupling 2%. Knowing that the laser linewidth is $\Delta\nu_0^* = 1.7$ GHz, calculate the frequency difference between these modes.

5.9. The data of Fig. 5.19 refer to the case of a ruby laser having a rod diameter of 6.3 mm, length 7.5 cm, and with the two mirrors directly attached to the rod end faces. The peak cross section of the laser transition is $\sigma = 2.5 \times 10^{-20}$ cm^2, the rod refractive index is $n = 1.76$, and the doping of the rod corresponds to an active ion concentration of $N_t = 1.6 \times 10^{19}$ ions/cm^3. From the steady-state values for $N_0 V_a$ and q_0 indicated in the figure, calculate the total losses γ and the amount x by which laser threshold is exceeded.

5.10. For the Nd:YAG Q-switched laser considered in Fig. 5.28, calculate the expected threshold, output energy, and pulse duration (for $E_{in} = 10$ J) when the output coupling is reduced to 20%.

5.11. A mode-locked He–Ne laser has an oscillation bandwidth of 0.6 GHz, and the spectrum can be approximately described by a Gaussian function. Calculate the corresponding output pulse duration when the mode-locking condition given by (5.67) is satisfied.

5.12. By approximating the sum over all modes in (5.74) with an integral, an important characteristic of the output behavior is lost. What is it?

5.13. The Fabry–Perot etalon of Fig. 5.8a has mirror electric field amplitude reflectivities r_1 and r_2 ($r_1^2 = R_1$ and $r_2^2 = R_2$ where R_1 and R_2 are the intensity reflectivities) and amplitude transmissions t_1 and t_2. Taking account of all the multiple reflections and also the phase delay of each reflection, find an expression for the overall amplitude reflectivity in the U direction.

5.14. Derive an expression for the birefringence ($n_{x'} - n_{y'}$) that the Pockels cell of Fig. 5.21 must have in order to convert a linearly polarized beam into a circularly polarized beam.

5.15. For a Pockels cell in the so-called longitudinal configuration, the birefringence $\delta n = n_{x'} - n_{y'}$ induced by an applied dc (longitudinal) voltage V to the cell is $\delta n = n_0^3 r_{63} V/l$, where n_0 is the (ordinary) refractive index, l is the length of the cell crystal and r_{63} is the appropriate nonlinear coefficient of the material. Using the result from the previous problem, derive an expression for the voltage required to keep the polarizer–Pockels-cell combination of Fig. 5.21 in the closed position.

5.16. For a Pockels cell made of KD_2PO_4 (deuterated potassium dihydrogen phosphate, also known as KD*P) the value of the r_{63} coefficient at $\lambda = 1.06$ μm is 26.4×10^{-12} m/V while $n_0 = 1.51$. From the result of Problem 5.15 calculate the voltage which needs to be applied for the closed position.

5.17. The Nd:YAG laser of Figs. 5.9 and 5.10 is pumped at $P_{in} = 10$ kW, and repetitively Q switched at a 10 kHz repetition rate by an acousto-optic cell (whose insertion losses are considered to be negligible). Calculate the output energy, pulse duration, and average power expected for this case.

5.18. Derive the expressions for output energy and pulse duration which apply for a Q-switched three-level laser.

5.19. Consider a ruby laser (three-level laser, $\lambda = 694.3$ nm) having a rod diameter of 6.3 mm and length of 7.5 cm. The rod is situated in a laser resonator consisting of two plane mirrors spaced by $L = 50$ cm and of transmission $T_2 = 0$ and $T_1 = 0.5$. Assume an internal loss per pass of $T_i = 10\%$ and use the values for N_t, n and σ given in Problem 5.9. Find the Q-switched output energy, peak power, and pulse duration when the laser is pumped with twice the threshold pump energy.

REFERENCES

1. M. Sargent, M. O. Scully, and W. E. Lamb, *Laser Physics* (Addison-Wesley Publishing Co., London, 1974).
2. H. Statz and G. De Mars, in *Quantum Electronics*, ed. by C. H. Townes (Columbia University Press, New York, 1960), p. 530.
3. R. Dunsmuir, *J. Electron. Control (GB)* **10**, 453 (1961).
4. R. H. Pantell and H. E. Puthoff, *Fundamentals of Quantum Electronics* (John Wiley and Sons, New York, 1969), Chapter 6, Section 6.4.2.
5. W. W. Rigrod, *J. Appl. Phys.* **36**, 2487 (1965).
6. A. Yariv, *Proc. IEEE* **51**, 1723 (1963).
7. C. L. Tang, H. Statz, and G. De Mars, *J. Appl. Phys.* **34**, 2289 (1963).
8. P. W. Smith, in *Lasers*, Vol. 4, ed. by A. K. Levine and A. J. De Maria (Marcel Dekker, New York, 1976), Chapter 2.
9. W. Koechner, *Solid-State Laser Engineering*, Springer Series in Optical Sciences, Vol. 1 (Springer-Verlag, New York, Berlin, 1976), Chapter 3, Section 3.6.3.
10. W. Koechner, *Solid-State Laser Engineering*, Springer Series in Optical Sciences, Vol. 1 (Springer-Verlag, New York, Berlin, 1976), Chapter 2, Section 2.3.
11. D. Findlay and D. W. Goodwin, in *Advances in Quantum Electronics*, ed. by D. W. Goodwin (Academic Press, New York, 1970), pp. 77–128.
12. W. Koechner, *Solid-State Laser Engineering*, Springer Series in Optical Sciences, Vol. 1 (Springer-Verlag, New York, Berlin, 1976), Chapter 3, Section 3.4.
13. Private communication, Istituto di Ricerca per le Tecnologie Meccaniche.
14. M. C. Fowler, *Appl. Phys. Lett.* **18**, 175 (1971).
15. E. Hoag *et al.*, *Appl. Opt.* **13**, 1959 (1974).
16. A. L. Schawlow and C. H. Townes, *Phys. Rev.* **112**, 1940 (1958).
17. C. H. Freed, *IEEE J. Quantum Electron.* **QE-3**, 203 (1967).
18. W. E. Lamb, *Phys. Rev.* **134**, 1429 (1964).

19. T. G. Polanyi and I. Tobias, in *Lasers*, Vol. 2, ed. by A. K. Levine and A. J. De Maria (Marcel Dekker, New York, 1976).

20. R. W. Hellwarth, in *Advances in Quantum Electronics*, ed. by J. R. Singer (Columbia University Press, New York, 1961), p. 334.

21. A. Yariv, *Introduction to Optical Electronics* (Holt, Rinehart and Winston Inc., New York, 1971), Chapter 9.

22. B. A. Lengyel, *Lasers*, 2nd ed. (Wiley-Interscience, New York, 1971), pp. 182–190.

23. W. Koechner, *Solid-State Laser Engineering* (Springer-Verlag, New York, Berlin, 1976), Chapter 8.

24. W. Koechner, *Solid-State Laser Engineering* (Springer-Verlag, New York, Berlin, 1976), Chapter 11, Section 11.5.

25. P. W. Smith, M. A. Duguay, and E. P. Ippen, in *Progress in Quantum Electronics*, ed. by J. H. Sanders and K. W. Stevens (Pergamon, Oxford, 1974), Vol. 3, pp. 107–229.

26. R. V. Ambartsumyan, N. G. Basov, P. G. Kriukov, and V. S. Lethokov, in *Progress in Quantum Electronics*, ed. by J. H. Sanders and K. W. Stevens (Pergamon Press, London, 1971), Vol. 1, pp. 107–185.

6

Types of Lasers

6.1 INTRODUCTION

This chapter contains miscellaneous data and practical information about a number of lasers. It should be pointed out that there are many more lasers in existence than just those described here. This chapter, however, concentrates on those types which are most commonly used and whose characteristics are representative of a whole category of lasers. It should also be noted that some of the data presented in this chapter (for example, on output powers and energies) are likely to be rapidly superseded. These data are, therefore, presented only as a rough guide.

We will consider the following types of lasers: (1) solid-state (crystal or glass) lasers, (2) gas lasers, (3) dye lasers, (4) chemical lasers, (5) semiconductor lasers, (6) color-center lasers, and (7) free-electron lasers.

6.2 SOLID-STATE LASERS

The term solid-state laser is usually reserved for those lasers having as their active medium either an insulating crystal or a glass. Semiconductor lasers will be considered in a separate section since the mechanisms for pumping and for laser action are quite different. Solid-state lasers often use as their active species impurity ions introduced into an ionic crystal. Usually the ion belongs to one of the series of transition elements in the periodic table (e.g., transition metal ions, notably Cr^{3+}, or rare earth ions, notably Nd^{3+} or Ho^{3+}). The transitions used for laser action involve states belonging to the inner unfilled shells. These transitions are therefore not so strongly influenced by the crystal field. This in turn means that the transitions are quite sharp (i.e., σ reasonably large) and the nonradiative channels are fairly weak (i.e., τ reasonably long). Consequently the threshold pump rate ($W_p \propto 1/\sigma\tau$ for a four-level laser) is low enough to allow laser action.

6.2.1 The Ruby Laser[1]

This type of laser was the first to be made to operate,[2,3] and still continues to be used. Ruby, which has been known for hundreds of years as a naturally occurring precious stone, is a crystal of Al_2O_3 (corundum) in which some of the Al^{3+} ions are replaced by Cr^{3+} ions. As a laser material, it is usually obtained by crystal growth from a molten mixture of Cr_2O_3 ($\sim0.05\%$ by weight) and Al_2O_3. The laser energy levels are those of the Cr^{3+} ion in the Al_2O_3 lattice, and the main levels of interest are indicated in Fig. 6.1. Laser action usually occurs on the $\bar{E}\rightarrow{}^4A_2$ transition (R_1 line, $\lambda_1 \simeq 694.3$ nm, red). Ruby has two main pump bands ${}^4F_1$ and ${}^4F_2$ centered at wavelengths of 0.55 μm (green) and 0.42 μm (violet), respectively. These bands are connected by a fast ($\sim10^{-7}$ s) nonradiative decay to both $2\bar{A}$ and $\bar{E}$ states. Since these last two states are also connected to each other by a very fast nonradiative decay ($\sim10^{-9}$ s), thermalization of their population occurs which results in the $\bar{E}$ level being the more heavily populated. The frequency separation between $2\bar{A}$ and $\bar{E}$ (~29 cm^{-1}) is small compared to (kT/h), and the $2\bar{A}$ population is comparable with the $\bar{E}$ level population. It is thus possible to obtain laser action on the $2\bar{A}\rightarrow{}^4A_2$ transition also (R_2 line, $\lambda_2 \simeq 0.6928$ μm) by using, for instance, the dispersive systems of Fig. 5.7. Despite the complication of having these two laser transitions, it is apparent that ruby operates as a three-level laser.

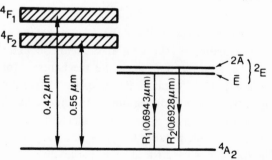

FIG. 6.1. Energy levels of ruby.

As already discussed in connection with Fig. 2.14, the R_1 transition is predominantly homogeneously broadened at room temperature, the broadening being the result of the interaction of Cr^{3+} ions with lattice phonons. The width of the transition (FWHM) is $\Delta\nu_0 = 11$ cm$^{-1} = 330$ GHz $(T = 300°K)$ and the $2\bar{A}$ and $\bar{E}$ levels have the same lifetime of $\sim 3 \times 10^{-3}$ s $(T = 300°K)$. At $T = 77°K$ the lifetime increases to 4.3×10^{-3} s, which shows that the room temperature lifetime also has a contribution from nonradiative decay. Note that the lifetime is in the millisecond range and it thus corresponds to an electric–dipole forbidden transition.

Ruby lasers are usually operated in a pulsed regime. For this, a medium-pressure (~ 500 Torr) xenon flashtube is used, either with the pump configuration as in Fig. 3.2b or, more often, as in Fig. 3.2a. Typical rod diameters range between 5 and 10 mm with a length between 5 and 20 cm. The output performance can be summarized as follows: (i) when Q switched, 10–50 MW in a single giant pulse of 10–20 ns duration, and (ii) when mode locked, a few gigawatts peak power in a pulse of ~ 10 ps duration. Ruby lasers can also run cw, pumped by a high-pressure mercury lamp.

Ruby lasers, once very popular, are now less widely used, since they have been superseded by their competitors, the Nd:YAG or Nd:glass lasers. Since, in fact, ruby works on a three-level scheme, the required threshold pump energy is about an order of magnitude larger than that for a Nd:YAG laser of comparable size. Ruby lasers are, however, still commonly used for a number of scientific applications, such as pulsed holography and ranging experiments (including military rangefinders).

6.2.2 Neodymium Lasers[4–6]

Neodymium lasers are the most popular type of solid state laser. The laser medium is usually either a crystal of $Y_3Al_5O_{12}$ (commonly called YAG, an acronym for *yttrium aluminum garnet*) in which some of the Y^{3+} ions are replaced by Nd^{3+} ions, or simply a glass which has been doped by Nd^{3+} ions. Neodymium lasers can oscillate on several lines, the strongest and thus the most commonly used one being at $\lambda = 1.06$ μm.

A simplified energy-level scheme for Nd:YAG is shown in Fig. 6.2. The energy-level scheme is much the same for Nd:glass since, as already mentioned, the energy levels involved are not strongly influenced by the crystal field. The $\lambda = 1.06$ μm laser transition is the strongest of the $^4F_{3/2} \rightarrow ^4I_{11/2}$ transitions. The two main pump bands occur at 0.73 and 0.8 μm respectively. These bands are coupled by a fast nonradiative decay to

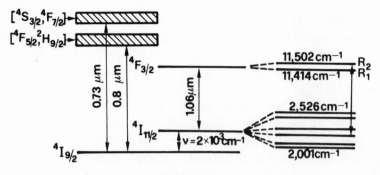

FIG. 6.2. Simplified energy levels of Nd:YAG.

the $^4F_{3/2}$ level, while the lower $^4I_{11/2}$ level is also coupled by a fast nonradiative decay to the $^4I_{9/2}$ ground level. Furthermore, the energy difference between the $^4I_{11/2}$ and $^4I_{9/2}$ levels is almost an order of magnitude larger than kT. Thus it follows that the Nd^{3+} laser works on the four-level scheme. As in the case of ruby, the laser transition is (predominantly) homogeneously broadened and the corresponding width is $\Delta\nu_0 = 6.5$ cm^{-1} = 195 GHz at $T = 300°K$. The lifetime of the upper laser level is in this case also very long ($\tau = 0.23$ ms) since the transition is forbidden for electric–dipole interaction.

Nd:YAG lasers can operate either cw or pulsed. For both cases, linear lamps in single-ellipse (Fig. 3.2b), close-coupling (Fig. 3.2c), or multiple-ellipse configurations (Fig. 3.3) are commonly used. Medium-pressure (500–1500 Torr) Xe lamps and high-pressure (4–6 atm) Kr lamps are used for the pulsed and cw cases respectively. The rod dimensions are typically the same as those given for ruby. The output performance can be summarized as follows: (i) output power up to 150 W from a single stage and up to 700 W from a cascade of amplifiers, for cw operation; (ii) output power up to 50 MW in Q-switched operation; (iii) pulse duration down to ~20 ps in mode-locked operation. The slope efficiency is about 1–3% for both cw and pulsed operation. Nd:YAG lasers are widely used in a variety of applications such as materials processing (in cw or repetitively pulsed operation), ranging, and laser surgery.

The rod dimensions for Nd:glass may be much larger than those for Nd:YAG (up to perhaps 1 m in length and a few tens of centimeters in diameter). Glass, due to its much lower melting temperature, can in fact be grown much more easily than YAG. Since, however, the thermal conductivity of glass is about an order of magnitude smaller than that of YAG, Nd:glass lasers are usually operated in a pulse regime. The output performance can be summarized as follows: (i) Output energy and peak power in Q-switched operation are comparable with those obtainable from a Nd: YAG rod of comparable dimensions. (ii) Since the laser transition is considerably broader than that of Nd:YAG (the additional, inhomogeneous broadening being due to the variation of ion environments within the glass matrix), pulsewidths as short as ~5 ps can be obtained in mode-locked operation. Nd:glass can be used instead of Nd:YAG in all those applications where the repetition rate of the laser is low enough not to cause thermal problems in the rod. A very important application of Nd:glass is as laser amplifiers in the very high energy systems used in laser fusion experiments. A system based on Nd:glass lasers, delivering pulses with a peak power of more than 20 TW and total energy of ~15 kJ, has already been built (the Shiva laser), and a system which should give an order of magnitude more power and energy is under construction (Nova laser, 200 kJ, 100–300 TW).

6.3 GAS LASERS

In general, for gases, the broadening of the energy levels is rather small (of the order of a few gigahertz or less), since the line-broadening mechanisms are weaker than in solids. For gases at the low pressures often used in lasers (a few Torr), the collision-induced broadening is very small, and the linewidths are therefore essentially determined by Doppler broadening. For this reason optical pumping with lamps of the type used for solid state lasers is not used in the case of gases. This would, in fact, be very inefficient since the emission spectrum of these lamps is more or less continuous, whereas there are no broad absorption bands in the active material. The only case in which laser action has been obtained in a gas by means of optical pumping of this type is that of Cs pumped by a linear lamp containing He. In this case the situation was quite favorable for optical pumping since some He emission lines are coincident with absorption lines of Cs. However, the importance of this laser lies more in its historical significance: Cs, which vaporizes at a temperature of 175°C, is a highly reactive substance.

Gas lasers are usually excited by electrical means, i.e., pumping is achieved by passing a sufficiently large current (dc or pulsed) through the gas. The principal pumping mechanisms occurring in gas lasers have already been discussed in Section 3.3. Other pumping mechanisms which are peculiar to certain lasers (e.g., Penning ionization and charge transfer) will be introduced in this chapter. We also wish to point out that some gas lasers can also be pumped by mechanisms other than electrical pumping. In particular, we mention pumping by gas-dynamic expansion, chemical pumping, and optical pumping by means of another laser.

Once a given species is in its excited state, it can decay to lower states, including the ground state, by four different processes: (i) collisions between an electron and the excited species, in which the latter gives up its energy to the electron (collision of the second kind), (ii) collisions between atoms (for a gas with more than one constituent), (iii) collisions with the walls of the container, and (iv) spontaneous emission. Regarding this last case, we should always consider the possibility (particularly for the usually very strong UV and VUV transitions) of radiation trapping. This process, already discussed in Section 2.3.4, slows down the effective rate of spontaneous emission.

For a given discharge current, these various processes of excitation and de-excitation lead eventually to some equilibrium distribution of population among the energy levels being established. Thus it can be seen that the production of a population inversion in a gas is a more complicated process than in a solid state laser, owing to the large number of phenomena involved. In general we can say that a population inversion between two given levels will occur when either (or both) of the following circumstances occur: (i) The excitation rate is greater for the upper laser level than for the lower laser level, and (ii) the decay of the upper laser level is slower than that of the lower laser level. We recall that the latter is a necessary condition for cw operation [see (5.26)]. If this condition is not satisfied, however, laser action can still occur under pulsed operation provided condition (i) is fulfilled (self-terminating lasers).

As far as their construction is concerned, many gas lasers have the arrangement illustrated schematically in Fig. 6.3. The gas is contained in a tube of suitable diameter (from a few millimeters to a few centimeters) which is terminated by two end windows inclined at Brewster's angle θ_β. We recall that, for this angle of incidence, a laser beam polarized in the plane of the figure suffers no reflection losses at the window surfaces, and consequently this is the direction of polarization that the laser output

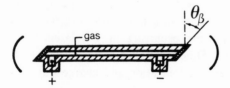

FIG. 6.3. Schematic diagram of a gas laser.

adopts. Spherical mirrors rather than plane mirrors are generally used since the former give better resonator stability (see discussion at the end of Section 4.4.2).

6.3.1 Neutral Atom Lasers

The He–Ne[7-10] laser may be considered to be a typical example (and in fact a particularly important one) of this category of laser. It can oscillate at any of the three following wavelengths: $\lambda_1 = 3.39$ μm, $\lambda_2 = 0.633$ μm, and $\lambda_3 = 1.15$ μm. It was the first gas laser to be made to oscillate (at $\lambda_3 = 1.15$ μm).[7] The 0.633-μm (red) He–Ne laser is one of the most popular and most widely used of lasers.

The energy-level schemes of He and Ne are shown in Fig. 6.4. Laser action occurs between energy levels of Ne, whereas the He is added to assist in the pumping process. In fact, as can be seen from the figure, the levels 2^3S and 2^1S of He are resonant with the levels $2s$ and $3s$, respectively, of Ne. Since the 2^3S and 2^1S levels are metastable, it is found that He proves very efficient in pumping the Ne $2s$ and $3s$ levels by resonant energy transfer. It has been confirmed that this process is the dominant one producing population inversion in the He–Ne laser, although direct electron–Ne collisions also contribute to the pumping. From what has been said earlier, it can be seen that the Ne $2s$ and $3s$ levels can build up their populations, and they are, therefore, likely candidates as upper levels for laser transitions. Taking account of the selection rules, we see that the possible transitions are those to p states. In addition, the decay time of the s states ($\tau_s \simeq 100$ ns) is an order of magnitude longer than the decay time of the p states ($\tau_p \simeq 10$ ns). So, the condition (5.26) for operation as a cw laser is satisfied. From these considerations it is seen that laser oscillation might be expected on any of the transitions a, b, and c of Fig. 6.4. Of the various transitions of type a, the strongest turns out to be that between sublevel $3s_2$ of the $3s$ group and sublevel $3p_4$ of the $3p$ group ($\lambda_1 = 3.39$ μm). Among the transitions of type b, it is the $3s_2 \rightarrow 2p_4$ transition ($\lambda_2 = 0.633$ μm, red) which features in the usual commercial He–Ne laser. The transition $2s_2 \rightarrow 2p_4$ (of type c) produces the wavelength $\lambda_3 = 1.15$ μm. The He–Ne laser will oscillate on transitions a, b, or c according to whether the maximum mirror reflectivity is at λ_1, λ_2, or λ_3. The multilayer dielectric mirrors are, therefore, made in such a way as to give a maximum reflectivity at the desired wavelength.

The laser transition is broadened predominantly by the Doppler effect. For instance, at $\lambda = 632.8$ nm, from (2.106) natural broadening can be

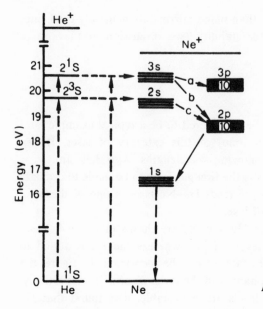

FIG. 6.4. Energy levels of He and Ne.

estimated to be $\Delta \nu_{nat} = 1/2\pi\tau \simeq 19$ MHz, where $\tau^{-1} = \tau_s^{-1} + \tau_p^{-1}$, and τ_s, τ_p are the lifetimes of the s and p states respectively. Collision broadening contributes even less than natural broadening [e.g., for pure Ne, $\Delta \nu_c \simeq 0.6$ MHz at the pressure $p \simeq 0.5$ Torr, see (2.105a)]. Finally it should be noted that the experimentally measured linewidth agrees well with the one calculated above, thus confirming that the effective temperature of the Ne atoms is the ambient temperature.

Early designs of He–Ne laser followed the general scheme of Fig. 6.3, but this has been superseded by an arrangement in which the two ends of the discharge tube are terminated by the two cavity mirrors, whose coated faces are actually in the discharge region. Because of the complicated processes which contribute to the excitation and de-excitation of its levels, the He–Ne laser has optimum values for a number of its operational parameters. In particular, these are (i) optimum value of the product of total gas pressure p and tube diameter D ($pD = 3.6$–4 Torr × mm), (ii) optimum value of the He:Ne ratio ($\sim 5:1$ at $\lambda = 632.7$ nm and $\sim 9:1$ at $\lambda = 1.15$ μm), and (iii) optimum value of the discharge current density J. The fact that there is an optimum value of pD indicates that it is the electron temperature that is being optimized. The elementary theory of a glow discharge in a positive column yields in fact a Maxwellian electron energy distribution whose temperature depends only on the product pD (see Section 3.3.2). The optimum value of current density (at least for the 3.39

and 0.6328 μm transitions) occurs because, at high current densities, de-excitation of the He (2^1S) metastable state occurs not only by diffusion to the walls but also by superelastic collision processes such as

$$\text{He}(2^1S) + e \rightarrow \text{He}(1^1S) + e \qquad (6.1)$$

Since the rate of this process is proportional to N_e, the electron density (and hence to J), the overall rate of de-excitation can be written as $k_2 + k_3J$, where k_2 represents diffusion to the walls and k_3J represents the super-elastic collision process (6.1). Since the excitation rate of the 2^1S level can be expressed as k_1J, the steady state 2^1S population will be given by $Nk_1J/(k_2 + k_3J)$, where N is the population of ground-state He atoms. Therefore the 2^1S population of He and hence that of the 3s state of Ne will saturate at high current densities, as indicated by the above relation. On the other hand, it has been found experimentally that the population of the lower laser level (3p or 2p) keeps increasing with J (due to direct pumping from ground-state Ne atoms and to radiative cascading from upper laser levels). As the discharge current density is increased, so the population difference increases to some maximum value and then decreases. Thus the laser gain and hence also the output power will have a maximum value for some particular current density. It should also be noted that the laser gain is found experimentally to vary as D^{-1}, provided pD is kept constant. This can be understood when it is realized that, at constant pD, the electron temperature is constant. Hence, all electron-collision excitation processes scale simply as the number of atoms available for excitation. Since both upper and lower laser levels are ultimately populated by electron-collision processes, their population and hence the laser gain is directly proportional to pressure or to D^{-1} at constant pD.

The preceding considerations indicate that, for a given laser tube, the range of possible current and pressure variation is rather limited. However, by increasing the tube diameter, at constant pD, one can increase the laser output. In this case, the gain decreases approximately as the inverse of the tube diameter, and the cross-sectional area of the discharge increases as the square of the diameter. The combined effect is to produce an output power that is roughly proportional to the tube diameter. Well above threshold, the output power increases linearly with length. A typical optimum output for a 100 cm × 6 mm cylindrical discharge might be 100 mW. Actually, most He–Ne lasers are operated with a bore diameter of 1–6 mm, for reasons of mode control. Since, as we have noted earlier, the linewidth $\Delta\nu$ (for the 633 nm transition) is about 1700 MHz, it is possible to obtain oscillation in a single longitudinal mode by using a cavity length which is short enough to give a longitudinal mode separation $(c/2L)$ comparable to $\Delta\nu$. In fact this implies $L < 15$–20 cm.

He–Ne lasers oscillating on the red transition are widely used for many applications where a low-power visible beam is needed (e.g., alignment, character reading, metrology, holography, video disk memories).

Besides the He–Ne laser there are other neutral atom gas lasers, covering most of the inert gases (He, Ne, Kr, Ar, Xe). In general, for all of these, one finds an energy-level scheme similar to the type shown for Ne in Fig. 6.4, apart from a change in scale. The first excited level ($1s$) is not usually used as a lower laser level since it is metastable. Hence, the levels used in obtaining laser action are higher than the first (or first two) excited levels. Because of this, neutral gas lasers usually operate in the red or near infrared (1–10 μm).

Finally we note that the neutral atom lasers are not represented solely by the inert gases and, in particular, we mention the class of metal vapor lasers (Pb, Cu, Au, Ca, Sr, and Mn). Of these, the most important at present is the Cu laser[10] which oscillates in the green (510.5 nm) where the efficiency is quite high ($\sim$1%), and in the yellow, at 578.2 nm. All metal vapor lasers are self-terminating and are therefore operated in a pulsed regime.

A general scheme for the relevant energy levels of this class of lasers is shown in Fig. 6.5. The $g \to 2$ transition is allowed while the $g \to 1$ transition is forbidden by electric–dipole interaction. Under the Born approximation we thus expect the electron-impact cross section of the $g \to 2$ transition to be appreciably larger than that of the $g \to 1$ transition. To accumulate sufficient population in the upper laser level, the $2 \to g$ radiative transition rate, which would usually be fast, must be slowed down to a value comparable to the $2 \to 1$ radiative rate. This means that a sufficient atomic density must be present to produce radiation trapping on the $2 \to g$ transi-

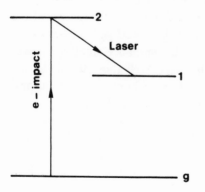

FIG. 6.5. General energy level scheme for a self-terminating metal vapor laser.

tion. Note that since the $1 \rightarrow g$ transition is forbidden, the laser can only operate on a pulsed basis with pulse duration of the order of or shorter than the lifetime of level 2. The $1 \rightarrow g$ decay usually occurs by collisions with the walls or by atom–atom deactivation. The corresponding decay rate sets an upper limit to the repetition rate of the laser.

The construction of a metal vapor laser follows the general scheme of Fig. 6.3. Two characteristic features are, however, worth noting: (i) Self-terminating lasers exhibit a very high gain per pass. Oscillation can therefore occur even without any mirrors through amplified spontaneous emission (see Section 2.3.4). However, a unidirectional output and a lower threshold are obtained by using a 100% reflecting mirror at one end of the tube and taking the output from the other end. (ii) To obtain the required vapor densities the laser must be operated at high temperature ($\sim 1500°C$). The laser tube is therefore usually made of alumina and the central region is held in an oven. A few Torr of Ne is also added to the laser tube to prevent deposition of metal vapor on the (cold) end windows. The problem of high operating temperatures can be considerably alleviated by using metal-halogen compounds (e.g., $CuBr$) rather than the pure metals. In this case the required temperature is considerably lower ($\sim 550°C$ for $CuBr$) and it can be achieved simply as the result of the heat generated by the discharge (when the laser is operated at sufficiently high repetition rate). However, the vapor then consists of $CuBr$ rather than Cu, and to produce atomic copper, a double discharge technique is used: the first discharge pulse dissociates the $CuBr$ molecule, while the second produces laser action.

Copper vapor lasers have been operated with average powers of ~ 40 W and repetition rates of ~ 15 kHz, and in fact provide the most efficient green laser source so far available. They are of interest for underwater communications and remote sensing of submerged objects (seawater is relatively transparent in the blue-green region), and also for some laser-photochemistry applications.

6.3.2 Ion Lasers

In the case of an ionized atom, the scale of energy levels is expanded. In this case, in fact, a given electron of the atom experiences the field due to the positive charge Ze of the nucleus (Z being the atomic number of the atom and e the electronic charge) screened by the negative charge $(Z - 2)e$ of the remaining electrons. The net effective charge is thus $2e$, whereas, for a neutral atom, it is only e. This expansion in energy scale means that ion lasers typically operate in the visible or ultraviolet regions. We will separate the ion lasers into two categories: (i) ion gas lasers and (ii) metal vapor lasers.

6.3.2.1 Ion Gas Lasers[10-12]

In an ion gas laser, the upper laser level becomes populated by two successive collisions with the electrons in the discharge. The first produces an ion from the neutral atom, while the second excites this ion. The pump process is therefore a two-step process involving the discharge current density J (i.e., it is proportional to J^2 or to higher powers of J, as we shall see later on). For this process to be efficient, a high current density is required. An ion gas laser thus requires a much higher current density than a neutral gas laser.

Of the various ion gas lasers, we will consider in some detail the Ar^+ laser. A level scheme showing the principal energy levels of Ar^+ is given in Fig. 6.6. Population of the upper level ($4p$) of the laser transition can be achieved by three distinct processes: (i) electron collisions with Ar^+ ions in their ground state [process (a)]; (ii) electron collisions with ions in metastable levels [process (b)]; (iii) radiative cascade from higher levels [process (c)]. If we let N_i be the density of Ar^+ ions in the ground state and N_e the electron density, and if we assume that the plasma as a whole is neutral, then we can say that $N_i \simeq N_e$. With this assumption, process (a) produces a pump rate per unit volume $(dN_2/dt)_p$ of the form

$$(dN_2/dt)_p \propto N_e N_i \propto N_e^2 \qquad (6.2)$$

Since the discharge reaches a condition in which the electric field is constant, the electron density N_e will be proportional to the discharge current density J. From (6.2) it follows that $(dN_2/dt)_p \propto J^2$. This quadratic dependence on current has been confirmed by observing the variation of spontaneously emitted power as a function of J. This would at first sight appear to be evidence in favor of process (a). However, processes (b) and (c) also give similar dependences of $(dN_2/dt)_p$ on J. This is immediately obvious in the case of process (c). In fact, the populations of those levels from which the cascade process originates will also be proportional to $N_e N_i$ and hence to N_e^2. In the case of process (b) the calculation is slightly more complicated. The population N_m of the metastable levels, which is determined by a balance between excitation and de-excitation processes, is given by

$$N_m \propto N_e N_i / (K + N_e) \qquad (6.3)$$

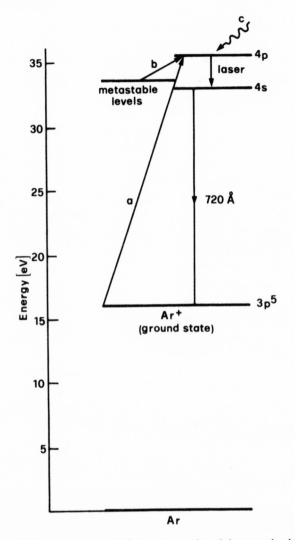

FIG. 6.6. Three different processes contributing to pumping of the upper level (4*p*) of an Ar$^+$ laser: (a) electron collisions with ground-state ions, (b) electron collisions with metastable-state ions, (c) radiative cascade from higher levels.

The term K in the denominator of (6.3) accounts for spontaneous de-excitation of the metastable level, while the term N_e accounts for de-excitation by electron collisions. From (6.3) one finds that process (b) produces a pumping rate

$$(dN_2/dt)_p \propto N_m N_e \propto N_e^3/(K + N_e) \qquad (6.4)$$

However, since the metastables are more likely to be de-excited by electron collisions than by spontaneous emission (i.e., $K \ll N_e$), it is seen from (6.4) that one again has the result that $(dN_2/dt)_p \propto N_e^2$. It is probable then that all three of the processes listed contribute to populating the laser level. It has been demonstrated, in fact, that a fraction, 23–50%, of the upper level population is due to the cascade process (c). Finally, we note that the lifetime of the upper laser level is $\sim 10^{-8}$ s, while the lower laser level ($4s$) is connected to the ground state by a radiative transition with a much shorter lifetime (10^{-9} s). So in this case also, condition (5.26) is satisfied. The Doppler linewidth $\Delta \nu_0^*$ is ~ 3500 MHz and from (2.113) it is seen that this implies a temperature $T \simeq 3000°\text{K}$. The ions are therefore very "hot" as a result of being accelerated by the electric field in the discharge.

A schematic diagram of an Ar^+ laser tube construction is given in Fig. 6.7. Because of the high current density there is a migration of Ar^+ ions towards the cathode (cataphoresis), and a return tube, as shown in the figure, is provided to compensate for this. Obviously, the return tube length must be greater than that of the laser tube to prevent the discharge passing along the return tube instead of the laser tube. At the high current densities involved, one of the most serious technological problems is damage to the tube caused by ions colliding with it ($T \simeq 3000°\text{K}$). Because of this, the tube is usually made of a ceramic material (beryllia) or of graphite. Also, a static magnetic field is applied, parallel to the tube axis, in the discharge region. With this arrangement the Lorentz force reduces the rate of diffusion of electrons towards the walls. This increases the number of free electrons at the center of the tube which leads to an increase in the pump rate and hence in the output power. By confining the discharge towards the center of the tube, the magnetic field also alleviates the problem of wall damage. Unlike the He–Ne laser, in this case, the gain does not depend on the internal diameter of the tube since an accumulation of population in the metastable levels does not decrease the population inversion. In commercial lasers, however, the tube diameter is kept small (a few millimeters) to confine oscillation to the TEM_{00} mode and to reduce the total current

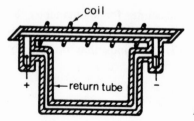

FIG. 6.7. Schematic diagram of an Ar^+ laser tube.

required. On the other hand, if one wants to increase the output power or reduce the problem of wall damage, there is an advantage to be gained from using substantially greater diameters.

The Ar^+ laser can oscillate at a number of wavelengths, the most intense being at $\lambda_1 = 488$ nm (blue) and $\lambda_2 = 514.5$ nm (green). It is possible to achieve oscillation on just a single line using the scheme of Fig. 5.7. An important characteristic of the Ar^+ laser (and of ion lasers in general) is that the output power increases rapidly with increasing discharge current. Unlike the He–Ne laser, the output power of the Ar^+ laser continues to increase with increasing excitation power. This is because the process of saturation of inversion (in this case due to resonance trapping of radiation on the 720 Å transition of Fig. 6.6) only becomes significant for current densities much higher than those that can be reached in practice. For the reasons given above, it has been possible to obtain very high output powers from Ar^+ lasers (up to 200 W continuous from a 1-cm-diameter tube). The laser efficiency is very low, however ($< 10^{-3}$). Argon lasers are widely used as a pump for cw dye lasers, for a variety of scientific applications (light–matter interactions), in laser printers, in laser surgery, and in the field of laser entertainment.

We conclude this section by mentioning the Kr^+ laser which, of the various other ion gas lasers, is the most widely used. It also oscillates at a number of wavelengths, of which the most powerful is in the red (647.1 nm).

6.3.2.2 Metal Vapor Lasers

The following metals have been used in vapor form to produce laser action: Sn, Pb, Zn, Cd, and Se. Of these lasers, the most widely used are those using Cd or Se vapors. Cd vapor produces strong cw laser action at the wavelengths $\lambda_1 = 441$ nm and $\lambda_2 = 325$ nm. The latter wavelength is particularly interesting for many applications since it falls in the UV region of the e.m. spectrum. Se vapor gives strong cw laser action on at least 19 wavelengths that cover most of the visible spectrum. Unlike ion gas lasers, in metal vapor lasers there are two quite different pump processes[†] that can be used: (i) Penning ionization, (ii) charge transfer ionization. Since these are single-step processes, the corresponding pump rate is now proportional to J rather than to J^2 (or J^3) as for the ion gas lasers. Much lower current

---

[†]The Ar^+ laser cannot use these processes because its laser levels are very high in energy (~35 eV, see Fig. 6.6).

density and electrical power per unit length are therefore required for metal vapor lasers compared to ion gas lasers.

The Penning ionization process can be written as

$$A^* + B \rightarrow A + B^+ + e \qquad (6.5)$$

where the ion B^+ in its final state may or may not be internally excited. Of course, this can only occur if the excitation energy of the excited atom A^* is greater than or equal to the energy required to ionize the other atom B. The surplus energy is transformed to kinetic energy of the electron. The process is most prominent when the excited species A^* is in a metastable state. Note that, unlike resonant energy transfer, Penning ionization is a nonresonant process: The excitation energy of A^* need only be greater than the ionization energy plus the excitation energy of atom B (if atom B is to be left in an excited state). Any surplus energy can in fact be removed as kinetic energy of the ejected electron. Charge transfer ionization, on the other hand, is a process of the type

$$A^+ + B \rightarrow A + (B^+)^* \qquad (6.6)$$

Here the ionization energy of atom A is transformed into ionization plus excitation energy of atom B. Since no electrons are ejected in this case, the process must be resonant: The ionization energy of A must equal the ionization plus excitation energy of B. The process is particularly effective if the A^+ ion is metastable (i.e., if it has a long lifetime).

After this brief discussion of the main pump mechanisms of metal vapor lasers, we will now describe the two most widely used lasers in this category: the He–Cd and He–Se lasers. The energy levels of the He–Cd system are shown in Fig. 6.8. The Cd laser is thus seen to be pumped predominantly by the Penning ionization process. The 2^1S and 2^3S metastable states of He can excite either the $^2D_{3/2}$ and $^2D_{5/2}$ states or the $^2P_{3/2}$ and $^2P_{1/2}$ states of Cd^+. Although the process is not resonant, it has been found that the cross section for excitation of the D states is about three times greater than that of the P states. What is more important, however, is that the lifetime of D states (10^{-7} s) is much longer than the lifetime of P states (10^{-9} s). Population inversion between the D and P states can, therefore, be produced readily and laser action has been achieved on the $^2D_{3/2} \rightarrow ^2P_{1/2}$ ($\lambda = 325$ nm) and the $^2D_{5/2} \rightarrow ^2P_{3/2}$ ($\lambda = 441.6$ nm) lines. The Cd^+ ions then drop to the $^2S_{1/2}$ ground state by radiative decay. In the case of the He–Se laser, the energy of the upper laser levels of the Se^+ ion (i.e., the sum of the ionization plus excitation energy of the Se atom) is ~ 25 eV (Fig. 6.9), i.e., greater than the excitation energy of the He metastable

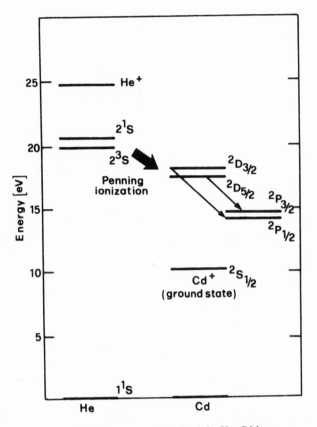

FIG. 6.8. Relevant energy levels of the He–Cd laser.

states. Therefore the upper laser levels can only be pumped by charge transfer ionization (the He^+ ion has in fact an energy of ~ 25 eV). This process is very effective since the He^+ ion has a long lifetime (it is determined only by electron recombination).

In its construction, a metal vapor laser is not very different from that shown in Fig. 6.3. In one possible configuration, however, the tube has a small reservoir near the anode to contain the metal. This reservoir is heated to a high enough temperature ($\sim 250°C$) to produce the desired vapor pressure in the tube. When the vapor reaches the discharge, some of the atoms are ionized and these migrate toward the cathode. The discharge produces enough heat to prevent condensation of the vapor on the walls of the tube. The vapor condenses, however, when it reaches the cathode region, where there is no discharge, and the temperature is low. The net

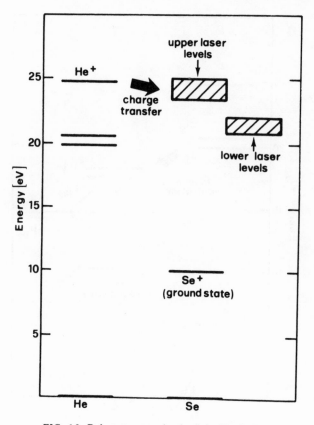

FIG. 6.9. Relevant energy levels of the He–Se laser.

result is a continuous flow of metal vapor from the anode to the cathode (this flow is called *cataphoresis*). Therefore a sufficient supply of Cd (1 g per 1000 h) must be provided for the life of the tube. He–Cd and He–Se lasers can give output powers of 50–100 mW, which places them in an intermediate position between red He–Ne lasers (a few milliwatts) and Ar^+ lasers (a few watts). He–Cd lasers are attractive for many applications where a blue or UV beam of moderate power is of interest (e.g., facsimile systems, reprographic systems, and Raman and fluorescence experiments).

6.3.3 Molecular Gas Lasers

These lasers exploit transitions between the energy levels of a molecule. Depending on the type of transition involved, molecular gas lasers can be placed in one of the three following categories: (i) Vibrational–rotational lasers. These lasers use transitions between vibrational levels of the same electronic state (the ground state). The energy difference between the levels involved in this type of transition (see Section 2.9.1) means that these lasers oscillate in the middle- and far-infrared (5–300 μm). (ii) Vibronic lasers. These lasers use transitions between vibrational levels of different electronic states (the word "vibronic" is a contraction from the words "vibrational" and "electronic"). In this case the oscillation wavelength falls in the visible/UV region. (iii) Pure rotational lasers which use transitions between different rotational levels of the same vibrational state. The corresponding wavelength falls in the far infrared (25 μm to 1 mm). Laser action is more difficult to achieve in this type of laser since the relaxation between rotational levels is generally very fast. These lasers are usually pumped optically,[13] using the output of another laser as the pump (commonly a CO_2 laser). Optical pumping excites the given molecule (e.g., CH_3F, $\lambda = 496$ μm) to a rotational level belonging to some vibrational state above the ground level. Laser action then takes place between rotational levels of this upper vibrational state.

6.3.3.1 Vibrational–Rotational Lasers

Of the various vibrational–rotational lasers, we will discuss the CO_2 laser in some detail.[14,15] This laser uses a mixture of CO_2, N_2, and He. Oscillation takes place between two vibrational levels in CO_2, while, as we shall see, the N_2 and He greatly improve the efficiency of laser action. The CO_2 laser is actually one of the most powerful lasers (output powers of ~80 kW have been demonstrated from a CO_2 gas-dynamic laser) and one of the most efficient (15–20%). Only the CO laser and the pulsed, electron-beam-initiated HF chemical laser have been reported with higher efficiencies.

Figure 6.10 shows the vibrational energy-level schemes for the electronic ground states of the CO_2 and N_2 molecules. N_2, being a diatomic molecule, has only one vibrational mode, and the lowest two vibrational levels ($v = 0$, $v = 1$) are indicated in the figure. The energy levels for CO_2 are more complicated since CO_2 is a linear triatomic molecule. In this case, there are three nondegenerate modes of vibration (Fig. 6.11): (1) symmetric stretching mode, (2) bending mode, and (3) asymmetric stretching mode. The oscillation behavior is therefore described by means of three quantum numbers n_1, n_2, and n_3, which give the number of quanta in each vibrational mode. The corresponding level is therefore designated by these three

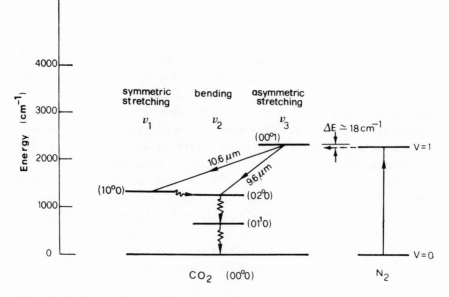

FIG. 6.10. The lowest vibrational levels of the ground electronic state of an N_2 molecule and a CO_2 molecule (for simplicity, the rotational levels are not shown).

quantum numbers written in the order n_1, n_2, n_3. For example, the 01^10 level[†] corresponds to an oscillation in which there is one vibrational quantum in mode 2. Since mode 2 has the smallest force constant of the three modes (the vibrational motion is transverse), it follows that this level will have the lowest energy. Laser action takes place between the 00^01 and 10^00 levels ($\lambda \simeq 10.6$ μm) although it is also possible to obtain oscillation between 00^01 and 02^00 ($\lambda \simeq 9.6$ μm). In fact, taking account of the rotational levels (which are not shown in Fig. 6.10), oscillation can take place on two sets of lines centered around $\lambda = 10.6$ μm and $\lambda = 9.6$ μm, respectively. The 00^01 level is very efficiently pumped by two processes:

    (i) Electron Collisions. $e + CO_2(00^00) \rightarrow e + CO_2(00^01)$. The electron

---

[†]The superscript (which we will denote by l) on the bending quantum number arises from the fact that the bending vibration is, in this case, doubly degenerate: it can occur both in the plane of Fig. 6.11 and in the plane orthogonal to it. A bending vibration therefore consists of a suitable combination of these two vibrations. The superscript l characterizes this combination; more precisely, $l\hbar$ gives the angular momentum of this vibration about the axis of the CO_2 molecule. For example, in the 02^00 state ($l = 0$) the two degenerate vibrations combine in such a way to give an angular momentum $l\hbar = 0$.

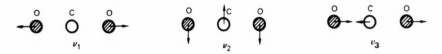

FIG. 6.11. The three fundamental modes of vibration for a CO_2 molecule: (v_1) symmetric stretching mode, (v_2) bending mode, (v_3) asymmetric stretching mode.

collision cross section for this process is very large. Electron collisions populate the 00^01 level preferentially (and not the lower laser levels 10^00 and 02^00), probably because the $00^01 \rightarrow 00^00$ transition is an allowed optical transition, whereas the $00^00 \rightarrow 10^00$ transition is not.

(ii) Resonant Energy Transfer from N_2 Molecule. This process is also very efficient because of the small energy difference between the two levels ($\Delta E = 18$ cm^{-1}). In addition the excitation of N_2 from the ground level to the $v = 1$ level by electron collisions is a very efficient process and the $v = 1$ level is metastable (the $1 \rightarrow 0$ transition is forbidden for electric–dipole radiation since, by virtue of its symmetry, a N–N molecule cannot have a net electric dipole moment). Finally the higher vibrational levels of N_2 are also closely resonant ($\Delta E < kT$) with the corresponding CO_2 levels (up to 00^04), and transitions between the excited levels $00n$ and the 001 are fast. In fact they occur very effectively through collisions with ground-state CO_2 molecules in the following (nearly) resonant process:

$$CO_2(0,0,n) + CO_2(0,0,0) \rightarrow CO_2(0,0,n-1) + CO_2(0,0,1) \quad (6.7)$$

This process tends to degrade all excited molecules to the $(0,0,1)$ state. Actually, level thermalization between the $(0,0,1)$ and the upper vibrational state is readily established in this way, and this vibrational system can be described by a vibrational temperature T_1. It can be seen then that, through these various processes, pumping of the upper laser level is a very efficient process, and this explains the high efficiency of the CO_2 laser.

The next point to consider is the decay of the upper laser level and how it compares with the decay rate of the lower laser level. Although the transitions $00^01 \rightarrow 10^00$, $00^01 \rightarrow 02^00$, $10^00 \rightarrow 01^10$, and $02^00 \rightarrow 01^00$ are optically allowed, the corresponding decay times τ_{sp} for spontaneous emission are very long (we recall that $\tau_{sp} \propto 1/\omega^3$). The decay of these various levels is therefore essentially determined by collisions. Accordingly, the decay time τ_s of the upper laser level can be obtained from a formula of the type

$$\frac{1}{\tau_s} = \sum a_i p_i \quad (6.8)$$

where the p_i are partial pressures and the a_i are constants which are characteristic of the gases in the discharge. Taking, for example, the case of a partial pressure of 1.5 Torr for CO_2, 1.5 Torr for N_2, and 12 Torr for He, one finds that the upper level has a lifetime $\tau_s \simeq 0.4$ ms. As far as the relaxation rate of the lower level is concerned, we begin by noting that the $100 \rightarrow 020$ transition is very fast and it occurs even in an isolated molecule. In fact the energy difference between the two levels is much smaller than kT. Furthermore, a coupling between the two states is present (Fermi resonance) because a bending vibration tends to induce a change of distance between the two oxygen atom (i.e., induce a symmetric stretching). Levels 10^00 and 02^00 are then effectively coupled to the 01^10 level by a near-resonant, collision process involving CO_2 molecules in the ground state:

$$CO_2(10^00) + CO_2(00^00) \rightarrow CO_2(01^10) + CO_2(01^10) + \Delta E \quad (6.9a)$$

$$CO_2(02^00) + CO_2(00^00) \rightarrow CO_2(01^10) + CO_2(01^10) + \Delta E' \quad (6.9b)$$

The two above processes have a very high probability since ΔE and $\Delta E'$ are much smaller than $kT^\dagger$. It follows, therefore, that the three levels 10^00, 02^00, and 01^10 reach thermal equilibrium in a very short time. This amounts to saying that the populations of these three levels can be described by a vibrational temperature T_2. Generally this temperature T_2 is not the same as T_1. We are therefore left with the decay from the 01^10 to the ground level 00^00. If this decay were slow, it would lead to an accumulation of molecules in the 01^10 level during laser action. This in turn would produce an accumulation in the 10^00 and 02^00 levels since these are in thermal equilibrium with the 01^10 level. Thus a slowing down of the decay process of all three levels would occur, i.e., the $01^10 \rightarrow 00^00$ transition would constitute a "bottleneck" in the overall decay process. It is, therefore, important to look into the question of the lifetime of the 01^10 level. This lifetime is also given by an expression of the type (6.8), and in this case the lifetime is greatly influenced by the presence of He (i.e., the coefficient a_i for He is very large). For the same partial pressures as considered in the example above, one obtains a lifetime of about 20 μs. It follows from the above discussion that this will also be the value of the lifetime of the lower laser level. Therefore condition (5.26) is easily satisfied in this case. Note that, since the $01^10 \rightarrow 00^00$ transition is the least energetic transition in any of the molecules in the discharge, relaxation from the 01^10 level can only occur by transferring this vibrational energy to translational energy of the colliding partners (V–T relaxation). Finally, we note that the presence of

---

†Relaxation processes in which vibrational energy is given up as vibrational energy of another like or unlike molecule are usually referred to as V–V relaxations.

He has another valuable effect. The He, because of its high thermal conductivity, helps to keep the CO_2 cool by conducting heat away to the walls. A low translational temperature for CO_2 is necessary to avoid population of the lower laser level by thermal excitation. The energy separation between the levels is, in fact, comparable to kT. In conclusion, the beneficial effects of the N_2 and He can be summarized as follows: The N_2 helps to produce a large population in the upper laser level while the He helps to empty population from the lower laser level.

From the point of view of their construction, CO_2 lasers can be separated into six categories: (i) lasers with longitudinal flow, (ii) sealed-off lasers, (iii) waveguide lasers, (iv) transverse-flow lasers, (v) transversely excited atmospheric pressure (TEA) lasers, and (vi) gas-dynamic lasers.

(i) *Lasers with Longitudinal Gas Flow.* The first CO_2 laser operation[16] was achieved in a laser of this type, and Fig. 6.12 shows one possible configuration. The mirrors can be internal (in contact with the gas), as in the figure, or external. In the latter case, the tube has a Brewster-angle window at each end (see Fig. 6.3). In the former case, at least one of the (metallic) mirrors needs to be kept at high voltage. The main reason for flowing the gas mixture is to remove the dissociation products, in particular CO, which would otherwise contaminate the laser. It is important to note that, except for very high flow velocities (supersonic flows), the heat dissipated in the discharge is removed by heat diffusion to the walls of the tube (which is water-cooled). In this case there is a maximum laser output power that can be obtained per unit length of the discharge (50–60 W/m) independent of the tube diameter. This comes about as a result of the

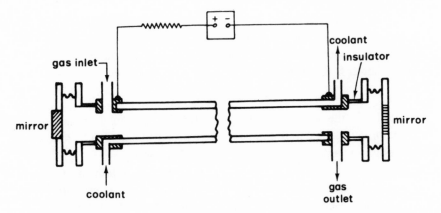

FIG. 6.12. Schematic diagram of a CO_2 laser with longitudinal gas flow.

following three circumstances: (1) If the tube diameter and the pressure are fixed, there will be some optimum value of current density. This is due to the fact that, at high current densities, there will be a rise in gas temperature with a consequent increase in population of the lower laser level. (2) If the diameter is fixed, there will be some optimum set of values for the partial pressures of the gases in the mixture and particularly of the CO_2. To explain this optimum CO_2 pressure, we begin by noting that from equations (5.17) and (5.18) one has the result that, at threshold, the number of atoms pumped per second into the upper laser level is

$$(dN_2/dt)_p = W_p(N_t - N_c) = (\gamma/\sigma l\tau) \propto \Delta\omega_0/\tau$$

where $\Delta\omega_0$ is the linewidth and τ is the lifetime of the upper level. Since this lifetime is determined by collisions, it is inversely proportional to the pressure p. The transition linewidth is the combined result of Doppler broadening and collision broadening. Therefore, $\Delta\omega_0$ increases with increasing pressure (for high pressures $\Delta\omega_0 \propto p$). Since the threshold electrical power P_e is proportional to $(dN_2/dt)_p$, it follows that P_e will increase with increasing pressure ($P_e \propto p^2$ at high pressures). The power dissipated in the gas therefore increases rapidly with increasing pressure. Above a certain pressure this will produce such a large temperature rise that the output power decreases. (3) The optimum values for current density J and pressure p are more or less inversely proportional to the laser tube diameter D (e.g., $p_{op} = 15$ Torr for $D = 1.5$ cm). This can be understood when one realizes that for larger diameters the generated heat has more difficulty in escaping to the walls. If we call σ_e the CO_2 cross section for electron impact excitation to the $00^0 1$ level, the number of molecules pumped into the upper level per second is given by (3.19):

$$\left(\frac{dN_2}{dt}\right)_p = \frac{J\sigma_e(N_t - N_c)}{e} \simeq \frac{J\sigma_e N_t}{e}$$

where e is the electron charge. For pump rates well above threshold, the output power is proportional to $(dN_2/dt)_p$ and therefore

$$P \propto JN_t V_a \propto JpD^2 l \tag{6.10}$$

where V_a is the volume of active material and l its length. Since the optimum values of J and p are inversely proportional to D, it follows that the optimum value of P depends only on the length l.

The total gas pressure in a longitudinal-flow CO_2 laser is of the order of 15 Torr (for $D = 1.5$ cm). At this pressure a major contribution to the laser linewidth comes from Doppler broadening (~ 50 MHz). The small value of Doppler linewidth (compared to visible gas lasers) is a result of the

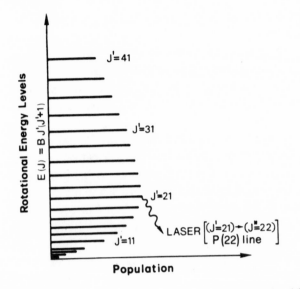

FIG. 6.13. Relative population of the rotational levels of a given vibrational level (e.g., the upper laser level of CO_2).

low frequency ω_0 of the transition. The small Doppler width also means that in this case collision broadening is not negligible. In fact it amounts to $\Delta\nu_c = 7.58(\psi_{CO_2} + 0.73\psi_{N_2} + 0.6\psi_{He})p(300/T)^{1/2}$ MHz, where the ψ are fractional partial pressures of the gas mixture and p the total pressure (in Torr). With this small linewidth, longitudinal-flow lasers automatically have their oscillation confined to a single longitudinal mode provided the resonator length is less than ~ 1 m. In this case, therefore, it becomes necessary to finely adjust the cavity length to ensure that a mode falls at the center of the gain curve. Actually, in our discussion so far we have ignored the fact that the upper laser level is made up from many rotational levels whose population is given by the Boltzmann distribution (see Fig. 6.13).[†] Accordingly, the laser transition consists of several equally spaced rotational–vibrational transitions (separated by ~ 2 cm^{-1}) belonging to both P and R branches (see Fig. 2.25). However, it is usually only the rotational transition with the largest gain, i.e., originating from the most heavily populated level, that actually oscillates [$P(22)$ transition]. This is because the rate of thermalization of the rotational levels ($\sim 10^7$ s^{-1} Torr^{-1}) is faster than the rate of decrease of population (due to spontaneous and stimulated emission) of the rotational level which is oscillating. Therefore, the entire

†Note that, for symmetry reasons, only levels with odd values of J are occupied.

population of rotational levels will contribute to laser action on the rotational level with highest gain. It was mentioned earlier that laser action can take place either on the $00^01 \rightarrow 10^00$ transition or on the $00^01 \rightarrow 02^00$ transition. Since the first of these has the greater gain and since both transitions have the same upper level, it follows that it is usually the $00^01 \rightarrow 10^00$ transition (10.6 μm) which oscillates. To summarize, we can say that oscillation usually takes place in a single rotational line of the $00^01 \rightarrow 10^00$ transition. To obtain oscillation on the 9.6-μm line or on a different rotational line, some appropriate frequency-selective device is placed in the cavity to suppress laser action on the line with highest gain. In fact the arrangement of Fig. 5.7b is commonly used. Finally we note that, with its long lifetime for the upper laser level ($\tau \simeq 0.4$ msec), the CO_2 laser lends itself particularly well to Q-switched operation. Repetitive Q switching is achieved by spinning one of the two mirrors at high speed while the gas is pumped by a continuous electrical discharge. However, the average power obtained in this way is only a small fraction ($\sim$5%) of that available from the same laser when operated cw. This is due to the fact that, when Q switched, the output pulse duration is comparable to the time taken for thermalization of the rotational levels. It is then no longer possible for the entire population of the rotational levels to contribute to laser action on the rotational line which is oscillating.

CO_2 lasers with longitudinal gas flow typically produce output powers of 50–500 W. Powers of 50–100 W are needed for laser surgery while powers up to 500 W are used in applications such as ceramic scribing, cutting of nonmetallic materials, resistor trimming, and welding of metals with a few millimeters thickness.

(ii) Sealed-off Lasers. If the flow of the gas mixture were stopped in the arrangement shown in Fig. 6.12, laser action would cease in a few minutes. This is because the chemical reaction products formed in the discharge (CO, in particular) would no longer be removed and would instead be absorbed in the walls of the tube or react with the electrodes, thus upsetting the CO_2–CO–O_2 equilibrium. Ultimately this would lead to dissociation of the CO_2. For a sealed-off laser, some kind of catalyst must be present in the gas tube to promote the regeneration of CO_2 from the CO. A simple way to achieve this is to add a small amount of H_2O (1%) to the gas mixture. This leads to regeneration of CO_2, probably through the reaction

$$CO^* + OH \rightarrow CO_2^* + H \tag{6.11}$$

involving vibrationally excited CO and CO_2 molecules. The relatively small amount of H_2O vapor required may be added in the form of hydrogen and oxygen gas. In fact, since oxygen is produced during the dissociation of CO_2, it is found that only hydrogen need be added. Another way of inducing the recombination reaction relies on the use of a hot (300°C) Ni cathode which acts as a catalyst. With these techniques, lifetimes for sealed-off tubes in excess of 10,000 h have been demonstrated.

Sealed-off lasers have produced output powers per unit length of ~60 W/m, comparable to those for longitudinal-flow lasers. Low power (~1 W) sealed-off lasers of short length and hence operating in a single mode are often used as local oscillators in optical heterodyne experiments. Sealed-off CO_2 lasers of somewhat higher power (~10 W) are attractive for laser microsurgery.

(iii) Waveguide Lasers.[17] If the diameter of the laser tube in Fig. 6.12 is reduced to about 1 mm, a situation is reached where the laser radiation is guided by the inner walls of the tube. Such waveguide CO_2 lasers have a low diffraction loss. Tubes of BeO or SiO_2 have been found to give the best performance. The power per unit length and efficiency of a waveguide laser are both somewhat smaller than the corresponding values for a conventional longitudinal-flow laser. The main advantage of a waveguide laser is the relatively large tuning range (~1 GHz) resulting from the increased optimum operating pressure (100–200 Torr). This is a very attractive feature where the laser is to be used as a local oscillator in an optical heterodyne experiment.

(iv) Transverse-Flow Lasers. We have seen that, for longitudinal-flow lasers (and also for sealed-off lasers) there is some maximum laser power that can be extracted. This is essentially due to a heating problem; with an efficiency of 20%, about 80% of the electrical power is dissipated in the discharge as heat. In those lasers the heat removal is effected simply by diffusion from the center of the tube towards the walls (which are cooled). A much more efficient way is to flow the gas perpendicular to the discharge (Fig. 6.14). If the flow is fast enough, the heat gets carried away by convection rather than by diffusion. The saturation of output power versus discharge current density now occurs at much higher values than in the longitudinal-flow arrangement. The optimum total pressure is now an order of magnitude greater (~100 Torr) and consequently, output powers well in excess of 1 kW can be obtained from these lasers with a length of ~1 m (see also Fig. 5.13). The increased total pressure p requires a corresponding increase of the electric field $\mathcal{E}$ in the discharge. In fact, for optimum operating conditions, the ratio $\mathcal{E}/p$ must remain approximately the same for all of these cases since this ratio determines the average energy of the

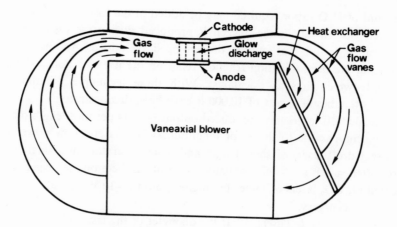

FIG. 6.14. Schematic diagram for a transverse-flow CO_2 laser.

discharge electrons [see (3.23)]. With this larger value of electric field, a longitudinal-discharge arrangement such as in Fig. 6.12 would be impractical since it would require a very high applied voltage (100–500 kV for a 1-m discharge). For this reason, the discharge is usually applied in a direction perpendicular to the resonator axis (lasers with transverse electric field, TE lasers). TE discharges can be classified into two types: self-sustained and non-self-sustained. In the self-sustained discharge the gas ionization is effected by the discharge. Therefore the value of $\mathcal{E}/p$ must be sufficiently large to promote avalanche ionization of the gas. In the non-self-sustained discharge, ionization is provided by auxiliary means such as a source of ionizing radiation. The $\mathcal{E}/p$ ratio can then be reduced to the value which maximizes the excitation of the upper laser level. In Fig. 6.15 an example is shown in which the main discharge is sustained by an auxiliary electron beam.

TE CO_2 lasers with fast transverse flow and high output powers (1–15 kW) are used in a great variety of metalworking applications (cutting, welding, surface hardening, surface metal alloying).

(v) Transversely Excited Atmospheric Pressure CO_2 Lasers.[18] In a cw TE CO_2 laser, it is not easy to increase the operating pressure above ~ 100 Torr. Above this pressure and at the current densities normally used, glow discharge instabilities set in and result in the formation of arcs within the discharge volume. To overcome this difficulty, the voltage can be applied to the transverse electrodes in the form of a pulse. If the pulse duration is sufficiently short (a fraction of a microsecond), the discharge instabilities

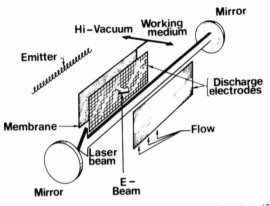

FIG. 6.15. E-beam sustained TE CO_2 laser [after Daugherty[37]].

have no time to develop and the operating pressure can then be increased up to and above atmospheric pressure. These lasers are therefore referred to as TEA lasers, the abbreviation standing for transversely excited atmospheric pressure. These lasers thus produce a pulsed output and are capable of large output energies per unit volume of the discharge (10–50 J/liter). To avoid arc formation, some form of ionization is also applied, which just precedes the voltage pulse producing the gas excitation (*pre-ionization*). A possible configuration is shown in Fig. 6.16, where the cathode structure consists of a trigger electrode in close proximity to a mesh and insulated from it by a dielectric sheet. A trigger pulse of high voltage is first applied between the trigger electrode and the mesh, and ions will thus be created near the cathode (corona effect). The main discharge pulse is then applied between the anode and the mesh cathode to excite the entire laser volume. This excitation method is often referred to as the double-discharge technique. Other preionization techniques include the use of pulsed *e*-beam guns (*e-beam pre-ionization*) or of suitable UV-emitting sparks to produce UV photoionization (*UV pre-ionization*). Since the transverse dimensions of

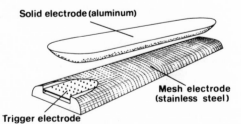

FIG. 6.16. Electrode assembly for double-discharge TEA CO_2 laser [after Richardson et al.[39]].

the laser are usually large, the two end mirrors are often chosen to give an unstable resonator configuration (positive-branch unstable confocal resonator, see Fig. 4.26). For low pulse repetition rates (~ 1 Hz), it proves unnecessary to flow the gas mixture. For higher repetition rates (up to a few kilohertz) the gas mixture is flowed transversely to the resonator axis and is cooled by a suitable heat exchanger. Another interesting characteristic of these lasers is their relatively broad linewidths (~ 4 GHz at $p = 1$ atm, due to collision broadening). Thus, by mode-locking TEA lasers, optical pulses with less than 1-ns duration have been produced. A very important application of TEA CO_2 lasers is in laser fusion experiments. A system based on TEA CO_2 lasers, delivering pulses with a peak power of 20 TW and total energy of 10 kJ, has already been built (the Helios laser). A system which should give an order of magnitude more power and energy is under construction (Antares laser, 100 kJ, 100–200 TW).

(vi) Gas-Dynamic CO_2 Laser.[19] The gas-dynamic CO_2 laser deserves a special mention since, in this laser, population inversion is not produced by an electrical discharge but by rapid expansion of a gas mixture (containing CO_2) which has initially been heated to a high temperature. Population inversion is produced downstream in the expansion region. Gas-dynamic CO_2 lasers have produced some of the largest powers so far reported in the unclassified literature.

The operating principle of a gas-dynamic laser can be summarized as follows (see Fig. 6.17). Suppose that the gas mixture is first held at high temperature (e.g., $T = 1400°K$) and pressure (e.g., $p = 17$ atm) in an appropriate container. Since the gas is initially in thermal equilibrium and at a high temperature, the population of the $00°1$ level of CO_2 will be appreciable ($\sim 10\%$ of the ground-state population, see Fig. 6.17b). The lower-level population is, of course, higher than this ($\sim 25\%$), and there is no population inversion. Now suppose the mixture is made to expand through some expansion nozzles (Fig. 6.17c). Since the expansion is adiabatic, the translational temperature of the mixture will be reduced to a much lower value. Due to V–T relaxation, the populations of both upper and lower laser levels will tend to relax to the new equilibrium values. However, since the lifetime of the upper state is longer than that of the lower state, the relaxation of the lower level will occur at an earlier stage in the expansion process (Fig. 6.17b). Thus there will exist a fairly extensive region downstream from the expansion zone where there will be a population inversion. The length L of this region is roughly determined by the time taken for the N_2 molecule to transfer its excitation to the CO_2 molecule. The two laser mirrors are thus chosen to have a rectangular

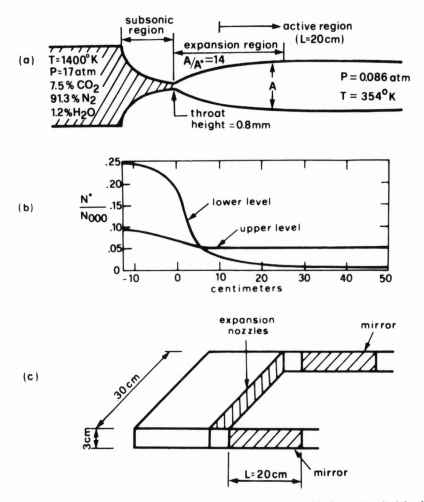

FIG. 6.17. Schematic illustration of the operation of a gas-dynamic CO_2 laser: (a) principle of the system, (b) spatial behavior of the population N^* for the upper and lower laser levels (normalized with respect to the population N_{000} of the ground level), (c) cavity geometry. Parts (a) and (b) have been reprinted by permission from *IEEE Spectrum*, Vol. 7, No. 11, November, 1970, pp. 51–58. Copyright 1970 by the Institute of Electrical and Electronics Engineers, Inc.

shape and are positioned as shown in Fig. 6.17c. This method of producing a population inversion will only work effectively if the expansion process reduces the temperature and pressure of the mixture in a time which is (i) short compared to the lifetime of the upper laser level and (ii) long compared to the lifetime of the lower laser level. To satisfy these conditions, expansion to supersonic velocities (Mach 4) is required. Finally we note that the gas mixture is usually raised to its initial high temperature by combustion of appropriate fuels (e.g., combustion of CO and H_2 or of benzene, C_6H_6, and nitrous oxide, N_2O, thus automatically supplying a CO_2/H_2O ratio of 2 : 1).

Gas-dynamic CO_2 lasers have been reported which produce output power up to 80 kW with a chemical efficiency[†] of 1%. So far, this type of laser can only be operated continuously for a short time (a few seconds) because of the heating produced by the laser beam in some of the components (particularly the mirrors).

The category of gas lasers using vibrational–rotational transitions is obviously not limited to the CO_2 laser. Other examples which should be mentioned are the CO ($\lambda \simeq 5$ μm) and the HCN laser. The latter can oscillate at wavelengths as long as $\lambda = 773$ μm = 0.773 mm, thus reaching the millimeter wave region. The CO laser has attracted considerable interest on account of its high power and efficiency. Output powers in excess of 100 kW and efficiencies in excess of 60% have been demonstrated.[20] However, to achieve this sort of performance the gas mixture must be kept at cryogenic temperature (77–100°K). Laser action, in the 5 μm region, arises from several rotational–vibrational transitions [e.g., from $v'(11) \rightarrow v(10)$, to $v'(7) \rightarrow v(6)$ at $T = 77°K$] of the highly excited CO molecule.

Pumping of the CO vibrational levels is achieved by electron-impact excitation. Like the isoelectronic N_2 molecule, the CO molecule has an unusually large cross section for electron-impact excitation of its vibrational levels. Thus nearly 90% of the electron energy in a discharge can be converted into vibrational energy of CO molecules. Another important feature of the CO molecule is that V–V relaxation proceeds at a much faster rate than V–T relaxation (which is unusually low). As a consequence of this a non-Boltzmann population buildup in higher vibrational levels by a process known as "anharmonic pumping" plays a very important role.[‡] Although this phenomenon does not allow a *total* inversion in the vibrational population of a CO molecule, a situation known as *partial* inversion occurs. This is illustrated in Fig. 6.18 in which the rotational populations of two neighboring vibrational states are indicated. Although the total population for the two vibrational states is equal, an inversion is seen to exist in the two P transitions [$(J' = 5) \rightarrow (J = 6), (J' = 4) \rightarrow (J = 5)$] and two R-

---

[†] *Chemical efficiency* is defined as the ratio of laser output energy to the total chemical energy that can be produced by combustion of the fuel.

[‡] Anharmonic pumping arises from the process $CO(v = n) + CO(v = m) \rightarrow CO(v = n + l) + CO(v = m - l)$ which, due to the vibration anharmonicity, is favored when $n > m$. This process allows the first CO molecule to climb up the ladder of the vibrational levels with a resulting non-Boltzmann distribution of the population among these levels.

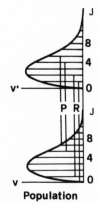

FIG. 6.18. Partial inversion between two vibrational transitions (V and V') having the same total population.

branch transitions indicated in the figure. Under conditions of partial inversion, laser action can thus take place and a new phenomenon, called *cascading*, plays an important role. The laser action depopulates a rotational level of the upper state and populates a rotational level of the lower vibrational state. The latter level can then accumulate enough population to result in population inversion with a rotational level of a still lower vibrational state. At the same time, the rotational level of the upper state may become sufficiently depopulated to result in population inversion with a rotational level of a still higher vibrational state. This process of cascading coupled with the very low V–T rate results in most of the vibrational energy being extracted as laser output energy. This, together with the very high excitation efficiency, accounts for the high efficiency of the CO laser. The low-temperature requirement arises from the need for very efficient anharmonic pumping. In fact, the overpopulation of the high vibrational levels compared to the Boltzmann distribution, and hence the degree of partial inversion, increases rapidly with decreasing translational temperature.

As in the case of the CO_2 laser, the CO laser has been operated with longitudinal flow, with *e*-beam pre-ionized pulsed TE, and with gas-dynamic excitation. The requirement of cryogenic temperatures has so far limited the commercial development of CO lasers.

6.3.3.2 Vibronic Lasers[21]

As a particularly relevant example of vibronic lasers, we will consider the N_2 laser. This laser has its most important oscillation at $\lambda = 337$ nm (UV), and belongs to the category of self-terminating lasers. The pulsed nitrogen laser is commonly used as a pump for dye lasers.

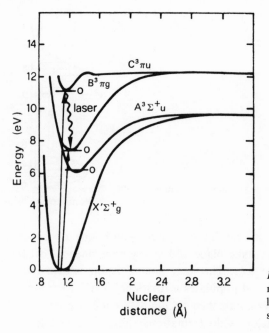

FIG. 6.19. Energy levels of the N_2 molecule. For simplicity only the lowest vibrational level ($v = 0$) is shown for each electronic state.

The relevant energy level scheme for the N_2 molecule is shown in Fig. 6.19. Laser action takes place in the so-called second positive system, i.e., in the transition from the $C^3\prod_u$ state (henceforth called the C state) to the $B^3\prod_g$ state (B state).[†] The excitation of the C state is believed to arise from electron-impact collisions with ground-state N_2 molecules. Since both C and B states are triplet states, transitions from the ground state are spin-forbidden. On the basis of the Franck–Condon principle, we can, however, expect that the excitation cross section to the $v = 0$ level of the C state will be larger than that to the $v = 0$ level of the B state. Compared to the ground state, the potential minimum of the B state is in fact shifted to larger internuclear separation than that of the C state. The lifetime (radiative) of the C state is 40 ns, while the lifetime of the B state is 10 μs. Clearly the laser cannot operate cw since condition (5.26) is not satisfied. It can, however, be excited on a pulsed basis provided the electrical pulse is appreciably shorter than 40 ns. Laser action takes place predominantly on several rotational lines of the $v''(0) \rightarrow v'(0)$ transition ($\lambda = 337.1$ nm). Besides being favored by the pumping process, as already mentioned, this transition in fact exhibits the largest Franck–Condon factor. Oscillation on

[†] Under different operating conditions laser action can also take place (in the near infrared, 0.74–1.23 μm) in the first positive system involving the $B^3\prod_g \rightarrow A^3\sum_u^+$ transition.

the $v''(1) \rightarrow v'(0)$ ($\lambda = 357.7$ nm) and on the $v''(0) \rightarrow v'(1)$ ($\lambda = 315.9$ nm) transitions also occur, although with lower intensity.

A possible configuration for a N_2 laser is shown schematically in Fig. 6.20a. Due to the high value of electric field required (~ 10 kV/cm at the typical operating pressure of $p \simeq 30$ Torr), a TE laser configuration is normally used. A fast discharge pulse (a few nanoseconds) is needed and a discharge circuit which achieves this, the so-called Blumlein configuration, is shown in Fig. 6.20. The transmission-line analog of this circuit is shown in Fig. 6.20b, where Z is the impedance of the discharge channel and Z_0 is the characteristic impedance of the line. If the line is initially charged to a voltage V, and if $Z = 2Z_0$, it can be shown that, upon closing the switch, a voltage pulse of value $V/2$ and duration $2L/c$ is produced across Z (c is the e.m. propagation velocity in the line). By making L short enough, the system of Fig. 6.20a can produce a short voltage pulse suitable for driving the N_2 laser.

Due to the high gain of this self-terminating transition, oscillation takes place in the form of amplified spontaneous emission. Thus the laser

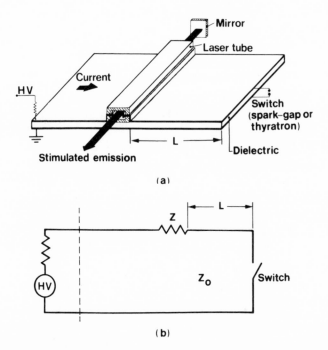

FIG. *6.20.* (a) Blumlein pulse generator using a flat transmission line. The discharge channel may typically be 2×0.5 cm, the large dimension being along the discharge direction. (b) Transmission line analog of the Blumlein generator above.

can be operated without mirrors. However, usually, a single mirror is placed at one end of the cavity, since this reduces the threshold power [see (2.91b) and (2.91d)] and also provides a unidirectional output. In this way, the output beam divergence is also reduced and is given by the ratio of the transverse dimension of the discharge to twice the cavity length. With this type of laser, it is possible to obtain peak power up to ~ 1 MW in pulses ~ 10 ns wide with a pulse repetition rate up to 100 Hz. The repetition rate is limited by heating effects. More recently N_2 lasers working at atmospheric pressure have been developed. The problem of arcing is alleviated by further reducing (to ~ 1 ns) the duration of the voltage pulse. Due to the increased gain per unit length and the fast discharge, this type of laser can give output pulses of duration 100–500 ps (100–200 kW peak power). No mirrors are used in this case. Such a laser, when used to pump dye lasers, can thus produce dye-laser pulses of duration well down into the subnanosecond range. These short pulses are useful for investigating relaxation processes in various materials.

There are, of course, other examples of vibronic lasers besides the N_2 laser. Of these other lasers we shall mention the H_2 laser. It oscillates on a series of lines around the wavelength $\lambda \simeq 160$ nm (Lyman band) and around $\lambda \simeq 116$ nm (Werner band). These wavelengths fall in the so-called vacuum ultraviolet (VUV). In fact, at these wavelengths, atmospheric absorption becomes so high that beam propagation must be done in a vacuum (or in a gas such as He). To provide the necessary fast discharge (~ 1 ns) a Blumlein configuration (Fig. 6.20a) is again used. This laser is also self-terminating, with its output being produced by amplified spontaneous emission.

It is interesting to note that the 116-nm wavelength is so far the shortest produced by laser action. The difficulty in getting to still shorter wavelengths (down to the x-ray region) is worth emphasizing at this point. From (3.25), (5.18), and (5.17) we find that the threshold pump power per unit volume is

$$\frac{dP_{th}}{dV} = \frac{1}{\eta_p} \hbar\omega_p W_{cp}(N_t - N_c) = \frac{\hbar\omega_p}{\eta_p} \frac{\gamma}{\sigma l \tau} \tag{6.12}$$

On the other hand, from (2.145) we find (for $\Delta\omega = 0$) that $(1/\sigma\tau)$ $\propto \omega_0^2/g_t(0) \propto \omega_0^2\Delta\omega_0$. At frequencies in the UV and at moderate pressures we may assume that the linewidth $\Delta\omega_0$ is determined by Doppler broadening. Hence [see (2.113)] $\Delta\omega_0 \propto \omega_0$ and (dP/dV) increases as ω_0^4 (if we take $\omega_p \simeq \omega_0$). At still higher frequencies (x-ray region) the linewidth is determined by natural broadening since the value of radiative lifetime becomes very short. In this case $\Delta\omega_0 \propto \omega_0^3$ and (dP/dV) increases as ω_0^6. Due to the rapid increase of (dP/dV) with frequency, the required threshold power becomes very large. This explains why, despite many attempts, no one has so far succeeded in making an x-ray laser work.[†]

6.3.3.3 Excimer Lasers[(22)]

An interesting and important class of molecular lasers involving transitions between different electronic states is that of the excimer lasers.

Consider a diatomic molecule A_2 with potential energy curves as in Fig. 6.21 for the ground and excited states, respectively. Since the ground state is repulsive, the molecule does not exist in this state (i.e., species A only exists in the monomer form A in the ground state). Since, however, the potential energy curve for the excited state has a minimum, the molecule A_2 does exist in the excited state (i.e., species A exists in the dimer form A_2 in the excited state). Such a molecule A_2^* is called an "excimer" from a contraction of the words "excited dimer."

Now suppose a large number of excimers has somehow been created in a given volume. Laser action can then be produced on the transition between the upper (bound) state and the lower (free) state (bound–free transition). This is called an excimer laser. An excimer laser has two

[†]Recently, the achievement of a pulsed x-ray laser at the wavelength of 14 Å has been announced. The laser was pumped by the x rays produced by a small nuclear detonation (an experimental condition not so easily duplicated in everyone's laboratory!).

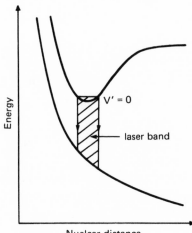

FIG. 6.21. Energy levels of an excimer laser.   Nuclear distance

peculiar but important properties, both due to the fact that the ground state
is repulsive: (i) Once the molecule, after undergoing the laser transition,
reaches the ground state, it immediately dissociates. This means that the
lower laser level will always be empty. (ii) No well-defined rotational-
vibrational transitions exist, and the transition is broad band. This allows
the possibility of tunable laser radiation over this broad-band transition.

As a particularly relevant class of excimer lasers we will consider those
in which a rare gas atom (e.g., Ar, Kr, Xe) is combined, in the excited state,
with a halogen atom (e.g., F, Cl) to form a rare-gas-halide excimer.[†]
Specific examples are ArF, (λ = 193 nm), KrF (λ = 248 nm), XeCl (λ = 308
nm), and XeF (λ = 351 nm), all oscillating in the UV. The reason why rare
gas halides are readily formed in the excited state can be seen when one
realizes that an excited rare gas becomes chemically similar to an alkali
atom, and these are known to react readily with halogens. This analogy also
indicates that the bonding in this excited state must be of ionic character:
In the bonding process, the excited electron is transferred from the rare gas
atom to the halogen atom. This bound state is therefore also referred to as a
charge-transfer state.

The pumping mechanisms in a rare-gas-halide laser are rather complex
since they involve several ionic species as well as excited atomic and
molecular species. In KrF, for example (using Kr, F_2, and a buffer gas in
the mixture), the following mechanisms play a very important role: (i)
direct reaction of the excited rare gas with the halogen, viz.,

$$Kr^* + F_2 \to KrF^* + F \tag{6.13}$$

and (ii) dissociative attachment of an electron to the halogen (6.14a)
followed by three-body recombination of the negative halogen ion (6.14b),
i.e.,

$$e + F_2 \to F^- + F \tag{6.14a}$$

and

$$F^- + Kr^+ + M \to KrF^* + M \tag{6.14b}$$

where M is an atom of the buffer gas (Ar or He).[‡]

Rare-gas-halide excimer lasers can be pumped either by an electron
beam or by an electrical discharge. In the latter case either e-beam or UV

[†]Strictly speaking these should not be called excimers since they involve unlike atoms. The
words hetero-excimer or *exciplex* (from *exci*ted state com*plex*) would perhaps be more
appropriate in these cases. However, the word excimer is now widely used in this context and
we will follow this usage.
[‡]The process in equation (6.14b) requires the presence of a buffer gas atom M since otherwise
the momentum and the energy of the reacting partners (F and Kr) cannot both be conserved.

preionization techniques are used, the laser is pulsed, and its design is in many respects similar to that of a TEA CO_2 laser. The laser pulse length is of the order of a few tens of nanoseconds, being limited by the onset of discharge instabilities (arc formation). Average output powers up to 100 W, pulse repetition rates up to 1 kHz, and electrical efficiencies of 1% have been obtained. Excimer lasers are very promising for sophisticated photochemical processes, such as isotope separation, and for numerous other applications in which a strong and efficient UV source is required.

6.4 LIQUID LASERS (DYE LASERS)[23,24]

The liquid lasers we will be considering are those in which the active medium consists of solutions of certain organic dye compounds in liquids such as ethyl alcohol, methyl alcohol, or water. These dyes usually belong to one of the following classes: (i) polymethine dyes (0.7–1 μm), (ii) xanthene dyes (0.5–0.7 μm), (iii) coumarin dyes (0.4–0.5 μm), and (iv) scintillator dyes ($\lambda < 0.4 \mu$m). By virtue of their wavelength tunability, wide spectral coverage, and simplicity, organic dye lasers are playing an increasingly important role in various fields of application (from spectroscopy to photochemistry).

6.4.1 Photophysical Properties of Organic Dyes

Organic dyes are large and complicated molecular systems[†] containing conjugated double bonds. Usually they have strong absorption bands in the UV or visible range of the spectrum, and when excited by light of appropriate wavelength, they display intense broad-band fluorescence spectra, such as that shown in Fig. 6.22 for Rhodamine 6G in ethanol solution.

A simple understanding of the energy levels of a dye molecule can be obtained using the so-called free-electron model. We will illustrate this by considering the case of the cyanine dye shown in Fig. 6.23a. The π-

---

[†]As an example, the structural formula for the widely-used dye Rhodamine 6G (xanthene dye) is

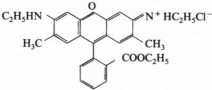

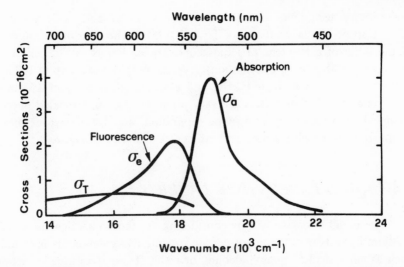

FIG. 6.22. Absorption cross section σ_a, stimulated-emission cross section σ_e (singlet–singlet transition), and absorption cross section σ_T (triplet–triplet transition) for an ethanol solution of Rhodamine 6G.

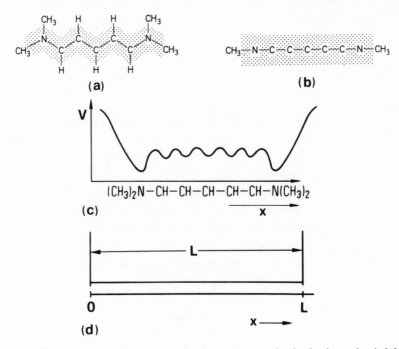

FIG. 6.23. Free-electron model to explain the electronic energy levels of a dye molecule [after Försterling and Kuhn[36]].

electrons of the carbon atoms form two planar distributions, one above and one below the plane of the molecule (Fig. 6.23b). The electronic states of the molecule originate from this π-electron distribution. In the free-electron model, these π-electrons are assumed to move freely within their planar distributions, limited only by the repulsive potential of the group at each end of the dye. The energy levels of the electrons are therefore simply those of a free electron in a potential well of the form shown in Fig. 6.23c. If this well is approximated by a rectangular one (Fig. 6.23d), the energy levels are well known and given by $E_n = h^2 n^2 / 8 m L^2$ where n is an integer, m is the electron mass, and L is the length of the well. It is important to note, at this point, that dye molecules have an even number of electrons in the π-electron cloud.[†] If we let the number of these electrons be $2N$, the lowest energy state of the molecule will correspond to the situation where these electrons are occupying the lowest N energy levels. Each level can in fact be occupied by two electrons with opposite spin. This molecular state will thus have zero spin angular momentums (singlet state) and is labeled S_0 in Fig. 6.24. In the same figure the highest occupied level and the next (empty) one above it are indicated by two squares one above the other. The first excited singlet state (labeled S_1 in the figure) is then obtained by

[†]Molecular systems with unpaired electrons are known as radicals and they tend to react readily, thus forming a system with paired electrons.

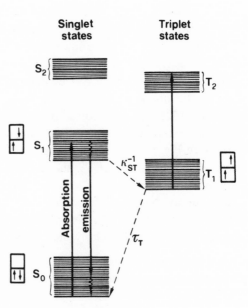

FIG. 6.24. Typical energy levels for a dye in solution. The singlet and triplet levels are shown in separate columns.

promoting one of the two highest-lying electrons, without flipping its spin, to the next level up. If the spin is flipped, the resulting state is a triplet state (total spin $S = 1$, labeled T_1 in the figure). Excited singlet (S_2) and triplet (T_2) states result when the electron is further promoted to the next level, and so on. Note that in Fig. 6.24 each electronic state is actually made up of a set of vibrational levels (the heavier lines in the figure) and rotational levels (the lighter lines). The separation between vibrational levels is typically 1400–1700 cm^{-1}, whereas the separation of rotational levels is typically 100 times less. Since the line broadening mechanisms are much more important in liquids than in solids, the rotational lines are not resolved and therefore give rise to a continuum of levels between the vibrational levels.

We now look at what happens when the molecule is subjected to electromagnetic radiation. First, we recall that the selection rules require that $\Delta S = 0$. Hence singlet–singlet transitions are allowed, whereas singlet–triplet transitions are forbidden. Therefore, the interaction with electromagnetic radiation can raise the molecule from the ground level S_0 to one of the vibrational levels of the S_1 level. Since the rotational and vibrational levels are unresolved, the absorption spectrum will show a broad and featureless transition (see, for example, Fig. 6.22). Note that an important characteristic of these dyes is that they have a very large dipole matrix element μ. This is because the π-electrons are free to move over a distance comparable to the molecular dimension a, and since a is quite large it follows that μ is large ($\mu \simeq ea$). It then follows that the absorption cross section σ, which is proportional to μ^2, is also large ($\sim 10^{-16}$ cm^2). Once in the excited state the molecule decays in a very short time (nonradiative decay, $\tau_{nr} \simeq 10^{-12}$ s) to the lowest vibrational state of the S_1 level.[†] From there it decays radiatively to some vibrational level of S_0 (*fluorescence*). The transition probability will be determined by the appropriate Franck–Condon factors. It is therefore clear from what has already been said (see also Fig. 2.22) that the fluorescent emission will take the form of a broad and featureless band shifted to the long-wavelength side of the absorption band (see Fig. 6.22). Having dropped to an excited rotational–vibrational state of the ground level S_0, the molecule will then return to the lowest vibrational level by another very fast (of the order of picoseconds) nonradiative decay. Note that, when the molecule is in the lowest level of S_1, it can also decay to level T_1. This process is called *intersystem crossing* and is caused by collisions. Similarly the transition $T_1 \rightarrow S_0$ takes place mainly by

---

[†]More precisely, thermalization among the rotational–vibrational levels of the S_1 state will occur.

way of collisions but partly also by a radiative process (the transition $T_1 \rightarrow S_0$ is prohibited radiatively as mentioned above). This radiation is called *phosphorescence*. We will characterize these three decay processes by the following three constants: (i) τ_{sp}, spontaneous-emission lifetime of level S_1, (ii) k_{ST} intersystem crossing rate (s^{-1}) between singlet and triplet systems, and (iii) τ_T, lifetime of the T_1 level. If we call τ the lifetime of the S_1 level, then we have [see (2.93)]

$$\frac{1}{\tau} = \frac{1}{\tau_{sp}} + k_{ST} \tag{6.15}$$

Due to the large value of the dipole matrix element μ, the radiative lifetime τ_{sp} is very short (a few nanoseconds). Since k_{ST}^{-1} is usually much longer (~ 100 ns), it follows that most of the molecules decay from level S_1 by fluorescence. The fluorescence quantum yield (number of photons emitted by fluorescence divided by number of atoms put into S_1) is therefore nearly unity. In fact, one has [see (2.96)]

$$\phi = \tau / \tau_{sp} \tag{6.16}$$

The triplet lifetime τ_T depends on the experimental conditions and particularly on the amount of dissolved oxygen in the solution. The lifetime can range from 10^{-7} s in an oxygen-saturated solution to 10^{-3} s or more in a solution which has been deoxygenated.

6.4.2 Characteristics of Dye Lasers

From what has been said it is quite reasonable to expect these materials to be capable of exhibiting laser action at the fluorescence wavelengths. In fact, the fast nonradiative decay within the excited single-state S_1 populates the upper laser level very effectively while the fast nonradiative decay within the ground state is effective in depopulating the lower laser level. Note also that the dye solution is quite transparent to the fluorescence wavelengths (i.e., the corresponding absorption cross section σ_a is very low, see for example Fig. 6.22). In fact it was quite late in the general development of laser devices before the first dye laser was operated (1966),[25] and we now look at some reasons for this. One problem which presents itself is the very short lifetime τ of the S_1 state since the required pump power is inversely proportional to τ. Although this is to some extent compensated by the comparatively large value of the cross section, the product $\sigma\tau$ [recall that threshold pump power is $\propto (\sigma\tau)^{-1}$, see (6.12)] is still about three orders of magnitude smaller than for solid-state lasers such as

Nd:YAG. A second problem arises as a consequence of intersystem cross-ing. If, in fact, τ_T is long compared to k_{ST}^{-1}, then molecules accumulate in the triplet state T_1. This introduces absorption through the $T_1 \rightarrow T_2$ transi-tion (which is optically allowed). Unfortunately, this absorption tends to occur in the same wavelength region as the fluorescence (see again, for example, Fig. 6.22), and is therefore a serious obstacle to laser action. It can be shown, in fact, that continuous laser action is possible only if τ_T is less than some particular value, depending on the characteristics of the dye material. To derive this result we first note that the fluorescence emission curve of the dye (see Fig. 6.22) can be described in terms of the stimulated emission cross section σ_e. Thus, if N_2 is the total population of the S_1 state, the corresponding (unsaturated) gain, at the given wavelength to which σ_e refers, is $\exp(N_2 \sigma_e l)$, where l is the length of the active material. If we now let N_T be the population in the triplet state T_1, a necessary condition for laser action is that the gain due to stimulated emission exceed the loss due to triplet–triplet absorption, i.e.,

$$\sigma_e N_2 > \sigma_T N_T \qquad (6.17)$$

In the steady state, the rate of decay of triplet population N_T/τ_T must equal the rate of increase due to intersystem crossing $k_{ST}N_2$, i.e.,

$$N_T = k_{ST}\tau_T N_2 \qquad (6.18)$$

Combining (6.17) and (6.18), we get

$$\tau_T < \sigma_e/\sigma_T k_{ST} \qquad (6.19)$$

which is a necessary condition for cw laser action [i.e., in a sense equivalent to (5.26)]. If this condition is not satisfied, the dye laser can only operate in a pulsed regime. In this case, the duration of the pump pulse must be short enough to ensure that an excessive population does not accumulate in the triplet state. Finally, a third crucial problem comes from the presence of thermal gradients produced in the liquid by the pump. These tend to produce refractive-index gradients which prevent laser action. These gradients produce effects which are similar in some respects to those due to intersystem crossing. Both of these processes tend to cause laser action to terminate after the pump has been applied for a certain length of time. Fortunately, however, as already mentioned, τ_T can be reduced if certain substances (e.g., oxygen) are added to the solution, while thermal effects can also be reduced with a suitable experimental arrangement.

Pulsed laser action has been obtained from very many different dyes by using one of the following pumping schemes: (i) fast flashlamps (with a risetime of $< 1\ \mu s$), (ii) a short light pulse from another laser. The N_2 laser in particular is very frequently used for this application. Its UV output is suitable for pumping many dyes that oscillate in the visible range. This pumping scheme is particularly efficient: Very high gains and a conversion efficiency (from UV to visible light) of the order of 10% have been achieved. The efficiency of the N_2 laser is rather low, however ($\sim$0.2%). For this reason, excimer lasers (in particular KrF and XeF) are being used increasingly as pumps for dye lasers. For both N_2 and excimer laser pumping, a transverse pump configuration (i.e., direction of the pump beam orthogonal to the resonator axis) is used (Fig. 6.25a). The telescope shown in the figure serves to enlarge the beam on the echelle grating (used as a wavelength-selective element, see Fig. 5.7), thus increasing its resolving power. The Fabry–Perot etalon (see also Fig. 5.8) is used for fine tuning of the output wavelength. Continuous laser action has been obtained in a number of laser dyes covering the entire visible range. Pumping is provided by another cw laser (usually an Ar^+ laser), and a longitudinal (or near longitudinal) pump configuration such as that in Fig. 6.25b is commonly used. Note the presence of the dispersive prism in the laser cavity which serves the double purpose of (i) tuning the laser wavelength (see again Fig. 5.7) and (ii) allowing the pump laser beam to be physically separated from the dye laser beam in the region indicated in the figure. Since the pump beam comes in around the side of the dye laser mirror, rather than through it, one avoids the need for special mirrors which are transparent to the pump and highly reflecting for the dye wavelengths. A rather interesting configuration for single-longitudinal-mode cw dye lasers is the ring cavity shown in Fig. 6.26. Pumping is again provided by an ion laser, and the dye

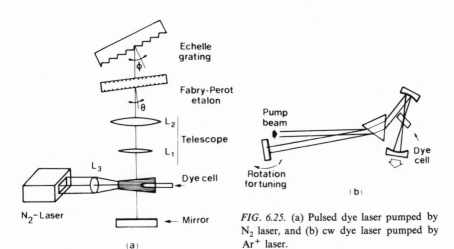

FIG. 6.25. (a) Pulsed dye laser pumped by N_2 laser, and (b) cw dye laser pumped by Ar^+ laser.

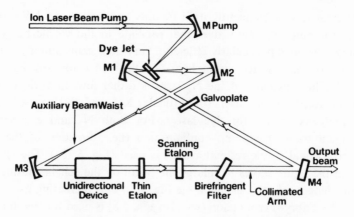

FIG. 6.26 High power single longitudinal mode ring dye laser.

is circulated by means of a liquid jet system. Single-longitudinal-mode oscillation and frequency scanning are achieved by the combination of birefringent filter, scanning etalon, and thin Fabry–Perot etalon. The special feature of this cavity is that, with the help of the unidirectional device, the laser beam can only travel in one sense around the ring cavity (indicated by the arrows in the figure). Thus there is no standing wave formed in the cavity and, in particular, within the dye medium. Therefore the phenomenon of spatial hole burning does not occur and this has two consequences: (i) Oscillation on a single longitudinal mode is much easier to obtain as can be understood by referring to the discussion in connection with Fig. 5.6. (ii) Higher output power is available in this single mode since now the whole of the active material (rather than just those regions around the maxima of the standing-wave pattern) contributes to the laser output. As a result of this, output powers have been obtained which are as much as an order of magnitude greater than those from conventional single-mode dye lasers (e.g., of the type of Fig. 6.25b).

Average output powers up to 100 W with an efficiency somewhat less than 1% have been obtained with flashlamp-pumped dye lasers. A very interesting property of dye lasers is their wide oscillation bandwidth ($\sim$10 nm). Tuning of the output wavelength over this bandwidth can therefore be achieved using wavelength-selecting cavities such as those of Fig. 5.7. A large oscillation bandwidth is also very important for mode-locked operation. Continuous-wave (Ar^+ laser-pumped) mode-locked dye lasers in a ring configuration have been operated with output pulse durations as short as $\sim$0.03 ps. These are the shortest pulses so far obtained from lasers.

Dye lasers are now widely used in many scientific and technological applications where wavelength tunability or pulses of short duration are required. Photodegradation of the dye due to the pump light remains an inconvenient feature of these lasers.

6.5 CHEMICAL LASERS [26,27]

A *chemical laser* is usually defined as one in which the population inversion is "directly" produced by a chemical reaction. According to this definition, the gas-dynamic type of CO_2 laser cannot be regarded as a chemical laser. Chemical lasers usually involve a chemical reaction between gaseous elements. In this case, a large proportion of the reaction energy is left in the form of vibrational energy of the molecules. The laser transitions are therefore often of vibrational–rotational type (the only notable exception being perhaps the photochemical-dissociation laser to be described later on), and the corresponding wavelengths lie at present between 3 and 10 μm. These lasers are interesting for two main reasons: (i) They provide an interesting example of direct conversion of chemical energy into electromagnetic energy. (ii) Since the amount of energy available in a chemical reaction is very large,[†] one can expect high output powers.

As an illustrative example of chemical lasers we will consider the HF laser. This laser oscillates over several rotational–vibrational lines in the 2.6 to 3.3 μm band and gives cw output powers up to 10 kW and pulsed energies up to a few kilojoules with a chemical efficiency up to $\sim$10%.

The main pumping mechanism for the HF laser comes from the chemical reaction

$$F + H_2 \rightarrow HF^* + H \qquad (6.20)$$

Since the heat of reaction is 31.6 kcal/mole, the HF molecule can be left in an excited state as high as the $v = 3$ vibrational level (see Fig. 6.27). As a consequence of the different rates of decay to the various vibrational levels, the $v = 2$ level has by far the greatest population, and a large population inversion builds up on the $(v' = 2) \rightarrow (v = 1)$ transition. It can be seen from the figure that more than 60% of the reaction energy is released as vibrational energy. The reason why, after chemical reaction, the HF molecule is left in an excited state can be understood in a simple way. Consider the reaction given in equation (6.20). On account of the high electron affinity of F, at large distances the F–H_2 interaction is strongly attractive and leads to a considerable polarization of the H_2 charge distribution. Since the electron is light, the HF bond can form before the proton has adjusted

---

[†]For example, a mixture of H_2, F_2, and other substances (16% of H_2 and F_2 at atmospheric pressure) has a heat of reaction equal to 2000 J/liter of which 1000 J is left as vibrational energy.

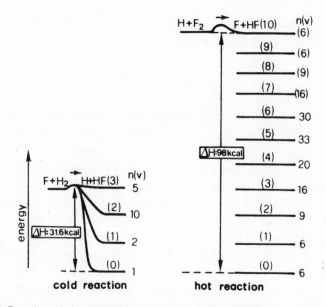

FIG. 6.27. Pumping of the vibrational levels of the HF molecule by the two reactions, $F + H_2 \rightarrow H + HF^*$ and $H + F_2 \rightarrow F + HF^*$. The relative populations $n(v)$ produced in this way are also shown.

to the internuclear separation appropriate to the HF ground electronic state. Thus, there is a considerable probability that the proton, after reaction, will be found at a distance which is greater than the equilibrium distance of the HF bond. This therefore leads, classically, to a vibrational motion.

Note that, for the chemical reaction in equation (6.20) to occur, atomic fluorine must be present. This is produced by dissociation of some suitable fluorine-donating molecule such as SF_6 or molecular F_2. Dissociation may be achieved in several ways, e.g., by electron collision in an electrical discharge ($SF_6 + e \rightarrow SF_5 + F + e$). When molecular fluorine is used, the undissociated F_2 molecules can react with atomic hydrogen [itself produced by reaction (6.20)] to yield

$$H + F_2 \rightarrow HF^* + F \qquad (6.21)$$

The atomic fluorine produced in this way can then take part again in reaction (6.20). This leads to a chain reaction in which the number of excited HF molecules can greatly exceed the number of fluorine atoms produced initially. Note that the chemical energy of reaction (6.21) (98 kcal/mole) is substantially larger than that of (6.20). This can result in excitation up to the $v = 10$ vibrational level of the HF molecule (Fig. 6.27). Reaction (6.21) therefore helps to establish a population inversion between various vibrational levels of the HF molecule. From what has just been said, it would seem that molecular fluorine might be more suitable than SF_6 for use in an HF laser. However, the $H_2 + F_2$ mixture is more difficult to handle than the $H_2 + SF_6$ mixture, and it may even become explosive.

HF lasers can be made to operate either pulsed or cw. In pulsed lasers, atomic fluorine is produced by collisions between the fluorine donors and electrons generated either by an electrical discharge or by an auxiliary electron-beam machine. Where an electrical discharge is used, the pump configuration is similar to that of a TEA CO_2 laser, and UV pre-ionization is often used to ensure a more uniform discharge. When molecular fluorine is used as a reactant, a chain reaction is established, and the laser output energy can appreciably exceed the energy of either the electrical discharge or *e*-beam. In a cw laser, fluorine is thermally dissociated by an arc jet heater and then expanded through supersonic nozzles (to ~Mach 4). Molecular hydrogen is mixed in downstream and reacts according to (6.20) (Fig. 6.28). For high-power or high-energy lasers, unstable resonators are often used.

Laser action takes place on several vibrational transitions, from $1 \to 0$ up to $6 \to 5$ ($\lambda = 2.7$–3.3 μm) and on several rotational lines within each vibrational transition. As already discussed in the case of the CO laser, there are two reasons why oscillation can occur on so many lines, namely: (i) The phenomenon of cascading. If in fact the $2 \to 1$ transition (usually the strongest) lases, the population of level 2 will be depleted and will accumulate in level 1. Consequently laser action on $3 \to 2$ and $1 \to 0$ transition may now occur. (ii) The phenomenon of partial inversion (see Fig. 6.18) in which there may be a population inversion between some rotational lines even when no inversion exists between the overall populations of the corresponding vibrational levels.

Besides the HF laser, mention should be made of the DF, HCl, and HBr lasers which operate on similar schemes to HF and oscillate in the 3.5–5 μm range. This range is interesting since it is a spectral region where

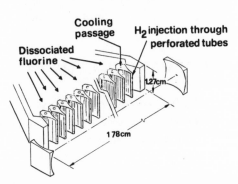

FIG. 6.28. Supersonic-diffusion HF chemical laser [after Chester[26]].

the atmospheric transmission is good. As already mentioned, chemical lasers of this type can give large output powers (or energies) with good chemical efficiency. Safety problems (F_2 is perhaps the most corrosive and reactive element known) greatly limit the applicability of these lasers, however. Although electrical-discharge lasers (using SF_6) are commercially available, the most important area of use for these lasers seems to be in high-power military applications.

As a second example of a chemical laser we will briefly mention the atomic iodine laser.[28] It belongs to the category of photochemical-dissociation (or photodissociation) lasers. Atomic iodine is in fact produced by photodissociation of either CH_3I or CF_3I or, more recently, C_3F_7I. When light ($\sim$300 nm) from a powerful flashlamp is absorbed by one of the above molecules, photodissociation leads to production of atomic iodine in the $^2P_{1/2}$ excited state at a greater rate than in the $^2P_{3/2}$ ground state. Thus laser oscillation takes place on the $^2P_{1/2} \rightarrow {}^2P_{3/2}$ line ($\lambda = 1.315$ μm). This line is forbidden as an electric–dipole transition but allowed as a magnetic–dipole transition. Since the corresponding spontaneous-emission lifetime is very long (in the millisecond range), the lifetime of the $^2P_{1/2}$ state is essentially governed by collisional deactivation. The lifetime of the $^2P_{3/2}$ ground state is governed by the three-body recombination process $I(^2P_{3/2}) + I(^2P_{3/2}) + M \rightarrow I_2 + M$, where M is another atom or molecule in the gas mixture (He, I_2). This lifetime is typically 100 μs. The characteristics of an iodine laser fall somewhat in between those typical of a gas laser and those typical of an optically pumped solid-state laser. The iodine, being a gas, must be contained in a glass tube (Fig. 6.1) just as for any other gas laser. However, the iodine laser is similar to solid-state lasers in two respects: (i) It is pumped by a flash in a geometrical configuration similar to those used for solid-state lasers (Fig. 3.2). (ii) As in the case of ruby and Nd^{3+} lasers, the laser line is a forbidden electric–dipole transition. This last property is particularly relevant. It means that the iodine laser has a long upper-state lifetime, and hence it can build up a large population inversion. This places the iodine laser (together with the Nd:glass and CO_2 lasers) among the most interesting systems for high-energy ($>$ 500 J) laser output.

6.6 SEMICONDUCTOR LASERS [29]

So far, we have only discussed atomic and molecular systems, whose energy levels are associated with localized wave functions, i.e., belong to single atoms or molecules. We will now consider the case of semiconductors for which it is no longer possible to talk about the wavefunction of an individual atom; it is necessary instead to deal with a wavefunction relating to the crystal as a whole. Likewise, one can no longer talk of energy levels of individual atoms.

6.6.1 Photophysical Properties of Semiconductor Lasers

The energy-level scheme for an idealized semiconductor is shown in Fig. 6.29. The energy-level spectrum consists of very broad bands: These are the valence band V and the conduction band C, separated by a region of forbidden energies (the band gap). Each band actually consists of a large number of very closely spaced energy states. According to the Pauli exclusion principle, there can be just two electrons (with opposite spin) occupying each energy state. Accordingly, the probability of occupation $f(E)$ of a given state of energy E is given by Fermi–Dirac statistics rather than by Maxwell–Boltzmann statistics. Thus

$$f(E) = \left\{ 1 + \exp\left[(E - F)/kT \right] \right\}^{-1} \tag{6.22}$$

where F is the energy of the so-called Fermi level. This level has the following physical significance: when $T \to 0$, one has

$$\begin{aligned} f &= 1 \quad \text{(for } E < F\text{)} \\ f &= 0 \quad \text{(for } E > F\text{)} \end{aligned} \tag{6.23}$$

so that this level represents the boundary between fully occupied and completely empty levels at $T = 0°$K. For nondegenerate semiconductors the Fermi level is situated inside the band gap (see Fig. 6.29). Therefore, at $T = 0°$K the valence band will be completely full, and the conduction band, completely empty. It can be shown that, under these conditions, the semiconductor will not conduct, and it is therefore an insulator. Note also that the Fermi level has also another physical meaning: At any temperature one has $f(F) = 1/2$.

Having made these preliminary remarks, we can now begin to describe the principles of operation of a semiconductor laser. For simplicity, we will

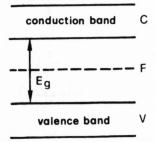

FIG. 6.29. Valence band, conduction band, and Fermi level of a semiconductor.

first assume that the semiconductor is at $T = 0°K$ (see Fig. 6.30a, in which the hatched area corresponds to completely full energy states). Now we suppose that electrons are somehow raised from the valence band to the conduction band. After a very short time ($\sim 10^{-13}$ s) the electrons in the conduction band will have dropped to the lowest levels in that band, and any electrons near the top of the valence band will also have dropped to the lowest unoccupied levels, thus leaving the top of the valence band full of "holes." This means that there is then a population inversion between the valence and conduction bands (Fig. 6.30b). The electrons in the conduction band fall back into the valence band (i.e., they recombine with holes), emitting a photon in the process (recombination radiation). Given a population inversion between conduction and valence bands as shown in Fig. 6.30b, the process of stimulated emission of recombination radiation will produce laser oscillation when the semiconductor is placed in a suitable resonator. From Fig. 6.30b it is seen that the frequency of the emitted radiation must satisfy the condition

$$E_g < h\nu < F_c - F_v \qquad (6.24)$$

which establishes the gain bandwidth of the semiconductor.

We now consider the situation where the semiconductor is held at a temperature $T > 0$. Referring again to Fig. 6.30b, we note that, although the semiconductor as a whole is not in thermal equilibrium, nevertheless equilibrium will be reached within a single band in a very short time. One can therefore talk of occupation probabilities f_v and f_c for the valence and conduction bands separately, where f_v and f_c are given by expressions of the same form as (6.22), namely,

$$f_v = \left\{ 1 + \exp\left[(E - F_v)/kT \right] \right\}^{-1} \qquad (6.25a)$$

$$f_c = \left\{ 1 + \exp\left[(E - F_c)/kT \right] \right\}^{-1} \qquad (6.25b)$$

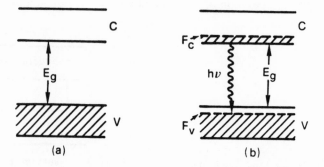

(a)   (b)

FIG. 6.30. Principle of operation of a semiconductor laser.

where F_v and F_c are the energies of the so-called quasi-Fermi levels of the valence and conduction bands respectively. From (6.25) and from our preliminary remarks, it is seen that when, for instance, $T = 0°K$, these levels separate the zones of fully occupied and completely empty levels for each band. Obviously the values of F_v and F_c depend on the number of electrons raised to the conduction band by the pumping process. Having introduced the concept of quasi-Fermi levels, we can readily obtain a necessary condition for laser action by imposing the requirement that the number of stimulated-emission events must be greater than the number of absorption events (the excess being necessary to overcome cavity losses). Both of these processes are proportional to the product of the number of photons present in the cavity and the B coefficient for the transition. On the other hand, the stimulated-emission rate will also be proportional to the product of the probability of occupation of the upper level with the probability of nonoccupation of the lower level, whereas the absorption rate will be proportional to the product of the occupation probability of the lower level with the probability of nonoccupation of the upper level. Therefore, to get stimulated emission, we must satisfy

$$Bq[\, f_c(1 - f_v) - f_v(1 - f_c)\,] > 0 \qquad (6.26)$$

This inequality means that $f_c > f_v$. From (6.25) this implies

$$F_c - F_v > E_2 - E_1 = h\nu \qquad (6.27)$$

where E_2 and E_1 are the upper- and lower-level energies respectively. We have thus rederived one of the two relations which were previously found with an intuitive approach for $T = 0°K$ [see (6.24)]. This derivation, however, shows that the relationship is valid for any temperature (as long as the concept of quasi-Fermi levels remains valid). Furthermore it has been shown that equation (6.27) is a consequence of the requirement that stimulated emission processes must exceed stimulated absorption processes. In this respect equation (6.27) is seen to be equivalent to the condition (5.26) established for a four-level laser.

6.6.2 Characteristics of Semiconductor Lasers

The pumping process in a semiconductor laser is usually achieved by preparing the semiconductor in the form of a $p-n$ junction diode with highly degenerate p-type and n-type regions, i.e., heavily doped (donor or acceptor concentration greater than 10^{18} atoms/cm^3). It will be seen that in this way the inversion is produced in the junction region.

As a first example of a junction laser we will consider the situation where the p-type and n-type materials are the same (e.g., GaAs) and are

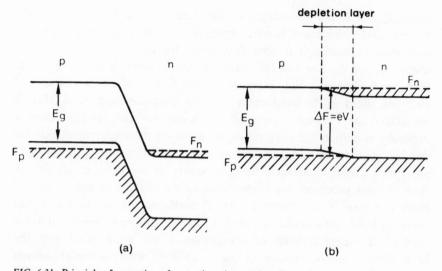

FIG. 6.31. Principle of operation of a *p–n* junction semiconductor laser with (a) zero bias and (b) forward bias.

joined directly to form the junction (which is therefore called a homojunction). The first semiconductor laser to work was of this type.[30,31] The operating principles of a diode constructed in this way are illustrated in Fig. 6.31. Since the materials are very heavily doped, the Fermi level F_p of the *p*-type semiconductor falls within the valence band, and the Fermi level F_n of the *n*-type semiconductor falls within the conduction band. It can be shown that, with no voltage applied, the two Fermi levels lie on the same horizontal line (Fig. 6.31a), i.e., they have the same energy. When a voltage V is applied, the two levels become separated by an amount given by

$$\Delta F = eV \qquad (6.28)$$

So, if the diode is forward biased, the energy levels will then be as shown in Fig. 6.31b. We see from the figure that a population inversion has been produced in the so-called "depletion layer" of the *p–n* junction. What the forward bias achieves essentially is the injection into the depletion layer of electrons from the conduction band of the *n*-type material and holes from the valence band of the *p*-type material. Finally, we note that since $\Delta F \simeq E_g$, where E_g is the band gap, it follows from (6.28) that $V \simeq E_g/e$. For a GaAs laser this means $V \simeq 1.5$ V.

Figure 6.32 shows a schematic diagram of a *p–n* junction laser, the shaded region being the depletion layer. It is seen that the diode has small dimensions. The thickness of the depletion region is usually very small (0.1

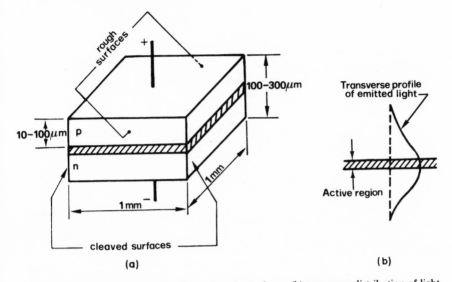

FIG. 6.32. (a) Schematic diagram of a semiconductor laser; (b) transverse distribution of light intensity.

μm). To obtain laser action, two end faces are made parallel, usually by cleavage along crystal planes. The other two are left with a rough finish to suppress oscillation in unwanted directions. Often the two surfaces are not provided with reflecting coatings: In fact, since the refractive index of a semiconductor is very large, there is already a sufficiently high reflectivity (~35%) for the semiconductor–air interface. The active region consists of a layer of thickness ~1 μm, i.e., somewhat wider than the depletion region. Due to diffraction the transverse dimension of the beam is in turn much greater (~40 μm) than the width of the active region (Fig. 6.32b). The laser beam therefore extends well into the p and n regions. However, since the transverse dimensions of the beam are still very small, the output beam ends up with a rather large divergence (a few degrees). Finally we point out that, at room temperature, the threshold current density for a homojunction laser is quite high (~10^5 A/cm^2 for GaAs). This is due to the high losses of the cavity mode since it extends far into the p and n regions (where absorption rather than gain is dominant). This current density, however, decreases rapidly with decreased operating temperature [approximately as $\exp(T/T_0)$, where the value of T_0 and the range of validity of the expression vary from one semiconductor to another]. This is a result of the fact that, as the temperature decreases, $f_c(1 - f_v)$ increases and $f_v(1 - f_c)$ decreases. Hence the gain [which depends on $f_c(1 - f_v) - f_v(1 - f_c)$, see

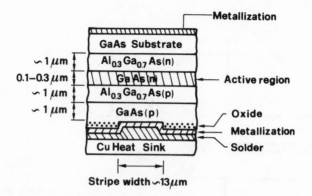

FIG. 6.33. Schematic diagram of a double-heterojunction semiconductor laser. The active region consists of the GaAs(n) layer (hatched area).

equation (6.26)] increases rapidly. As a consequence of this, homojunction lasers can operate cw only at cryogenic temperatures. This constitutes a serious limitation of this type of laser.

To overcome this difficulty, heterojunction lasers have been used. Figure 6.33 shows an example of a double heterojunction GaAs laser. In this diode there are two junctions [$Al_{0.3}Ga_{0.7}As(p)$–GaAs and GaAs–$Al_{0.3}Ga_{0.7}(n)$] between different materials. The active region consists of a thin layer of GaAs (0.1–0.3 μm). With such a diode, the threshold current density for room-temperature operation can be reduced by about two orders of magnitude (i.e., to $\sim 10^3$ A/cm^2) compared with the homojunction device. Thus cw operation at room temperature is made possible. The reduction of threshold current density is due to the combined effect of three circumstances: (i) The refractive index of GaAs ($n \simeq 3.6$) is significantly larger than that of $Al_{0.3}Ga_{0.7}As$ ($n \simeq 3.4$), thus providing an optical-waveguide structure. This means that the laser mode will now be confined in the GaAs layer, i.e., in the region where the gain is, and, unlike the situation in the homojunction diode, the wings of the field distribution no longer extend into the unpumped (and therefore absorbing) regions. (ii) The band gap of $Al_{0.3}Ga_{0.7}As$ (~ 1.8 eV) is significantly larger than that of GaAs (~ 1.5 eV). Energy barriers are therefore formed at the two junctions which effectively confine injected holes and electrons in the active layer (Fig. 6.34). For a given current density, the concentration of holes and electrons in the active layer is thus increased, and therefore the gain is also increased. (iii) The heat dissipation capability of the diode has been considerably improved. This has been achieved by cementing the GaAs(p) substrate to a copper (or tin) plate which, because of its mass and thermal conductivity, acts as a heat sink.

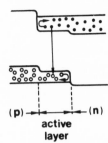

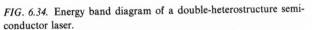

FIG. 6.34. Energy band diagram of a double-heterostructure semiconductor laser.

Semiconductor lasers cover a broad wavelength range from around 0.7 to ~30 μm. At present the most important semiconductor laser is perhaps the GaAs laser ($\lambda = 0.84$ μm). Continuous output powers up to a few milliwatts (5–10 mW) at room temperature with an overall slope efficiency of about 10% have been obtained. The internal quantum efficiency (fraction of injected carriers which recombine radiatively) is even higher (~70%). Semiconductor lasers are therefore among the most efficient of lasers. We note that, due to the large oscillation bandwidth (~10^{11} Hz for GaAs), the possibilities for mode-locked operation are attractive. Pulses of about 5 ps duration have indeed been obtained with passively mode-locked GaAs lasers. Note also that ternary compounds such as $Ga(As_{1-x}P_x)$ can also be used. The oscillating wavelength ranges from $\lambda = 0.84$ ($x = 0$) to 0.64 μm ($x = 0.4$). Thus, by varying the composition, it is possible to continuously vary the output wavelength. Gallium arsenide lasers are attractive as sources in optical communication links using optical fibers as the transmitting medium. Operating lifetimes in excess of 10^6 h have already been demonstrated with double-heterojunction GaAs lasers. The GaAs laser is also very interesting in a number of applications requiring only a low-power laser (such as optical reading) where it is no disadvantage to use infrared rather than visible light. Double-heterojunction semiconductor lasers operating at either $\lambda \simeq 1.3$ or $\lambda \simeq 1.6$ μm, where two minima for the loss of a quartz optical fiber occur, are now being vigorously developed. Here, the most interesting semiconductor for the active region seems to be the quaternary alloy $In_{1-x}Ga_xAs_yP_{1-y}$, while the p and n sides of the junctions may be made of the simple binary compound InP. If $y = 2.2x$, the quaternary alloy is lattice matched to InP, and by the appropriate choice of x the emission wavelength can be tuned from 0.92 to 1.5 μm.

Of the various other semiconductor lasers, mention should be made of the lead salt lasers,[38] all oscillating in the middle-to-far infrared, and in particular the ternary compounds $PbS_{1-x}Se_x$(4–8.5 μm), $Pb_{1-x}Sn_xTe$(6.5–32 μm), and $Pb_{1-x}Sn_xSe$(9–30 μm). Laser operation in these cases requires cryogenic temperatures ($T \simeq 77°K$ for cw operation). For a given index of composition x, the wavelength of the emitted radiation can be tuned by applying a magnetic field, by applying hydrostatic pressure, or by changing the diode current (heating effect). Typical applications of these lead salt lasers are found in the field of infrared spectroscopy, particularly in high-resolution spectroscopy. The linewidth of the emitted radiation can indeed be made very narrow (e.g., ~50 kHz for PbSnTe).

6.7 COLOR-CENTER LASERS [32]

A number of different types of color centers in alkali halide crystals are now being used as the basis of efficient, optically pumped lasers with broad tunability in the near infrared. At present color-center lasers allow operation over the wavelength range 0.8–3.3 μm. On a scale of increasing wavelength, these lasers thus take over just where the organic dyes give up.

Figure 6.35 indicates the structure of some color centers which are of interest to the present discussion. Of these only the F_A and F_2^+ centers have been made to lase. The ordinary F center can be regarded as an archetype of the other varieties of F-like centers: It consists of an electron trapped in an anion vacancy of the crystal. On the other hand, if one of the six metal ions immediately surrounding the vacancy is foreign (indicated by the smaller circle in the figure; e.g., Li^+ in a potassium halide), the defect is known as an F_A center. Two adjacent F centers along a (110) direction constitute a so-called F_2 center, and F_2^+ is its singly ionized counterpart. The general energy level scheme of an F center is shown in Fig. 6.36. After

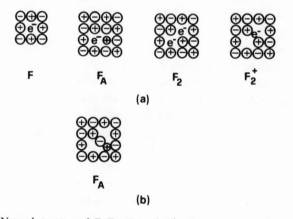

(a)

(b)

FIG. 6.35. (a) Normal structure of F, F_A, F_2, and F_2^+ color centers. (b) Relaxed structure of the F_A center. The electron (not shown) shares the two empty spaces of the lattice.

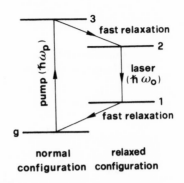

FIG. 6.36. The pumping cycle of an *F*-center laser.

it has been raised to its excited state 3, the *F* center rapidly (of the order of picoseconds) relaxes to its state 2. The configuration of this so-called relaxed state is also indicated in Fig. 6.35 for the F_A center. Relaxation consists of a simple expansion of the vacancy (or of the double vacancy) for F, F_2, and F_2^+ centers. The F center then decays (radiatively) to its relaxed ground state (state 1 in Fig. 6.36) and then from there it rapidly decays to the unrelaxed ground state g. Since both the excitation ($g \rightarrow 3$) and emission ($2 \rightarrow 1$) transitions are rather broad (~ 4000 Å), these spectra are reminiscent of dye lasers (see Fig. 6.22), the emission spectrum being Stokes shifted relative to the absorption spectrum. F centers are thus seen to fulfill the requirement of a four-level laser rather well. Not all F centers are good candidates for laser action, however, since some of them (e.g., the ordinary F center) have a very low fluorescence quantum efficiency. Of the F_A-type lasers, we mention the KCl:Li ($\lambda = 2.5$–2.9 μm) and RbCl:Li ($\lambda = 2.7$–3.3 μm) lasers. Of the F_2^+-type lasers we mention the NaF ($\lambda = 0.88$–1 μm), KF ($\lambda = 1.25$–1.45 μm), and LiF ($\lambda = 0.84$–1.04 μm) lasers. It should be noted that the preparation of laser crystals based on these F_A and F_2^+ types of color centers requires considerable care and skill.

Figure 6.37 illustrates the sort of layout used in a color-center laser. The laser is longitudinally pumped by another laser (usually a Kr^+ laser oscillating on its red, 647 nm, transition or a Nd:YAG laser) in a configuration similar to that used for cw dye lasers (compare with Fig. 6.25b). Here the pump beam passes through the input mirror, which has a high reflectivity at the laser wavelength and a high transmission at the pump wavelength. Coarse tuning of the laser is usually achieved by means of a dispersive optical system such as a prism or a grating (they are not included in Fig. 6.37; however, see Fig. 5.7). The fine tuning and selection of a single mode is achieved by using one or more Fabry–Perot etalons in the cavity (see Fig. 5.8). A complicating feature of color-center lasers is the need to keep the laser crystal at cryogenic temperatures (usually $T \simeq 77°K$). There are

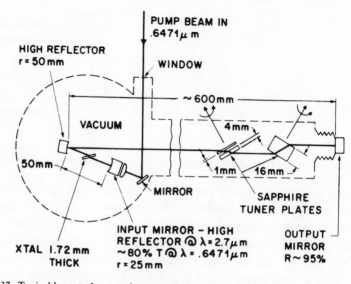

FIG. 6.37. Typical layout of a cw color-center laser. The data indicated in the figure refer to a KCl:Li laser longitudinally pumped by a Kr⁺ laser [after Mollenauer[32]].

two reasons for this: (i) The lifetime τ of the upper laser level decreases roughly as $1/T$ for the F_A center. Thus the laser threshold [$\propto 1/\sigma\tau$, see (6.12)] is expected to increase linearly with T. (ii) Both F_A and F_2^+ centers fade (in ~1 day) if the temperature is increased beyond ~200°K. This last point poses a problem as far as the shelf life of color-center lasers is concerned.[†] Note finally that the whole laser cavity is usually kept in a vacuum enclosure (dashed line in Fig. 6.37) for two reasons: (i) on account of the low temperature requirement of the laser crystal, and (ii) to prevent losses due to atmospheric absorption (especially by H_2O) from interfering with laser action.

The performance of color-center lasers can be summarized as follows. Typical threshold pump powers are of the order of a few tens of milliwatts (when the pump beam is focused down to a 20-μm-diameter spot in the crystal). Continuous wave output powers up to 1 W with slope efficiency up to 7% for F_A and up to 60% for F_2^+ have been obtained. The difference of almost an order of magnitude in slope efficiency between the two types of laser requires a comment. It arises from the fact that the pump quantum efficiency ($\hbar\omega_0/\hbar\omega_p$, see Fig. 6.36) is ~80% for F_2^+ centers while it is only ~10% for F_A centers. The slope efficiency in this case is essentially determined by the quantum efficiency since the slope efficiency is the

[†]Some new classes of color-center lasers (e.g., Tl⁺-doped alkali halides) have recently been shown to be both optically and thermally stable, however.

product of pump quantum efficiency and coupling efficiency [see (5.24b)] if we assume $\eta_p = 1$ (all pump photons absorbed) and $\eta_d = 1$ (in view of the very fast decay of the lower laser level). We finally note that some color-center lasers (LiF and KF, F_2^+ centers) have also been mode locked using the same technique of synchronous pumping as for dye lasers. Pulses as short as 5 ps and tunable over the laser's emission range have been obtained in this way.

By virtue of their broad tunability, their very narrow oscillation linewidth, and their picosecond pulse capabilities, color-center lasers look very interesting for applications in many areas such as molecular spectroscopy, chemical dynamics, and evaluation of optical fibers.

6.8 THE FREE-ELECTRON LASER [33]

In the previous sections our discussion has progressed from situations in which electrons are bound to a single atom or molecule, to situations in which the electron is free to move along the chain of atoms in a conjugated double-bonded molecule (dye lasers), and then to situations in which the electron is free to move through the entire volume of a semiconductor crystal (semiconductor lasers). In this last topic of the chapter, we describe one of the most recent and interesting types of laser, in which the electrons are even more free than in the cases considered so far: the free-electron laser. In fact in this laser the electrons move freely (they are in a vacuum) through a periodic magnetic field and the stimulated emission process comes about through the interaction of the e.m. field of the laser beam with the electrons moving in this periodic structure. Free-electron lasers can in principle generate stimulated radiation of high peak intensity ($\sim$a few MW/cm^2) at any wavelength from the IR to the UV and perhaps even in the x-ray region of the e.m. spectrum. So far, however, this type of laser has only been operated at $\lambda = 3.4$ μm.[34]

A schematic diagram of a free-electron laser is shown in Fig. 6.38. A beam of relativistic electrons is passed through a periodic transverse magnetic field (the "wiggler"). Stimulated emission along the direction of the

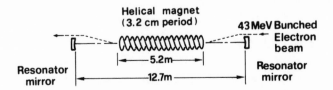

FIG. 6.38. Schematic diagram of a free-electron laser [after Deacon *et al.*[34]].

electron beam is fed back, just as in any other laser, by means of two mirrors of appropriate reflectivity. The stimulated emission process has its origin in the interaction of the e.m. wave with the electrons moving in the periodic magnetic structure. This effect may be properly termed magnetic bremsstrahlung. Radiation can also be absorbed through the process of inverse bremsstrahlung. The wavelengths for emission (plus sign) and absorption (minus sign) are given by[35]

$$\lambda = \frac{\lambda_q}{2\gamma^2}\left[1 + \left(\frac{1}{2\pi}\right)^2 \frac{\lambda_q^2 r_0}{mc^2} B^2\right]\left(1 \pm \frac{h\nu}{\gamma mc^2}\right) \qquad (6.29)$$

where λ_q is the period of the magnetic field, γmc^2 is the electron energy (ergs), m is the electron mass, r_0 is the classical electron radius (cm), and B is the transverse magnetic field (G). Since the emission wavelength is slightly longer than that for absorption, one observes gain on the long-wavelength side and loss on the short-wavelength side of the transition. The minimum theoretical value for the linewidth arises from spontaneous emission and is determined by the length of the magnet assembly. There are also inhomogeneous effects, however, arising, for instance, from the spread in electron energy, the angular divergence of the electron beam, and the variation in magnetic field over the cross section of the beam.

With the arrangement shown in Fig. 6.38, laser action has been obtained at $\lambda \simeq 3.4$ μm with an average output power of 0.36 W and peak power (the e beam was pulsed) of 7 kW. The light was circularly polarized with the same polarization as the helical magnet.

From (6.29) it is seen that the transition wavelength can be varied by changing the electron energy and/or the magnet period. Since, however, the gain of the transition scales as $\lambda^{3/2}J$,[35] where J is the e-beam current density, it follows that higher electron beam currents will be required for laser operation in the visible and ultraviolet. The facilities afforded by present-day e-beam machines offer the possibility of producing laser action to wavelengths at least as short as 100 nm. The need for a sophisticated e-beam machine and the relatively low efficiency of this laser ($< 0.5\%$) seem at present to be the most serious limitations. The future of free-electron lasers appears therefore to be dependent on the development of relatively large national facilities.

6.9 SUMMARY OF PERFORMANCE DATA

By way of a summary, a selection of the laser performance data given in the previous sections has been gathered together in Table 6.1. Although

TABLE 6.1. Performance Data of Some of the Lasers Described in This Chapter

Laser type	Mode of operation[a]	λ, nm	Average power, W	Peak power, kW	Pulse duration	Slope efficiency, %
Ruby	P	694.3	1	10–10^4	1 ms–10 ns	< 0.1
Nd:YAG	cw	1064	150			1–3
Nd:YAG	P	1064	400	10	1–5 ms	1–3
Nd:YAG	P	1064	4	2×10^4	10–20 ns	1–3
He–Ne	cw	632.8	10^{-3}–10^{-2}			
Cu	P	510.5	40	100	20–40 ns	1
Ar$^+$	cw	514.5	10–150			< 0.1
He–Cd	cw	325 441.6	0.1			
CO_2	cw	10.6 μm	$(1$–$15) \times 10^3$			10–20
CO_2	P	10.6 μm	10^3	10^4	0.1–0.5 μs	10
N_2	P	337.1	0.1	10^3	10 ns	< 0.1
KrF	P	248	100	5×10^3	10 ns	1
Rhodamine 6G	P	590	100	100	10 μs	0.5
HF	cw	2.6–3.3 μm	10^4			
HF	P	2.6–3.3 μm		10^3		
GaAs	cw	840	10^{-2}			10

[a] P = pulsed; cw = continuous wave.

the list of lasers covered in Table 6.1 is already quite numerous, it must be realized that this only represents a tiny fraction of the lasers that have been operated so far. By way of an illustration of this, Fig. 6.39 shows the wavelength ranges over which the various types of lasers have actually been operated. The same figure also shows the potential ranges for the three different types of gas-laser transitions which can be used: (i) transitions between electronic state, (ii) vibrational–rotational transitions, and (iii) rotational transitions. However, it should be noted that these ranges cannot, in general, be covered continuously by existing lasers. Dye lasers and color-center lasers, however, are exceptions to this, and the ranges shown for them can be covered continuously.

The wide range of wavelengths which can now be covered with lasers is worth emphasizing (roughly 0.1 to 10^3 μm, i.e., a factor of 10^4 between the extremes of the range). Besides the wavelength, there are other laser parameters which can span a wide range. In fact we have seen that the output power ranges from the milliwatt level of low-power cw lasers up to 100 kW (and possibly much more, but this information is classified) for high-power cw lasers and up to 100 TW for pulsed lasers. Likewise, laser pulsewidths can range from milliseconds (for pulsed solid-state lasers) to

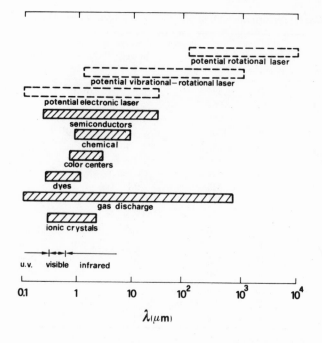

FIG. 6.39. Wavelength ranges for existing lasers in six of the categories considered in the text. The figure also shows the potential ranges for the three types of transitions used in gas lasers.

picoseconds (for mode-locked lasers). The physical dimensions of different types of laser also cover an extraordinary range: this goes from a few micrometers to a few tens of meters (one of the longest lasers was used for geodesic studies and had a length of 6.5 km!). This enormous variety of laser types and laser performance parameters is perhaps one of the most fascinating features of the laser field. It also means that, with such a range of lasers available, there is a considerable variety of applications, and these will be discussed at some length in Chapter 9.

PROBLEMS

6.1. Draw a scale covering the wavelength range of visible e.m. waves. Where do ruby, He–Ne, Ar^+, and Rhodamine 6G lasers fall within this range? What is the corresponding color of the emitted light?

6.2. List at least four lasers whose wavelengths fall in the infrared.

6.3. List at least three lasers whose wavelengths fall in the UV or VUV. What are the problems in getting laser action in the UV or VUV?

6.4. Estimate the width of the Lamb dip for a red He–Ne laser. Compare this with the Doppler width.

6.5. Estimate the width of the Lamb dip for an Ar^+ laser and compare it with the Doppler width.

6.6. Estimate the width of the Lamb dip for a longitudinal-flow CO_2 laser and compare it with the Doppler width.

6.7. For possible surgical applications, a laser with cw output power > 20 W is needed. What lasers satisfy this condition?

6.8. For metalworking applications, a laser with cw output power > 1 kW is needed. Which lasers meet this requirement?

6.9. Assume that the bond between the two nitrogen atoms of the N_2 molecule can be simulated by a spring of suitable elastic constant. Knowing the vibrational frequency (Fig. 6.10) and the atomic mass, calculate the elastic constant. Compare this constant with that obtainable from the ground-state curve of Fig. 6.19.

6.10. Assume that each of the two oxygen–carbon bonds of the CO_2 molecule can be simulated by a spring with force constant k_1. Calculate this constant from a knowledge of the ν_1 frequency (assume $\nu_1 = 1337$ cm^{-1}). With the further assumption that there is no interaction between the two oxygen atoms, calculate the expected frequency ν_3 of the asymmetric stretching mode.

6.11. Show that the bonds of the CO_2 molecule cannot be simulated by elastic springs connecting the three atoms if the harmonic oscillation corresponding to the bending mode of frequency ν_2 has to be calculated.

6.12. Show that, if the elastic constant of the N–N bond is taken to be the same as that of the isoelectronic CO molecule, the $(v' = 1) \rightarrow (v = 0)$ transition wavelength of the N_2 molecule is approximately the same as that of the CO molecule.

6.13. From the knowledge that the maximum population of the upper laser level of a CO_2 molecule occurs for the rotational quantum number $J' = 21$ (see Fig. 6.13), calculate the rotational constant B [assume $T = 400°$K, which corresponds to an energy kT such that the corresponding frequency (kT/h) is ~ 280 cm^{-1}].

6.14. Using the result of the previous problem, calculate the frequency spacing (in cm^{-1}) between the rotational lines of the CO_2 laser transition (assume that the rotational constant of the lower laser level is the same as that of the upper laser level, and remember that only levels with odd values of J are occupied).

6.15. The collision broadening of the CO_2 laser transition is $\Delta\nu_c = 7.58$ (ψ_{CO_2} + $0.73\psi_{N_2} + 0.6\psi_{He})p(300/T)^{1/2}$ MHz, where the ψ are the fractional partial pressures of the gas mixture and p is the total pressure (Torr). If the ratio of partial pressures of CO_2, N_2, and He molecules is $1:1:8$, calculate the total gas pressure needed to make all the rotational lines merge together. What would the width of the gain curve then be?

6.16. If a CO_2 laser with high enough pressure to have all its rotational lines merged together were mode locked, what would be the order of magnitude of the corresponding laser pulse width?

REFERENCES

1. V. Evthuhov and J. K. Neeland, in *Lasers*, ed. by A. K. Levine (Marcel Dekker, New York, 1966), Vol. 1, Chapter 1.
2. T. H. Maiman, *Nature* **187**, 493 (1960).
3. T. H. Maiman, *Br. Commun. Electron.* **7**, 674 (1960).
4. D. Findlay and D. W. Goodwin, in *Advances in Quantum Electronics*, ed. by D. W. Goodwin (Academic Press, New York, 1970), pp. 77–128.
5. H. G. Danielmeyer, in *Lasers*, ed. by A. K. Levine and A. J. DeMaria (Marcel Dekker, New York, 1976), Vol. 4, Chapter 1.
6. W. Koechner, *Solid State Laser Engineering* (Springer Verlag, New York, 1976), Chapter 2.
7. A. Javan, W. R. Bennett, and D. R. Herriott, *Phys. Rev. Lett.* **6**, 106 (1961).
8. C. S. Willet, *An Introduction to Gas Lasers: Population Inversion Mechanisms* (Pergamon Press, Oxford, 1974), Chapter 4, Section 4.1.1.
9. R. Arrathoon, in *Lasers*, ed. by A. K. Levine and A. J. De Maria (Marcel Dekker, New York, 1976), Vol. 4, Chapter 3.
10. W. B. Bridges, in *Methods of Experimental Physics*, Vol. 15, *Quantum Electronics*, ed. by C. L. Tang (Academic Press, New York, 1979), Part A, pp. 33–151.
11. C. C. Davis and T. A. King, in *Advances in Quantum Electronics* ed. by D. W. Goodwin (Academic Press, New York, 1975), Vol. 3, pp. 170–437.
12. M. H. Dunn and J. N. Ross, in *Progress in Quantum Electronics*, ed. by J. H. Sanders and S. Stenholm (Pergamon Press, London, 1977), Vol. 4, pp. 233–270.
13. Y. Y. Chang, in *Nonlinear Infrared Generation*, ed. by Y. R. Shen (Springer-Verlag, Berlin, 1977), Chapter 6.
14. P. K. Cheo in *Lasers* ed. by A. K. Levine and A. J. De Maria (Marcel Dekker, New York, 1974), Vol. 3, Chapter 2.
15. A. J. De Maria, in *Principles of Laser Plasmas*, ed. by G. Bekefi (Wiley-Interscience, New York, 1976), Chapter 8.
16. C. K. N. Patel, *Phys. Rev. Lett.* **12**, 588 (1964).
17. R. L. Abrams, in *Laser Handbook*, ed. by M. L. Stitch (North Holland, Amsterdam, 1979), Vol. 3, Part 2A.
18. O. R. Wood, *Proc. IEEE* **62**, 355 (1974).
19. J. D. Anderson, *Gasdynamic Lasers: An Introduction* (Academic Press, New York, 1971).
20. M. L. Bhaumik, in *High-Power Gas Lasers*, ed. by E. R. Pike (The Institute of Physics, Bristol and London, 1975), pp. 122–147.
21. C. S. Willet, *An Introduction to Gas Lasers: Population Inversion Mechanisms* (Pergamon Press, Oxford, 1974), Sections 6.2.1 and 6.2.3.
22. C. A. Brau, in *Excimer Lasers*, ed. by C. K. Rhodes (Springer-Verlag, Berlin, 1979), Part 4.
23. F. P. Schäfer, in *Dye Lasers*, 2nd edn. ed. by F. P. Schäfer (Springer-Verlag, Berlin, 1977).
24. O. G. Peterson, in *Methods of Experimental Physics*, Vol. 15, ed. by C. L. Tang (Academic Press, New York, 1979), Part A, pp. 251–355.
25. P. P. Sorokin and J. R. Lankard, *IBM J. Res. Dev.* **10**, 162 (1966).
26. A. N. Chester, in *High-Power Gas Lasers*, ed. by E. R. Pike (The Institute of Physics, Bristol and London, 1975) pp. 162–221.

27. C. J. Ultee, in *Laser Handbook*, Vol. 3, ed. by M. L. Stitch (North-Holland, Amsterdam, 1979), pp. 199–287.
28. K. Hohla and K. L. Kompa, in *Handbook of Chemical Lasers*, ed. by R. W. F. Gross and J. F. Bott (John Wiley and Sons, New York, 1976), pp. 667–702.
29. H. Kressel, in *Methods of Experimental Physics*, Vol. 15, ed. by C. L. Tang (Academic Press, New York, 1979), Part A, pp. 209–250.
30. R. N. Hall *et al.*, *Phys. Rev. Lett.* **9**, 366 (1962).
31. M. I. Nathan *et al.*, *Appl. Phys. Lett.* **1**, 62 (1962).
32. L. F. Mollenauer, in *Methods of Experimental Physics*, Vol. 15, ed. by C. L. Tang (Academic Press, New York, 1979), Part B, pp. 1–54.
33. *Physics of Quantum Electronics*, Vol. 7, ed. by W. Lamb, M. Sargent, and M. Scully (Benjamin, New York, 1980).
34. D. A. G. Deacon *et al.*, *Phys. Rev. Lett.* **38**, 892 (1977).
35. L. Elias *et al.*, *Phys. Rev. Lett.* **36**, 717 (1976).
36. H. D. Försterling and H. Kuhn, *Physikalische Chemie in Experimenten*, Ein Praktikum (Verlag Chemie, Weinheim/Bergstr., 1971).
37. J. D. Daugherty, in *Principles of Laser Plasmas*, ed. by G. Bekefi (John Wiley and Sons, New York, 1976), pp. 369–419.
38. I. Melngailis and A. Mooradian, in *Laser Applications in Optics and Spectroscopy*, ed. by S. Jacobs, M. Sargent, M. Scully, and J. Scott, (Addison-Wesley Company, 1975).
39. M. C. Richardson *et. al.*, *IEEE J. Quantum Electr.* **QE-9**, 236 (1973).

7

Properties of Laser Beams

7.1 INTRODUCTION

In Chapter 1 it was stated that the most characteristic properties of laser beams are (i) monochromaticity, (ii) coherence (spatial and temporal), (iii) directionality, (iv) brightness. The material presented in earlier chapters allows us to now examine these properties in more detail and compare them with the properties of conventional light sources (thermal sources). The discussion presented in this chapter is essential for an understanding of the application possibilities of laser beams (Chapter 9).

7.2 MONOCHROMATICITY

If the laser is oscillating on a single mode and if the output is constant in time, the theoretical limit of monochromaticity arises from zero-point fluctuations and is given by (5.45). This limit, however, gives a very low value for the oscillating bandwidth $\Delta\nu_{osc}$ (a value of $\Delta\nu_{osc}/\nu_{osc} \simeq 10^{-15}$ was calculated in Section 5.3.7 for a laser power of 1 mW) and is rarely reached. In practice, vibrations and thermal expansion of the cavity limit $\Delta\nu_{osc}$ to much higher values. If a sufficiently massive structure made of material with a low expansion coefficient (e.g., Invar) is used to support the laser cavity, $\Delta\nu_{osc}$ can be reduced to a value in the range of 1–10 kHz. For a low-pressure gas laser (e.g., He–Ne) locked to the center of the absorption line of an appropriate gas, one can obtain [1] $\Delta\nu_{osc} = 50$–500 Hz (i.e.,

$\Delta \nu_{osc}/\nu = 10^{-12}$–$10^{-13}$). In pulsed operation the minimum linewidth is obviously limited by the inverse of the pulse duration τ_p. For example, for a single-mode giant pulse laser, assuming $\tau_p \simeq 10^{-8}$ s, one has $\Delta \nu_{osc} \simeq 100$ MHz.

In the case of a laser oscillating on many modes, the monochromaticity is obviously related to the number of oscillating modes. For a solid-state laser (ruby, neodymium, semiconductor), where it is usually difficult to obtain single-mode oscillation (because of the large linewidth $\Delta \nu$) the oscillation bandwidth is often of the order of gigahertz. Of course, one does not always want to have a very narrow oscillating bandwidth. We recall, for example, that in order to get very short pulses of light (mode locking), it is desirable to obtain oscillation over as wide a bandwidth as possible.

7.3 FIRST-ORDER COHERENCE [2,3]

In Chapter 1 the concept of coherence of an e.m. wave was introduced in an intuitive fashion, with two types of coherence being distinguished: (i) spatial and (ii) temporal coherence. In this section we intend to give a more thorough discussion of these types of coherence. In fact, as we shall see better at the end of this chapter, spatial and temporal coherence describe the coherence properties of an e.m. wave only to first order.

7.3.1 Complex Representation of Polychromatic Fields

Before introducing the concepts of spatial and temporal coherence it is worth starting with a short description of a very useful complex representation for polychromatic fields (due to Gabor[4]). For the sake of simplicity we will consider a linearly polarized e.m. wave. This can then be specified by a single real scalar quantity $V^{(r)}(\mathbf{r}, t)$ (e.g., $|\mathbf{E}|$ or $|\mathbf{H}|$ or the modulus of the vector potential $|\mathbf{A}|$). This quantity, which is a function of position $\mathbf{r}$ and time t can be expressed as a Fourier integral, i.e.,

$$V^{(r)}(\mathbf{r}, t) = \frac{1}{2\pi} \int_{-\infty}^{+\infty} V(\mathbf{r}, \omega) \exp(-i\omega t)\, d\omega \qquad (7.1)$$

Equation (7.1) has the well-known inverse relationship

$$V(\mathbf{r}, \omega) = \int_{-\infty}^{+\infty} V^{(r)}(\mathbf{r}, t) \exp(i\omega t)\, dt \qquad (7.2)$$

Since $V^{(r)}$ is real, we see from (7.2) that $V(\mathbf{r}, -\omega) = V^*(\mathbf{r}, \omega)$, and hence the negative frequency spectrum does not add any further information about the field to that already contained in the spectrum of positive frequencies. So instead of $V^{(r)}$, we can consider the complex quantity $V(\mathbf{r}, t)$ defined by

$$V(\mathbf{r}, t) = \frac{1}{2\pi} \int_0^\infty V(\mathbf{r}, \omega) \exp(-i\omega t) \, d\omega \qquad (7.3)$$

$V(\mathbf{r}, t)$ is called the complex analytic signal associated with $V^{(r)}$. Obviously there is a unique relation between the two functions. In fact, given V, we find from (7.1) and (7.3) that

$$V^{(r)} = 2\text{Re}(V) \qquad (7.4)$$

Conversely, it is easy to see that, if $V^{(r)}$ is given, then V is uniquely determined. In fact, given $V^{(r)}$, then from (7.2) we obtain $V(\mathbf{r}, \omega)$. With the help of (7.3), we then get $V(\mathbf{r}, t)$.

The analytic signal V, introduced by Gabor,[4] proves much more convenient than the real signal as a way of representing the e.m. field. If, for example, the real signal is monochromatic, we can write $V^{(r)} = A\cos\omega t$. Hence, from (7.2) and (7.3) we have $V = A\exp(-i\omega t)/2$. In this case the analytic signal representation is just the well-known exponential representation for sinusoidal functions whose advantages are well known. Frequently, in a practical situation, the spectrum of the analytic signal has an appreciable value only in an interval $\Delta\omega$ which is very small compared to the mean frequency $\langle\omega\rangle$ of the spectrum (quasi-monochromatic wave). In this case we can write

$$V(t) = A(t) \exp\left\{i\left[\psi(t) - \langle\omega\rangle t\right]\right\} \qquad (7.5)$$

where $A(t)$ and $\psi(t)$ are both slowly varying, i.e.,

$$\left[\frac{dA}{A\, dt}, \frac{d\psi}{dt}\right] \ll \langle\omega\rangle \qquad (7.6)$$

Note that the analytic signal takes on a much deeper significance when one makes use of a quantum description of the e.m. field. It is then found that V and V^* have a close correspondence to the so-called photon creation and annihilation operators. We note finally that other quantities of the e.m. field can be expressed as functions of the analytic signal. For example, one can define the intensity $I(\mathbf{r}, t)$ of the beam by the relation

$$I(\mathbf{r}, t) = V(\mathbf{r}, t)V^*(\mathbf{r}, t) \qquad (7.7)$$

In fact this definition means that I is not, strictly speaking, proportional to $(V^{(r)})^2$. However, if the light is quasi-monochromatic, it can be shown from (7.5) that $I(\mathbf{r}, t)$ is equal to the mean value of $(V^{(r)})^2/2$ taken over a few optical cycles.

7.3.2 Degree of Spatial and Temporal Coherence

In order to describe the beam properties, we can now introduce a whole class of correlation functions for the analytic signal. For the moment, however, we will limit ourselves to looking at the first-order functions.

For a point r_1, we can define a first-order[†] correlation function $\Gamma^{(1)}$ as follows

$$\Gamma^{(1)}(r_1, r_1, \tau) = \lim_{T \to \infty} \frac{1}{2T} \int_{-T}^{T} V(r_1, t + \tau) V^*(r_1, t) \, dt \qquad (7.8)$$

This function is therefore just the autocorrelation function of $V(r, t)$ or, in other words, the mean value of the product $V(r_1, t + \tau) V^*(r_1, t)$. One can therefore write, more simply,

$$\Gamma^{(1)}(r_1, r_1, \tau) = \langle V(r_1, t + \tau), V(r_1, t) \rangle \qquad (7.9)$$

We can also define a normalized function $\gamma^{(1)}(r_1, r_1, \tau)$ as follows:

$$\gamma^{(1)} = \frac{\Gamma^{(1)}}{\langle V(r_1, t), V(r_1, t) \rangle} = \frac{\Gamma^{(1)}}{\langle I(r_1, t) \rangle} \qquad (7.10)$$

From the Schwarz inequality it can immediately be seen that $|\gamma^{(1)}(r_1, r_1, \tau)| \leqslant 1$. It can also be shown from (7.8) that $\gamma^{(1)}(r_1, r_1, -\tau) = \gamma^{(1)*}(r_1, r_1, \tau)$. The function $\gamma^{(1)}(r_1, r_1, \tau)$ is called the *complex degree of temporal coherence*. Its modulus $|\gamma^{(1)}|$ is called the *degree of temporal coherence*. It is evident that $\Gamma^{(1)}$ and therefore $\gamma^{(1)}$ are a measure of how much correlation there is between the analytic signal at the same point at two instants separated by a time τ. In the case where there is a complete absence of temporal coherence one finds from (7.8) and (7.10) that $\gamma^{(1)} = 0$ for $\tau > 0 (\gamma^{(1)} = 1$ for $\tau = 0)$. In the case of perfect temporal coherence [e.g., a sinusoidal wave, i.e., $V = A(r_1) \exp(-i\omega t)$] one gets $|\gamma^{(1)}| = 1$ for any τ. The quantity $|\gamma^{(1)}|$ is therefore a function whose value lies between 0 and 1 and describes the degree of temporal coherence of the wave. In general the function $|\gamma^{(1)}(\tau)|$ will be of the form shown in Fig. 7.1 [recall that $|\gamma^{(1)}(-\tau)| = |\gamma^{(1)}(\tau)|$]. One can therefore define a characteristic time τ_{co} (called the coherence time) as, for instance, the time for which $|\gamma^{(1)}| = 1/2$. For a perfectly coherent wave we obviously have $\tau_{co} = \infty$ while for a completely incoherent wave we have $\tau_{co} = 0$. Note that we can also define a temporal coherence length L_c as $L_c = c\tau_{co}$.

In a similar way, we can define a first-order correlation function

---

[†] Some authors[2] call this a second-order correlation function and, correspondingly, they talk of second-order coherence.

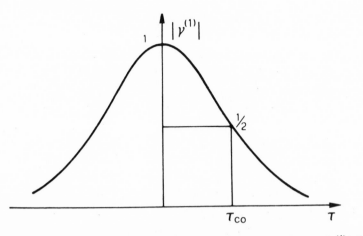

FIG. 7.1. Example of possible behavior of the degree of temporal coherence $|\gamma^{(1)}(\tau)|$. The coherence time can be defined as the half-width of the curve at half-height.

between two different points $\mathbf{r}_1$ and $\mathbf{r}_2$ at the same instant as

$$\Gamma^{(1)}(\mathbf{r}_1, \mathbf{r}_2, 0) = \lim_{T \to \infty} \frac{1}{2T} \int_{-\infty}^{+\infty} V(\mathbf{r}_1, t) V^*(\mathbf{r}_2, t) \, dt = \langle V(\mathbf{r}_1, t), V(\mathbf{r}_2, t) \rangle$$

$$(7.11)$$

We can also define the corresponding normalized function $\gamma^{(1)}(\mathbf{r}_1, \mathbf{r}_2, 0)$ as

$$\gamma^{(1)} = \frac{\Gamma^{(1)}(\mathbf{r}_1, \mathbf{r}_2, 0)}{\left[\Gamma^{(1)}(\mathbf{r}_1, \mathbf{r}_1, 0)\Gamma^{(1)}(\mathbf{r}_2, \mathbf{r}_2, 0)\right]^{1/2}}$$

$$(7.12)$$

From the Schwarz inequality one can again see that $|\gamma^{(1)}| \leqslant 1$. The quantity $\gamma^{(1)}(\mathbf{r}_1, \mathbf{r}_2, 0)$ is called the *complex degree of spatial coherence*, and its modulus, *the degree of spatial coherence*. By analogy with what has been said before, if $\mathbf{r}_1$ is fixed, it is clear that $\gamma^{(1)}$ as a function of $\mathbf{r}_2$ decreases from the value 1 (which occurs for $\mathbf{r}_2 = \mathbf{r}_1$) to zero as $|\mathbf{r}_2 - \mathbf{r}_1|$ increases. Thus $\gamma^{(1)}$ will be greater than some prescribed value (e.g., $1/2$) over a certain characteristic area on the wavefront around the point P_1 described by the vector $\mathbf{r}_1$. This will be called the *coherence area* of the wave at point P_1 of the wavefront.

The concepts of spatial and temporal coherence can be combined by means of the so-called mutual coherence function, defined as

$$\Gamma^{(1)}(\mathbf{r}_1, \mathbf{r}_2, \tau) = \langle V(\mathbf{r}_1, t + \tau), V(\mathbf{r}_2, t) \rangle$$

$$(7.13)$$

which can also be normalized as follows:

$$\gamma^{(1)}(\mathbf{r}_1,\mathbf{r}_2,\tau) = \frac{\Gamma^{(1)}(\mathbf{r}_1,\mathbf{r}_2,\tau)}{\left[\Gamma^{(1)}(\mathbf{r}_1,\mathbf{r}_1,0)\Gamma^{(1)}(\mathbf{r}_2,\mathbf{r}_2,0)\right]^{1/2}} \qquad (7.14)$$

This function, called the *complex degree of coherence*, provides a measure of the coherence between two different points of the wave at different instants of time. For a quasi-monochromatic wave, it follows from (7.5) and (7.14) that we can write

$$\gamma^{(1)}(\tau) = |\gamma^{(1)}|\exp\{i[\psi(\tau) - \langle\omega\rangle\tau]\} \qquad (7.15)$$

where $|\gamma^{(1)}|$ and $\psi(\tau)$ are both slowly varying, i.e.,

$$\left(\frac{d|\gamma^{(1)}|}{|\gamma^{(1)}|\,d\tau}, \frac{d\psi}{d\tau}\right) \ll \langle\omega\rangle \qquad (7.16)$$

7.3.3 Measurement of Spatial and Temporal Coherence

One very simple way of measuring the degree of spatial coherence between two points of a light wave is by using Young's interferometer (Fig. 7.2). This consists simply of a screen 1 in which two small holes have been made at positions corresponding to the points x_1 and x_2 and a screen 2 on which an interference pattern is produced by light passing through the two holes. More precisely, the interference at point P and at time t will arise from the superposition of the waves emitted from points x_1 and x_2 at times $[t - (L_1/c)]$ and $[t - (L_2/c)]$ respectively. One will therefore see interference fringes on screen 2 around point P which are more distinct the better the correlation between the two analytic signals of the light wave, $V[x_1, t - (L_1/c)]$ and $V[x_2, t - (L_2/c)]$, during the time taken for the measurement of the fringes (e.g., the exposure time of a photographic plate). If now the

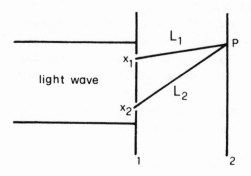

FIG. 7.2. The use of Young's interferometer for the measurement of the degree of spatial coherence of an e.m. wave.

point P on the screen is chosen so that $L_1 = L_2$, the visibility of the fringes around P will give a measure of the degree of spatial coherence between points x_1 and x_2. To be more precise we define the visibility $V_{(P)}$ of the fringes at point P as

$$V_{(P)} = \frac{I_{\max} - I_{\min}}{I_{\max} + I_{\min}} \qquad (7.17)$$

where $I_{\max}$ and $I_{\min}$ are, respectively, the maximum intensity of a bright fringe and the minimum intensity of a dark fringe in the region of P. If the two holes 1 and 2 produce the same illumination at point P and if the wave has perfect spatial coherence, then $I_{\min} = 0$ and therefore $V_{(P)} = 1$. For the case in which the signals at the two points x_1 and x_2 are completely uncorrelated (i.e., incoherent), the fringes disappear (i.e., $I_{\max} = I_{\min}$) and therefore $V_{(P)} = 0$. From what we have said in the previous section, it seems clear that V_P must be related to the modulus of the function $\gamma^{(1)}(x_1, x_2, 0)$. More generally, for any point P on the screen, we expect V_P to be related to the modulus of the function $\gamma^{(1)}(x_1, x_2, \tau)$ where $\tau = (L_2 - L_1)/c$. At the end of this section we will indeed show that, if the two holes produce the same illumination at point P, we have

$$V_{(P)}(\tau) = |\gamma^{(1)}(x_1, x_2, \tau)| \qquad (7.18)$$

Thus, by measuring the fringe visibility $V_{(P)}$ at a point P such that $L_1 = L_2$, the degree of spatial coherence between points x_1 and x_2 is obtained.

The Michelson interferometer (Fig. 7.3) provides a very simple method of measuring the temporal coherence. Let P be the point where the temporal coherence of the wave is to be measured. A combination of a small hole placed at P and a lens with its focus at P transforms the incident wave into a plane wave (see Fig. 7.8 also). This wave then falls on a partially reflecting mirror S_1 (reflectivity $R = 50\%$) which splits it into two waves A and B. These waves are reflected back by mirrors S_2 and S_3 ($R = 1$) and recombine to form the wave C. Since the waves A and B interfere, the illumination in the direction of C will be either light or dark according to whether $2(L_3 - L_2)$ is an even or odd number of half-wavelengths. Obviously this interference will only be observed as long as the difference $L_3 - L_2$ does not become so large that the two beams A and B are uncorrelated in phase. For a partially coherent wave the intensity I_c of beam C as a function of $2(L_3 - L_2)$ will behave as shown in Fig. 7.3b. In this case we can again define a fringe visibility exactly as in (7.17), where now $I_{\max}$ and $I_{\min}$ are as shown in Fig. 7.3b. Just as in the case of Young's

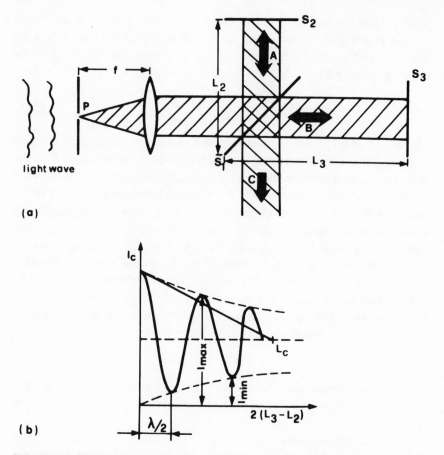

FIG. 7.3. (a) Michelson interferometer for the measurement of the degree of temporal coherence of an e.m. wave at point P; (b) behavior of the light output in the direction C as a function of the difference $L_3 - L_2$ between the lengths of the interferometer arms.

interferometer, it can now be shown that

$$V_{(P)}(\tau) = |\gamma^{(1)}(P, P, \tau)| \qquad (7.19)$$

where $\tau = 2(L_3 - L_2)/c_0$. Therefore the measurement of fringe visibility in this case gives a value for the degree of temporal coherence of the wave at point P. Once $V_{(P)}(\tau)$ is known, we can obtain, from this function, the value of the coherence time τ_{co} and hence the coherence length $L_c = c_0 \tau_{co}$. Note that L_c is equal to twice the difference $L_3 - L_2$ between interferometer arms at which the visibility falls to $V_{(P)} = 1/2$.

We conclude this section with a proof of (7.18) which also serves as an exercise in the use of the analytic signal. A similar sort of argument can be used to prove (7.19). Let us call $V(t')$ the analytic signal at point P of Fig. 7.2 at time t'. Since it is due to the superposition of signals coming from each of the two holes of Fig. 7.2, it can be written as

$$V = K_1 V(x_1, t' - t_1) + K_2 V(x_2, t' - t_2) \qquad (7.20)$$

where $t_1 = L_1/c, t_2 = L_2/c$. The factors K_1 and K_2 are inversely proportional to L_1 and L_2 and also depend on the hole dimensions and the angle between the incident wave and the wave diffracted to x_1 and x_2. Since the diffracted secondary wavelets are always a quarter of a period out of phase with the incident wave,[5] it follows that

$$K_1 = |K_1| \exp(-i\pi/2) \qquad (7.21a)$$

$$K_2 = |K_2| \exp(-i\pi/2) \qquad (7.21b)$$

If we now define $t = t' - t_2$ and $\tau = t_2 - t_1$, equation (7.20) can be written as

$$V = K_1 V(x_1, t + \tau) + K_2 V(x_2, t) \qquad (7.22)$$

The intensity at the point P therefore has the value

$$I = VV^* = I_1(t + \tau) + I_2(t) + 2\text{Re}\left[K_1 K_2^* V(x_1, t + \tau) V^*(x_2, t)\right] \quad (7.23)$$

where I_1 and I_2 are the intensities at point P due to the emission from point x_1 alone and point x_2 alone, respectively, and are given by

$$I_1 = |K_1|^2 |V(x_1, t + \tau)|^2 = |K_1|^2 I(x_1, t + \tau) \qquad (7.24a)$$

$$I_2 = |K_2|^2 |V(x_2, t)|^2 = |K_2|^2 I(x_2, t) \qquad (7.24b)$$

where $I(x_1, t + \tau)$ and $I(x_2, t)$ are the intensities at points x_1 and x_2. Taking the time average of both sides of (7.23) and using equation (7.13), we find

$$\langle I \rangle = \langle I_1 \rangle + \langle I_2 \rangle + 2|K_1||K_2|\text{Re}\left[\Gamma^{(1)}(x_1, x_2, \tau)\right] \qquad (7.25)$$

where equations (7.21) have also been used. Equation (7.25) can be expressed in terms of $\gamma^{(1)}$ by noting that from (7.14) we have

$$\Gamma^{(1)} = \gamma^{(1)}\left[\langle I(x_1, t + \tau)\rangle\langle I(x_2, t)\rangle\right]^{1/2} \qquad (7.26)$$

Substituting (7.26) in (7.25) and using (7.24) we get

$$\langle I \rangle = \langle I_1 \rangle + \langle I_2 \rangle + 2(\langle I_1\rangle\langle I_2\rangle)^{1/2}\text{Re}\left[\gamma^{(1)}(x_1, x_2, \tau)\right]$$

$$= \langle I_1 \rangle + \langle I_2 \rangle + 2(\langle I_1\rangle\langle I_2\rangle)^{1/2}|\gamma^{(1)}|\cos\left[\psi(\tau) - \langle\omega\rangle\tau\right] \quad (7.27)$$

where we have used (7.15). Now, since both $|\gamma^{(1)}|$ and $\psi(\tau)$ are slowly varying, it follows that the variation of intensity $\langle I \rangle$ as P is changed is due to the rapid variation of the cosine term with its argument $\langle\omega\rangle\tau$. So in the region of P, we have

$$I_{max} = \langle I_1 \rangle + \langle I_2 \rangle + 2(\langle I_1 \rangle \langle I_2 \rangle)^{1/2} |\gamma^{(1)}| \qquad (7.28a)$$

$$I_{min} = \langle I_1 \rangle + \langle I_2 \rangle - 2(\langle I_1 \rangle \langle I_2 \rangle)^{1/2} |\gamma^{(1)}| \qquad (7.28b)$$

and therefore, from equation (7.17)

$$V_{(P)} = \frac{2(\langle I_1 \rangle \langle I_2 \rangle)^{1/2}}{\langle I_1 \rangle + \langle I_2 \rangle} |\gamma^{(1)}(x_1, x_2, \tau)| \qquad (7.29)$$

For the case where $\langle I_1 \rangle = \langle I_2 \rangle$ equation (7.29) reduces to (7.18).

7.3.4 Relation between Temporal Coherence and Monochromaticity

From the previous paragraphs it is clear that the concept of temporal coherence is intimately connected with the monochromaticity. For example, the more monochromatic the wave is, the greater its temporal coherence. So it is clear that the coherence time must be inversely proportional to the oscillation bandwidth. In this section we wish to discuss this relationship in more depth.

We start by noting that the spectrum of an e.m. wave as measured by a spectrograph is proportional to the power spectrum $W(\mathbf{r}, \omega)$ of the signal $V(\mathbf{r}, t)$. Since the power spectrum W is equal to the Fourier transform of the autocorrelation function $\Gamma^{(1)}$, either one of these quantities can be obtained once the other is known. To give a precise expression for the relation between τ_{co} and $\Delta\nu_{osc}$ we need to redefine these two quantities in an appropriate way. So we will define τ_{co} as the mean square width of the function $|\Gamma^{(1)}(\tau)|^2$ i.e., such that $(\tau_{co})^2 = \int_{-\infty}^{+\infty}(\tau - \langle\tau\rangle)^2 |\Gamma(\tau)|^2 d\tau / \int_{-\infty}^{+\infty} |\Gamma(\tau)|^2 d\tau$. As a short-hand notation for the above expression, we will write

$$(\tau_{co})^2 = \langle (\tau - \langle\tau\rangle)^2 \rangle \qquad (7.30)$$

where the mean value $\langle\tau\rangle$ is defined by $\langle\tau\rangle = \int \tau |\Gamma(\tau)|^2 d\tau / \int |\Gamma(\tau)|^2 d\tau$. Since $|\Gamma(-\tau)| = |\Gamma(\tau)|$, we see from this definition that $\langle\tau\rangle = 0$ and (7.30) reduces to

$$(\tau_{co})^2 = \langle \tau^2 \rangle \qquad (7.31)$$

The coherence time defined in this way is conceptually simpler (although sometimes more lengthy to calculate) than that defined earlier (i.e., the half-width at half-height of the curve $|\Gamma(\tau)|$, see Fig. 7.1). If the curve in Fig. 7.1 were oscillatory, τ_{co}, as we first defined it, would not be uniquely determined. Similarly we define the oscillation bandwidth $\Delta\nu_{osc}$ as the mean square width of $W^2(\nu)$, i.e.,

$$(\Delta\nu_{osc})^2 = \left\langle (\nu - \langle\nu\rangle)^2 \right\rangle \tag{7.32}$$

where $\langle\nu\rangle$, the mean frequency of the spectrum, is given by $\langle\nu\rangle = \int \nu W^2 d\nu / \int W^2 d\nu$. Now, since W and Γ are related by a Fourier transform, it can be shown that $\Delta\nu_{osc}$ and τ_{co}, as we have just defined them, satisfy the condition

$$\tau_{co}\Delta\nu_{osc} \geqslant 1/4\pi \tag{7.33}$$

The relation (7.33) is closely analogous to the Heisenberg uncertainty relation and can be proved using the same procedure as used in the derivation of the uncertainty relation.[6] The equality sign in (7.33) applies when $|\Gamma^{(1)}(\tau)|$ [and hence $W(\omega)$] are Gaussian functions. This case is obviously the analogue of the minimum uncertainty wave packet.[6]

7.3.5 Some Numerical Examples

A laser oscillating on a single transverse mode has perfect spatial coherence [i.e., $|\gamma^{(1)}(\mathbf{r}_1, \mathbf{r}_2, 0)| = 1$ for any points $\mathbf{r}_1$, and $\mathbf{r}_2$]. The temporal coherence of course depends on $\Delta\nu_{osc}$. For the example considered in Section 7.2 of a continuous He–Ne laser ($\Delta\nu_{osc} = 50$–500 Hz), one would have $L_c = 60$–600 km! Conventional light sources (e.g., a sodium lamp) have coherence times $\tau_{co} \simeq 10^{-10}$ s, for which $L_c \simeq 3$ cm. With such a light source one would be able to observe fringes in a Michelson interferometer for a difference $L_3 - L_2$ between the interferometer arms of up to a few centimeters.

7.4 DIRECTIONALITY

Let us first consider a wave with perfect spatial coherence consisting of a plane wave beam of circular cross section having constant intensity over the cross section (Fig. 7.4a). As a result of diffraction, this beam has an intrinsic divergence θ_d. This can be understood with the help of Fig. 7.4a, which shows the wavefront $A'B'$ obtained from AB by applying the Fresnel–Huygens principle. It can be shown that the divergence θ_d is given by

$$\theta_d = 1.22 \frac{\lambda}{D} \tag{7.34}$$

where D is the beam diameter. To see what is meant by this divergence, let

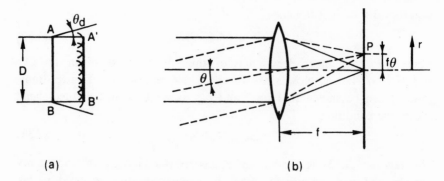

FIG. 7.4. (a) Divergence (due to diffraction) of a plane e.m. wave of constant intensity across its circular cross section; (b) method of measuring the divergence of the plane wave of (a).

us see what happens when the beam we are considering is focused by a lens (Fig. 7.4b). Since, as we have seen, the beam has some spread, it can be shown that the beam can be decomposed into a set of plane waves propagating in slightly different directions. One of these, inclined at an angle θ, is indicated by dotted lines in the figure. Now, as we can see, this wave will be focused to the point P in the focal plane and (for small θ) at a distance

$$r = f\theta \tag{7.35}$$

from the beam axis. Hence, a knowledge of the intensity distribution $I(r)$ in the focal plane gives the angular distribution of the original beam. Now, it is known from diffraction theory[7] that the function $I(r)$ is given by the Airy formula

$$I = \left[\frac{2J_1(krD/2f)}{krD/2f} \right]^2 I_0 \tag{7.36}$$

where $k = 2\pi/\lambda$, J_1 is the first-order Bessel function, and I_0 (the intensity at the center of the focal spot) has the value

$$I_0 = P_i \left(\frac{\pi D^2}{4\lambda^2 f^2} \right) \tag{7.37}$$

where P_i is the power of the beam incident on the lens.

The behavior of the intensity I is shown in Fig. 7.5 as a function of

$$x = \frac{krD}{2f} \tag{7.38}$$

Consequently, the diffraction pattern formed at the focal plane of the lens consists of a circular central zone (the Airy disc) surrounded by a series of

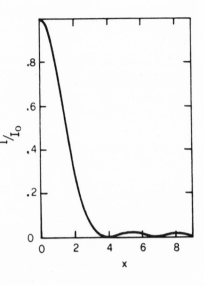

FIG. 7.5. Distribution of light intensity in the focal plane of Fig. 7.4b as a function of radial distance r (normalized, i.e., $x = krD/2f$).

rings of rapidly decreasing intensity. Now the divergence θ_d of the original beam is conventionally defined to correspond to the radius of the first minimum shown in Fig. 7.5. So, from Fig. 7.5, with the help of (7.38) and (7.35), we obtain (7.34). It can be seen then that the expression (7.34) for θ_d has a certain arbitrariness.

As a second example of the propagation of a spatially coherent beam, we consider the case of a Gaussian beam (TEM_{00}) such as can be obtained from a stable laser cavity consisting of two spherical mirrors. If we call w_0 the spot size at the beam waist, the spot size w and the radius of curvature R of the equiphase surface at a distance z from the waist can, according to (4.34), (4.35), and (4.36), be written as

$$w^2(z) = w_0^2 \left[1 + \left(\frac{\lambda z}{\pi w_0^2} \right)^2 \right] \tag{7.39}$$

$$R(z) = z \left[1 + \left(\frac{\pi w_0^2}{\lambda z} \right)^2 \right] \tag{7.40}$$

Note that, for given λ and z, both w and R (and hence the field distribution) depend only on w_0. This can be readily understood when we note that, at the plane corresponding to the beam waist, we know both the amplitude (since we know w_0 and we know that the field distribution is Gaussian) and the phase (since we know that $R = \infty$ at the waist) distribution of the field. The field distribution at some other point in space can then be obtained by means, for instance, of the Kirchhoff integral (4.10). This field is thus uniquely determined once w_0 is known.

To calculate the divergence properties of a Gaussian beam, we consider both (7.39) and (7.40) at a large distance from the waist (i.e., for $\lambda z / \pi w_0^2 \gg 1$). We have $w = \lambda z / \pi w_0$ and $R = z$. Since, at large distances, both w and R linearly increase with distance, the wave can be considered to be a spherical wave originating from the center of the waist. Its divergence can then be obtained as

$$\theta_d = \frac{w}{z} = \frac{\lambda}{\pi w_0} \qquad (7.41)$$

We can now compare (7.41) and (7.34). If, for the purpose of comparison, we put $D = 2w_0$, we see that, for the same diameter, a Gaussian beam has a divergence about half that of a plane beam.

To sum up the results obtained so far, we can say that the divergence θ_d of a spatially coherent wave can be written as

$$\theta_d = \beta \frac{\lambda}{D} \qquad (7.42)$$

where β is a numerical factor of the order of unity, whose exact value depends on the field amplitude distribution as well as on the way in which both θ_d and D are defined. Such a beam is commonly referred to as being diffraction-limited. A more satisfactory formulation of both θ_d and D can be obtained with the help of the arrangement shown in Fig. 7.6 where the beam divergence is measured through the field distribution $V_2(r_2)$ in the plane P_2 (see also Fig. 7.4b) while the beam diameter is measured through the field distribution $V_1(r_1)$ in the plane P_1. In a similar way to the treatment in Section 7.3.4, we can now define (for a circularly symmetric beam) the beam radii $\langle r_1 \rangle$ and $\langle r_2 \rangle$ in the planes P_2 and P_1 respectively as

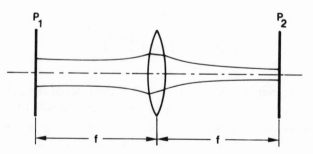

FIG. 7.6. Possible configuration to define both beam diameter and divergence.

follows:

$$\langle r_1 \rangle^2 = \langle r_1^2 \rangle \qquad (7.43a)$$

$$\langle r_2 \rangle^2 = \langle r_2^2 \rangle \qquad (7.43b)$$

where the average is taken over the square of the respective field distributions. The beam diameter and divergence can then be defined as

$$D = 2\langle r_1 \rangle \qquad (7.44a)$$

$$\theta_d = \langle r_2 \rangle / f \qquad (7.44b)$$

This provides a more general and satisfactory definition of both D and θ_d. Note that, since $V_1(r_1)$ and $V_2(kr_2/f)$ can be shown to be related by a Fourier transform,[14] we have [compare with (7.33)] $[k\langle r_1 \rangle\langle r_2 \rangle/f]$ $\geqslant 1/4\pi$. From (7.44) we then get

$$D\theta_d \geqslant \lambda/(2\pi)^2 \qquad (7.45)$$

and the equality sign holds when both V_1 and V_2 are Gaussian functions. We thus see that a Gaussian beam has the minimum possible divergence.

What has been said so far applies only to a spatially coherent wave. For an e.m. wave which does not have perfect spatial coherence the divergence is greater than for a spatially coherent wave having the same intensity distribution. This can, for example, be understood from Fig. 7.4a: if the wave is not spatially coherent, the secondary wavelets emitted over the cross section AB would no longer be in phase and the wavefront produced by diffraction would have a larger divergence than that given by equation (7.34). A rigorous treatment of this problem (i.e., the propagation of partially coherent waves) is beyond the scope of this book, and the reader is therefore referred to more specialized texts.[8] We will limit ourselves to considering first a particularly simple case of a beam of diameter D (Fig. 7.7a) which is made up of many smaller beams (shaded in

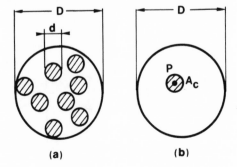

FIG. 7.7. Simple examples to illustrate the different divergence properties of coherent and partially coherent waves.

(a)   (b)

the figure) of diameter d. We will assume that each of these smaller beams is diffraction limited (i.e., spatially coherent). Now, if the various beams are mutually uncorrelated, the divergence of the beam as a whole will be equal to $\theta_d = \beta\lambda/d$. If, on the other hand, the various beams were correlated, the divergence would be $\theta_d = \beta\lambda/D$. This last case is actually equivalent to a number of antennae (the small beams) all emitting in phase with each other. After this simple case we can consider the general case in which the partially coherent beam has a given intensity distribution over its diameter D and a given coherence area A_c at each point P (Fig. 7.7b). By analogy with the previous case we can readily understand that in this case $\theta_d = \beta\lambda/[A_c]^{1/2}$, where the numerical factor β depends both on the particular intensity distribution and on the way in which A_c is defined. The concept of directionality is thus intimately related to that of spatial coherence.

Given an incoherent lamp S, one can obtain a spatially coherent wave from it by using the arrangement of Fig. 7.8. The light from S is imaged on a pinhole of diameter d situated in the focal plane of the lens L'. Since the beam produced by diffraction from the hole has a divergence $\theta = 1.22\lambda/d$, one might expect that, if the aperture D of the lens L' satisfied the condition $D = 2\theta f = 2.44\lambda f/d$, the lens would collect just the light resulting from diffraction from the hole (shown shaded in Fig. 7.8) thus producing a coherent output beam. However, this argument is rather oversimplified because it uses equation (7.34), which is only valid when the hole is illuminated by light that is already coherent. A more rigorous treatment of this problem requires a study of the propagation of partially coherent e.m. waves.[8] Let us suppose for simplicity (and also because this is frequently the case in practice) that the wave arriving at the hole has no spatial coherence. Then, for this case, it follows from the well-known Van Cittert-Zernike theorem[8] that, if the exit beam from the lens L' of Fig. 7.8 is to have some particular value of spatial coherence, the diameter D of the lens must have the value $D = \beta\lambda f/d$, where β is a numerical factor which

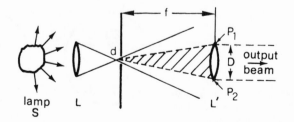

FIG. 7.8 Method for obtaining a coherent output beam from an incoherent lamp.

depends on the degree of coherence we stipulate. If, for instance, we require that the degree of spatial coherence between two extreme points P_1 and P_2 on the lens circumference have the value $|\gamma(P_1, P_2, 0)| = 0.88$, we get $\beta = 0.32$. This gives

$$D = \frac{0.32\lambda f}{d} \qquad (7.46)$$

which is of the same form as that established by the earlier simplified argument but with a different (and in fact significantly smaller) numerical factor.

7.5 *LASER SPECKLE*[9,10]

Following what has been said in the previous sections about first-order coherence, we now mention a very striking phenomenon characteristic of laser light, known as laser speckle. Laser speckle is apparent when one observes laser light scattered from a wall or transparent diffuser. The scattered light is seen to consist of a random collection of alternatively bright and dark spots (or speckles) (Fig. 7.9a). Despite the randomness, one can distinguish an average speckle (or grain) size. This phenomenon was quickly recognized by early workers in the field as being due to constructive and destructive interference of radiation coming from the small scattering centers on the surface of the wall or of the transparent diffuser. Since the phenomenon depends on there being a high degree of first-order coherence, it is an inherent feature of laser light.

The physical origin of the observed granularity can be readily understood for both free-space propagation (Fig. 7.9b) and for an imaging system (Fig. 7.9c), when it is realized that the surfaces of most materials are extremely rough on the scale of an optical wavelength. For free-space propagation, the resulting optical wave at any moderately distant point from the scattering surface consists of many coherent components or wavelets each arising from a different microscopic element of the surface. Referring to Fig. 7.9b, we note that the distances traveled by these various wavelets may differ by many wavelengths. Interference of the phase-shifted but coherent wavelets results in the granular intensity (or speckle pattern, as it is called). When the optical arrangement is that of an imaging system (Fig. 7.9c), an explanation of the observed pattern must take account of diffraction as well as interference. Even for a perfectly corrected imaging system, the intensity at a given image point can result from the coherent addition of contributions from many independent parts of the surface. It is only necessary that the point-spread function of the imaging system be

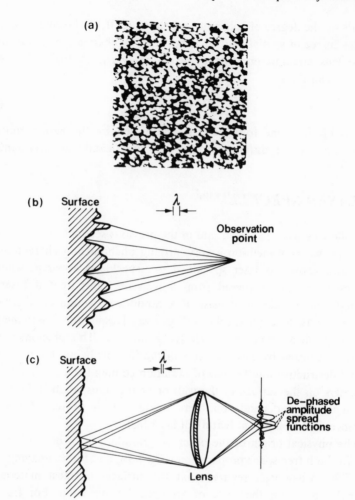

FIG. 7.9 Speckle pattern (a) and its physical origin for free-space propagation (b) and for an image-forming system (c).

broad in comparison to the microscopic surface variations to ensure that many phase-shifted coherent contributions add at each image point. Thus speckle can arise either from free-space propagation or from an imaging arrangement.

We can readily obtain an order-of-magnitude estimate for the grain size d_g (i.e., the average size of the spots in the speckle pattern) for the two cases just considered. In the first case (Fig. 7.10a) the scattered light is recorded on photographic film at a distance L from the diffuser with no lens between film and diffuser. Suppose that, at some point P in the

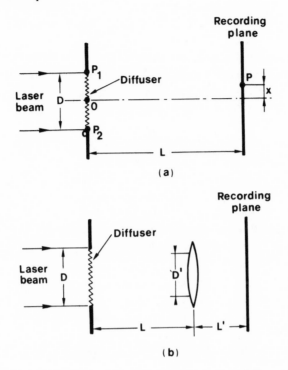

FIG. 7.10 Grain-size calculation for free-space propagation (a) and for an image-forming system (b).

recording plane, there is a bright speckle. This means that the light diffracted by the diffuser (e.g., by points P_1, P_2, and O) will interfere at point P predominantly in a constructive way so as to give an overall nonzero value for the field amplitude. We can now ask how far the point P must be moved along the x axis in the recording plane in order to destroy this constructive interference. This means that we now require the contributions of the diffracted waves from points P_1 and O at the new point P to interfere destructively rather than constructively. This implies that the change δx of the x coordinate must be such that the corresponding change $\delta(OP - P_1P)$ of the difference in lengths $OP - P_1P$ be equal to $\lambda/2$.[†] Since $OP = (x^2 + L^2)^{1/2}$ and $P_1P = \{[(D/2) - x]^2 + L^2\}^{1/2}$, we get (for $D \ll L$) $\delta(OP - P_1P) = (D/2L)\delta x$. If we require that $\delta(OP - P_1P) = \lambda/2$, we get $\delta x = \lambda L/D$ and hence

$$d_g = 2\delta x = 2\lambda L/D \qquad (7.47)$$

[†]Note that a similar argument can be applied at the lens aperture of Fig. 7.4b to get an estimate of the spot diameter ($\sim 2.44\lambda f/D$) of the beam in the focal plane.

We stress again that the size of the grain is thus a consequence of the fact that the whole beam acts coherently in its contribution of diffracted light to each individual spot. The above expression holds provided that: (i) the size d_s of the individual scatterers is much smaller than the aperture D; (ii) there is an appreciable overlap, at the recording plane, between wavelets diffracted from various scattering centers. This implies that the dimension of each of these wavelets at the recording plane $(\sim\lambda L/d_s)$ is larger than their mean separation (D). The length L must therefore be such that $L > d_s D/\lambda$. For $d_s = 10$ μm and $\lambda = 0.5$ μm, for instance, $L > 20D$.

The second case we will consider is that of scattered light recorded on a photographic film after it passes through a lens which images the diffuser on the film. An aperture of diameter D' is placed in front of the lens (Fig. 7.9c). If the length L is again such that $L > d_s D/\lambda$, the grain size d_g on the lens will be given by (7.47). As in the previous case, we will assume that (i) this grain size d_g is much smaller than the aperture D'; (ii) there is an appreciable overlap, at the recording plane, of wavelets diffracting from these various grains. This implies that the dimension of each of these wavelets at the recording plane $(\lambda L'/d_g)$ is larger than their mean separation (D'). By use of (7.47) this is seen to imply $D' < D(L'/L)$. Under the above two assumptions, the grain size d_g' at the recording plane is given by

$$d_g' = 2\lambda L'/D' \qquad (7.48)$$

Now it is the whole beam of aperture D' which acts coherently in its contribution of diffracted light to each individual spot. Note that the arrangement of Fig. 7.10b also describes the case where one looks directly at a diffusing surface. In this case the lens and the recording plane correspond to the lens of the eye and the retina. Accordingly, d_g' given by (7.48) is the grain size on the retina. Note that the apparent grain size on the diffuser d_{ag} is then given by $d_{ag} = d_g'(L/L') = 2\lambda L/D'$. This increases with increasing L, i.e., with increasing distance between the observer and the diffuser. It also decreases with increased aperture of the iris (i.e., when the eye is dark-adapted). Both these predictions are indeed confirmed by experimental observations.

Speckle noise often constitutes an undesirable feature of coherent light. The spatial resolution of an object illuminated by laser light is in fact often limited by speckle noise. Speckle noise is also apparent in the reconstructed image of a hologram, again limiting the spatial resolution of this image. Some techniques have therefore been developed to reduce speckle in coherently illuminated objects.[9] Speckle noise is not always a nuisance, however. In fact techniques have been developed which exploit the speckle behavior (speckle interferometry) to show up, in a rather simple way, the deformation of large objects due, for instance, to stresses or vibrations.[9]

7.6 BRIGHTNESS

The brightness B at a given point of a light source for a given direction of emission has already been defined in Chapter 1 [see Fig. 1.7 and equation (1.13)]. It is important to note that the most significant parameter of a laser beam (and in general of any light source) is neither power nor intensity, but brightness. In fact let us compare, for example, two lasers 1 and 2 having the same diameter and output power, one having a beam divergence θ_1, the other θ_2, where $\theta_2 > \theta_1$. From what was said in connection with Fig. 7.4b it can be seen that the first of these beams produces the higher intensity at the focus of a lens. Since the solid angle of emission Ω is proportional to the square of the divergence, the first beam is brighter than the second. It follows therefore that the intensity that can be produced at the focus of a lens is proportional to the beam brightness. Since, in most applications, one is interested in the beam intensity that can be produced by focusing with a lens, it follows that brightness is the significant quantity. This is further demonstrated by the fact that although the intensity of a beam can be increased, its brightness cannot. The simple arrangement of confocal lenses shown in Fig. 7.11 can be used to decrease the beam diameter if $f_2 < f_1$. The intensity of the exit beam is therefore greater than that of the entrance beam. However, the divergence of the exit beam $(\sim\lambda/D_2)$ is also greater than that $(\sim\lambda/D_1)$ of the entrance beam, and so one can see that the brightness remains invariant. This property, seen here for a particular case, is of general validity (for incoherent sources also): Given some light source and an optical imaging system, the image cannot

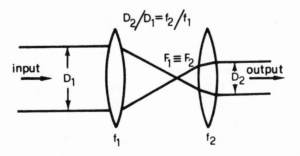

FIG. 7.11 Method for increasing the intensity of a plane wave.

be brighter than the original source[11] (this is true provided the source and image are surrounded by media of the same refractive index).

The brightness of lasers is several orders of magnitude greater than that of the most powerful incoherent sources. This is due to the extreme directionality of a laser beam. Let us compare, for example, a He–Ne laser oscillating on a single mode at a wavelength $\lambda \simeq 0.63$ μm with an output power of 1 mW with what is probably the brightest conventional source. This would be a high-pressure mercury vapor lamp (PEK Labs type 107/109), with an output power of ~ 100 W and a brightness B of ~ 95 W/cm^{2}sr in its most intense green line ($\lambda = 546$ nm, $\Delta\lambda = 10$ nm). To obtain a diffraction-limited beam we can use the arrangement of Fig. 7.8. The emission solid angle is $\Omega = \pi D^2/4f^2$, and the emitting area is $A = \pi d^2/4$. Since the brightness of the image at the pinhole cannot be greater than that of the lamp, the output beam power is at most

$$P = B\Omega A \simeq (\lambda/4)^2 B \simeq 1.7 \times 10^{-8} W \tag{7.49}$$

where we have used (7.46). The output power turns out to be about five orders of magnitude less than that of the He–Ne laser. We also note from (7.49) that the diffraction-limited power obtainable from a lamp depends only on its brightness. This further illustrates the importance of the concept of brightness.

Using the system of Fig. 7.8 we have thus arranged so that the two beams (from the laser and from the mercury lamp) have the same degree of spatial coherence. To obtain the same degree of temporal coherence it is necessary to insert a filter in the arrangement of Fig. 7.8 so as to pass a very narrow band, i.e., equal to the oscillation bandwidth $\Delta\nu_{osc}$ of the He–Ne laser. Assuming $\Delta\nu_{osc} \simeq 500$ Hz, since the linewidth of the mercury lamp under consideration is $\Delta\nu = 10^{13}$ Hz, it follows that this second operation further reduces the output power by more than ten orders of magnitude ($P \simeq 10^{-18}$ W). We recall that the lamp power that we started with was 100 W! This also shows how much more difficult it is to produce interference phenomena (which require sources with good coherence) starting with incoherent sources.

This output beam from the mercury lamp now has the same spatial and temporal coherence characteristics as a He–Ne laser. It is therefore natural to ask whether this light now has exactly the same coherence characteristics as a laser beam. The answer, however, is negative. Despite having paid such a heavy penalty in terms of output power, the laser light is still more coherent than the "filtered" light from the lamp. However, to understand this difference, some further discussion of the coherence properties of light beams is necessary.

7.7 HIGHER-ORDER COHERENCE [2,12]

It has been seen that, if the analytic signal $V(\mathbf{r}, t)$ is given, then the correlation function $\Gamma^{(1)}(\mathbf{r}_1, \mathbf{r}_2, \tau)$ is uniquely determined. However, the converse is not generally true. In other words, given $\Gamma^{(1)}$, it is not possible to find V. This means that one can define higher-order correlation functions which will generally be mutually independent and also independent of the first-order correlation function. For the sake of brevity we will use the symbol $x_i = (\mathbf{r}_i, t_i)$ for the space and time coordinates of a point, and we can then write the definition of the nth-order correlation function as follows:

$$\Gamma^{(n)}(x_1, x_2, \ldots, x_{2n}) = \langle V(x_1) V(x_2) \cdots V(x_{2n}) \rangle \qquad (7.50)$$

which involves the product of $2n$ terms, these being the function V evaluated at $2n$ space–time points $x_1, x_2, \ldots, x_{2n}$. The corresponding normalized quantity is then given by

$$\gamma^{(n)}(x_1, x_2, \ldots, x_{2n}) = \frac{\Gamma^{(n)}(x_1, x_2, \ldots, x_{2n})}{\prod_{r=1}^{2n} \left[\Gamma^{(1)}(x_r, x_r) \right]^{1/2}} \qquad (7.51)$$

where $\prod$ stands for product. Obviously these expressions reduce to (7.13) and (7.14) for the case $n = 1$. In general, to determine the field $V(\mathbf{r}, t)$ it is necessary to know all the correlation functions $\Gamma^{(n)}$.

From what has been said earlier, we see that if two light waves (e.g., light from a laser and from a mercury lamp) have the same degree of spatial and temporal coherence, then they have the same correlation functions to first order. This does not mean, however, that their corresponding analytic signals have the same characteristics. In other words, the two waves can still have different correlation functions for higher orders. It is therefore natural to ask what is meant by completely coherent or completely incoherent light. We will not consider this question in any detail since a correct understanding involves the use of quantized e.m. fields. This is beyond the scope of this book, so we will limit ourselves to a few remarks and refer the reader elsewhere[2,12] for a more comprehensive study of this subject.

If the wave is perfectly coherent in first order (i.e., if $|\gamma^{(1)}(x_1, x_2)| = 1$), it can be shown that

$$\Gamma^{(1)}(x_1, x_2) = V(x_1) V^*(x_2) \qquad (7.52)$$

i.e., $\Gamma^{(1)}$ can be separated into a product of the analytic signal at x_1 with the signal at x_2. By analogy one defines a perfectly coherent e.m. wave as one for which $\Gamma^{(n)}$ factorizes for all n. This means

$$\Gamma^{(n)}(x_1, x_2, \ldots, x_{2n}) = \prod_{r=1}^{n} V(x_r) \prod_{k=n+1}^{2n} V^*(x_k) \qquad (7.53)$$

From (7.51), (7.52), and (7.53) one finds, that, for this case

$$|\gamma^{(n)}(x_1, x_2, \ldots, x_{2n})| = 1 \qquad (7.54)$$

for all orders of n. For the particular case where $x_1 = x_2 = \cdots = x_{2n} = x$, one finds from (7.53) that

$$\Gamma^{(n)}(x, x, \ldots, x) = \left[\Gamma^{(1)}(x, x) \right]^n \qquad (7.55)$$

A wave is said to be completely incoherent when it is Gaussian in nature. To understand what this means, let us suppose that T is the time taken for the measurement of $V^{(r)}(t)$ (or one of its correlation functions). In this case $V^{(r)}(t)$ at some point $\mathbf{r}$ in space can be expanded as a series in frequency $\omega_k = k\omega_0$, where ω_0 is the fundamental frequency $\omega_0 = 2\pi/T$. Therefore, in place of (7.1) we will write

$$V^{(r)}(t) = \sum_{k=0}^{\infty} (a_k \cos \omega_k t + b_k \sin \omega_k t) \qquad (7.56)$$

For the case of incoherent (or partially coherent) light, a_k and b_k are statistical variables. Let us define a probability distribution $p(a'_1, a'_2, \ldots ; b'_1, b'_2, \ldots)$ representing the probability that during the measurement time T the values a'_i and b'_i were found for the coefficients a_i and b_i. The condition for light to be incoherent is that the distribution should factorize into a product of Gaussian functions, as follows:

$$p(a_1, a_2, \ldots ; b_1, b_2, \ldots) = \prod_k \frac{1}{(2\pi\sigma_k)^{1/2}} \exp\left(-\frac{a_k^2 + b_k^2}{2\sigma_k^2} \right) \qquad (7.57)$$

where σ_k is arbitrary. In this case the total probability is the product of probabilities of the type $p_k(a_k) = [\exp(-a_k^2/2\sigma_k^2)]/(2\pi\sigma_k)^{1/2}$ which implies that all the variables a_k and b_k are statistically independent. An interesting property of Gaussian light is that the correlation function $\Gamma^{(n)}$ can be written as

$$\Gamma^{(n)}(x_1, \ldots, x_n, x_{n+1}, \ldots, x_{2n}) = \sum_{\Pi} \Gamma^{(1)}(x_1, x_{n+1}) \cdots \Gamma^{(1)}(x_n, x_{2n})$$

$$\qquad (7.58)$$

where the sum extends over all possible permutations $(n!)$ of the indexes from 1 to n. In the case where $x_1 = x_2 = \cdots = x_{2n} = x$ it then follows that

$$\Gamma^n(x, \ldots, x, \ldots, x) = n! \left[\Gamma^{(1)}(x, x) \right]^n \qquad (7.59)$$

A laser oscillating in a single mode and well above threshold is essentially coherent in all orders. A thermal light source (e.g., a mercury vapor lamp) gives almost completely incoherent light because the light output consists of contributions from a large number of uncorrelated emitters. We note that [see equation (7.59)] completely incoherent light can satisfy the coherence condition (7.55) only in first order ($n = 1$). It follows that one can at most arrange that a thermal light source has spatial and temporal coherence, as described in the previous section.

It now remains to be seen how one might measure the quantity $\Gamma^{(n)}$. We saw that in first order this can be done with Young's and Michelson's interferometers. The function $\Gamma^{(2)}$ can be measured with the so-called intensity interferometer of Hanbury-Brown and Twiss.[13] A very elegant method for obtaining $\Gamma^{(n)}(x, x, \ldots, x)$ is provided by the technique of photon counting. Suppose we count the number q of photons arriving in a

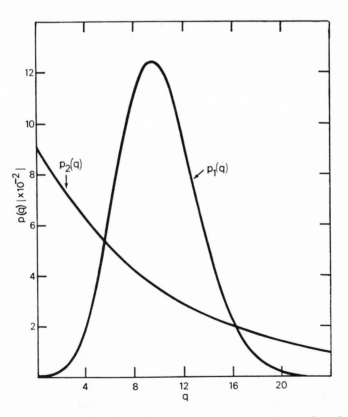

FIG. 7.12 Probability $p(q)$ that, in a photon counting experiment, the number of photons counted is q when the mean value $\langle q \rangle$ is 10: $p_1(q)$ = coherent case, $p_2(q)$ = incoherent case.

time T at the photocathode of a phototube. By repeating this measurement very many times we can measure the probability distribution $p(q)$. The nth-order moments of this distribution $[M_n = \sum q^n p(q)]$ are connected by simple algebraic relations with $\Gamma^{(n)}(x, x, \ldots, x)$. Measurement of $p(q)$ thus gives complete information about the field, within the limits of accuracy of the measurement operation. It can be shown that, for a completely coherent field,

$$p_1(q) = \frac{\langle q \rangle^q}{q!} \exp(-\langle q \rangle) \tag{7.60}$$

where $\langle q \rangle = \sum q p(q)$ is the mean value of the number of photons counted. The distribution in this case is the so-called Poisson distribution. For the case of Gaussian light, the distribution follows from Bose–Einstein statistics, and is given by

$$p_2(q) = \frac{1}{1 + \langle q \rangle}\left(\frac{\langle q \rangle}{1 + \langle q \rangle}\right)^q = \frac{\exp[-\alpha q]}{[1 + \langle q \rangle]} \tag{7.61}$$

where $\alpha = \ln[(1 + \langle q \rangle)/\langle q \rangle]$. The distribution is exponential in this case. The two distributions are shown in Fig. 7.12.

PROBLEMS

7.1. Show that, for a quasi-monochromatic e.m. wave, the relationship between the intensity $I(\mathbf{r}, t)$ as defined by (7.7) and $V^{(r)}$ is given by $2I = \langle V^{(r)2} \rangle$, where the average is taken over a few optical cycles [hint: use (7.5)].

7.2. Calculate $\Gamma^{(1)}(\mathbf{r}_1, \mathbf{r}_1, \tau)$ for a sinusoidal wave.

7.3. Calculate $\Gamma^{(1)}(\mathbf{r}_1, \mathbf{r}_1, \tau)$ for a sinusoidal wave undergoing phase jumps as in Fig. 2.6 with a probability p_τ as in (2.49). Plot the corresponding $\gamma^{(1)}(\mathbf{r}_1, \mathbf{r}_1, \tau)$ versus τ and compare this curve with that of Fig. 7.1.

7.4. Prove equation (7.19).

7.5. For a Michelson interferometer, find the analytical relation between I_c and $2(L_3 - L_2)$ for the e.m. wave of Problem 7.3. Calculate the corresponding fringe visibility $V_P(\tau)$.

7.6. A laser operating at $\lambda = 10.6$ μm produces an output having a Gaussian line shape with a bandwidth of 10 kHz [$\Delta \nu_{osc}$ is defined according to (7.32)]. With reference to Fig. 7.3b, calculate both the distance ΔL between two successive maxima of the intensity curve and the coherence length L_c.

7.7. A plane e.m. wave of circular cross section, uniform intensity, and perfect spatial coherence is focused by a lens. What is the increase in intensity at the focus compared to that of the incident wave?

7.8. A ruby laser beam with a diameter $D = 6$ mm and a constant intensity distribution over its cross section has a divergence $\theta = 7 \times 10^{-3}$ rad. Show that the beam is not diffraction limited and estimate the corresponding coherence area A_c.

7.9. How would you measure the beam divergence of the laser in the previous problem?

7.10. Suppose the beam of Problem 7.8 passes through an attenuator whose (power) transmission T varies with radial distance r according to the law $T = \exp \cdot [-(r/w_1)^2]$ with $w_1 = 0.5$ mm. Thus the beam, after the attenuator, has a Gaussian intensity profile. Does this mean that the beam is now a Gaussian beam of (intensity) spot size w_1?

7.11. The laser beam of Problem 7.8 is passed through a telescope as in Fig. 7.11. Calculate the diameter of the pinhole which must be inserted at the common focus $F_1 = F_2$ in order to produce a diffraction-limited output beam. Note that, since the beam already has a fairly good spatial coherence, one should use the equation appropriate to a coherent beam, rather than that for an incoherent beam [i.e., (7.46))].

7.12. Show that (7.52) holds for perfect sinusoidal waves {i.e., when $V(x_1) = A_1\exp[i\omega(t + \tau)]$ and $V(x_2) = A_2\exp(i\omega t)$, with $A_1 = A_2 = \text{const}$}.

7.13. Show that (7.53) holds for perfect sinusoidal waves.

REFERENCES

1. J. L. Hall, *Science*, **202**, 147(1978).
2. L. Mandel and E. Wolf, *Rev. Mod. Phys.* **37**, 321 (1965).
3. M. Born and E. Wolf, *Principles of Optics*, 4th ed. (Pergamon Press, London, 1970), pp. 491–544.
4. D. Gabor, *J. Inst. Elec. Eng.* **93**, 429 (1946).
5. M. Born and E. Wolf, *Principles of Optics*, 4th ed. (Pergamon Press, London, 1970), pp. 370–375.
6. W. H. Louisell, *Radiation and Noise in Quantum Electronics* (McGraw-Hill Book Co., New York, 1964), pp. 47–53.
7. M. Born and E. Wolf, *Principles of Optics*, 4th ed. (Pergamon Press, London, 1970), pp. 395–398.
8. M. Born and E. Wolf, *Principles of Optics*, 4th ed. (Pergamon Press, London, 1970), pp. 508–518.
9. *Laser Speckle and Related Phenomena*, ed. by J. C. Dainty (Springer-Verlag, Berlin, 1975).
10. M. Françon, *Laser Speckle and Applications in Optics* (Academic Press, New York, 1979).
11. M. Born and E. Wolf, *Principles of Optics*, 4th ed. (Pergamon Press, London, 1970), pp. 189, 190.
12. R. J. Glauber, in *Quantum Optics*, ed. by S. M. Kay and A. Maitland (Academic Press, New York, 1970), pp. 53–125.
13. R. Hanbury-Brown and R. Q. Twiss, *Nature*, **177**, 27 (1956).
14. J. W. Goodman, *Introduction to Fourier Optics* (McGraw-Hill Book Co., New York, 1968), Chapter 5.

8

Laser Beam Transformation

8.1 INTRODUCTION

Before it is put to use, a laser beam is generally transformed in some way. The most common type of transformation is that which occurs when the beam is made to propagate in free space or through a suitable optical system. Since this produces a change in the spatial distribution of the beam (e.g., the beam may be focused or expanded), we shall refer to this as a *spatial transformation* of the laser beam. A second type of transformation, also rather frequently encountered, is that which occurs when the beam is passed through an amplifier or chain of amplifiers. Since the main effect here is to alter the beam amplitude, we shall refer to this as *amplitude transformation*. A third, less trivial, case occurs when the wavelength of the beam is changed as a result of propagating through a suitable nonlinear optical material (*wavelength transformation*). Finally, the temporal behavior of the laser beam can be modified (e.g., the time variation of the output from a pulsed laser may be changed) by a suitable electro-optical or nonlinear optical element. This fourth and last case will be called *time transformation*. It should be noted that these four types of beam transformation are often interrelated. For instance, amplitude and wavelength transformation often result in spatial and time transformations occurring as well.

In this chapter the cases of spatial, amplitude, and wavelength transformation will be briefly considered. In the case of wavelength transformation, of the various nonlinear optical effects which can be used[1] to achieve this, only the so-called parametric effects will be considered here. These in fact provide some of the most useful techniques so far developed for producing new sources of coherent light. Time transformations will not

297

be considered at all here. We also leave out some amplitude and time transformations arising from the nonlinear phenomena of self-focusing and self-phase-modulation,[2] although it should be noted that they can play a very important role in limiting the performance of laser amplifiers.

8.2 TRANSFORMATION IN SPACE: GAUSSIAN BEAM PROPAGATION

In this section we will limit ourselves to considering the propagation of a lowest-order Gaussian beam (TEM_{00} mode). The important topics of propagation of coherent beams having non-Gaussian transverse distributions (for which the Kirchhoff integral or equation (8.10) can still be used) and propagation of partially coherent beams[3] will not be considered.

The case of free-space propagation of a TEM_{00} Gaussian beam has already been considered in Chapter 7 (see Section 7.4). For convenience we repeat here the expressions for beam spot size w and radius of curvature R of the equiphase surfaces, viz.,

$$w^2 = w_0^2 \left[1 + \left(\frac{\lambda z}{\pi w_0^2} \right)^2 \right]$$

(8.1a)

$$R = z \left[1 + \left(\frac{\pi w_0^2}{\lambda z} \right)^2 \right]$$

(8.1b)

where w_0 is the spot size at the beam waist and the z coordinate is measured along the propagation direction with its origin at the waist.[†] Figure 8.1 shows the behavior of the beam spot size and equiphase surfaces with distance z. We re-emphasize that the propagation properties of this beam depend only on the wavelength and the value w_0 of the spot size at the beam waist. We also recall that this can be understood by noting that, once w_0 is known, both the amplitude and phase are known at the waist (the wavefront is plane at the waist). Since the field distribution is thus known over the entire plane $z = 0$, diffraction theory [e.g., the Kirchhoff integral (4.10)] can be used to calculate the field amplitude at any given point in space. We shall not carry out such a calculation here and we limit ourselves to noting that equation (8.1a) shows that the square of the beam spot size at a distance z from the waist is given by the sum of the squares of the spot

---

[†]We recall that the sign convention for radius of curvature is that $R(z)$ is taken as positive when the center of curvature is to the left of the wavefront.

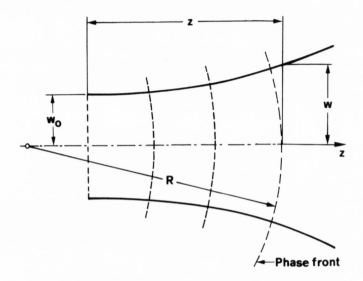

FIG. 8.1. Propagation of a Gaussian beam.

size at the waist, w_0^2, and the contribution $[(\lambda/\pi w_0)z]^2$ arising from diffraction. At the end of this section, as an exercise, equations (8.1) will be derived directly from Maxwell's equations rather than using the Kirchhoff integral.

We now turn our attention to the propagation of a TEM_{00} Gaussian beam through a lens system. Figure 8.2 depicts the behavior of the beam as a result of passing through a lens of focal length f. We begin by noting that, just before the lens, the spot size w_1 and radius of curvature R_1 of the beam can, according to (8.1), be written as

$$w_1^2 = w_{01}^2\left[1 + \left(\frac{\lambda L_1}{\pi w_{01}^2}\right)^2\right] \tag{8.2a}$$

$$R_1 = L_1\left[1 + \left(\frac{\pi w_{01}^2}{\lambda L_1}\right)^2\right] \tag{8.2b}$$

We also note that, for a thin lens, the amplitude distribution must remain unchanged upon passing through the lens, i.e., there cannot be a discontinuous change of spot size. Thus we can write

$$w_2 = w_1 \tag{8.3a}$$

for the beam spot size after the lens. To calculate the corresponding wavefront curvature we first consider the case of propagation of a spherical

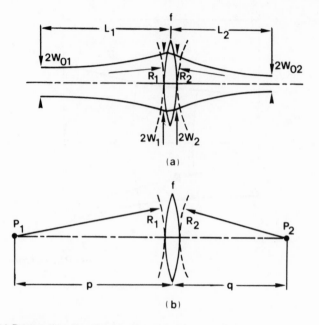

FIG. 8.2. (a) Propagation of a Gaussian beam through a lens; (b) propagation of a spherical wave through a lens.

wave through the same lens (Fig. 8.2*b*). Here a spherical wave originating from a source point P_1 is focused by the lens to the image point P_2. From geometrical optics there follows the well-known result that $p^{-1} + q^{-1} = f^{-1}$. Since the radii of curvature R_1 and R_2 of the two spherical waves just before and after the lens are equal to p and $-q$ respectively[†] we can also write

$$\frac{1}{R_1} - \frac{1}{R_2} = \frac{1}{f} \qquad (8.3b)$$

A spherical lens can then be seen to transform the radius of curvature R_1 of an incoming wave to a radius R_2 of the outgoing wave according to (8.3*b*). Similarly, the radius of curvature R_2 of the outgoing Gaussian beam of Fig. 8.2a will also be given by (8.3*b*), and so we now have both the amplitude [through (8.3*a*)] and phase [through (8.3*b*)] distribution of the outgoing wave. This wave therefore has a Gaussian amplitude distribution and spherical wavefront, i.e., a Gaussian beam remains a Gaussian beam after passing through a (thin) lens system. This result applies also to a thick lens

[†]Note the application of the sign convention referred to earlier.

system, as can be seen by considering a thick lens as a sequence of thin lenses. Once the spot size and radius of curvature of the outgoing wave are known just after the lens, we can calculate the corresponding values at any points in space. For instance, the spot size w_{02} at the new beam waist and the distance L_2 of this waist from the lens can be obtained by using equations (8.1) in reverse. After some straightforward manipulation we arrive at the following two equations:

$$L_1 = f \pm \left(\frac{w_{01}}{w_{02}} \right)(f^2 - f_0^2)^{1/2} \qquad (8.4a)$$

$$L_2 = f \pm \left(\frac{w_{02}}{w_{01}} \right)(f^2 - f_0^2)^{1/2} \qquad (8.4b)$$

from which both w_{02} and L_2 can be obtained. The quantity f_0 in equations (8.4) is given by

$$f_0 = \pi w_{01} w_{02}/\lambda \qquad (8.5)$$

and either the two plus or the two minus signs can be chosen. These equations prove very useful for solving a variety of problems which arise in Gaussian beam propagation (see Problems 8.2 and 8.3). We limit ourselves to pointing out here that, when the first waist is coincident with the first focal plane ($L_1 = f$), the second waist also coincides with the second focal plane of the lens ($L_2 = f$). We also note that, in general, the planes of the two waists are not conjugated in accordance with the geometrical optics result (i.e., $L_1^{-1} + L_2^{-1} \neq f^{-1}$).

Before ending this section we show, as an exercise, how (8.1) can be derived through Maxwell's equations rather than using the Kirchhoff integral. In the scalar case, Maxwell's equations lead to the wave equation[†]

$$\nabla^2 E - \frac{1}{c^2} \frac{\partial^2 E}{\partial t^2} = 0 \qquad (8.6)$$

For a monochromatic wave we write $E(x, y, z, t) = E(x, y, z) \exp(i\omega t)$, and equation (8.6) gives (Helmholtz equation)

$$\nabla^2 E(x, y, z) + k^2 E(x, y, z) = 0 \qquad (8.7)$$

with $k = \omega/c$. For a radially symmetric beam, expressing (8.7) in cylindrical coordinates gives

$$\left(\frac{\partial^2}{\partial r^2} + \frac{1}{r} \frac{\partial}{\partial r} + \frac{\partial^2}{\partial z^2} \right)E + k^2 E = 0 \qquad (8.8)$$

[†]It has been pointed out[15] that some care is needed to derive this equation in a rigorous fashion.

We now look for a solution of the form

$$E(r,z) = U(r,z)\exp(-ikz) \tag{8.9}$$

in which $U(r,z)$, as a function of z, is assumed to be slowly varying on the scale of a wavelength ($\lambda = 2\pi/k$). Substituting (8.9) into (8.8), and using this slowly varying amplitude approximation (i.e., putting $\partial^2 U/\partial^2 z \ll k\partial U/\partial z$) gives

$$\left(\frac{\partial^2}{\partial r^2} + \frac{1}{r}\frac{\partial}{\partial r}\right)U - 2ik\frac{\partial U}{\partial z} = 0 \tag{8.10}$$

This is the fundamental equation we require (known as the quasi-optical equation) and is widely used in diffraction theory. It has to be solved with the appropriate boundary conditions.

To solve (8.10) in our case, we set the boundary condition (see Fig. 8.1)

$$U(r,0) = \exp(-r/w_0)^2 \tag{8.11}$$

Accordingly, for $z > 0$, we look for a solution of the general Gaussian form

$$U(r,z) = \exp(\alpha - \beta r^2) \tag{8.12}$$

where both α and β are taken to be complex functions of z. Before proceeding it is appropriate to point out the physical significance of both α and β. The real part of α gives the change in amplitude on the beam axis (where $r = 0$) due to beam propagation. The imaginary part of α gives a phase shift which is additional to the plane wave phase shift kz already included in (8.9). The real part, β_r, of β is obviously related to the beam spot size w by the equation

$$\beta_r = 1/w^2 \tag{8.13}$$

To understand the meaning of the imaginary part, β_i, of β, consider a uniform spherical wave emitted from a point source P located at $z = 0$ (Fig. 8.3). The field $U(P_2)$ of this wave at point P_2 lying on a plane which is perpendicular to the z axis and which is placed at a distance R from point

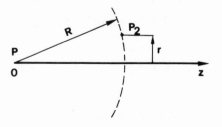

FIG. 8.3. Phase value at point $P_2(z = R)$ for a spherical wave originating from point P ($z = 0$).

P is

$$U(P_2) \propto \frac{\exp\left[-ik(r^2 + R^2)^{1/2}\right]}{\left[r^2 + R^2\right]^{1/2}}$$

$$\simeq \frac{\exp(-ikR)}{R} \exp\left[-(ikr^2/2R)\right] \qquad \text{for } r \ll R \qquad (8.14)$$

Note that R is also the radius of curvature of the spherical wave at the plane considered (dashed arc in Fig. 8.3). We thus recognize that a phase term of the form $kr^2/2R$ must represent a spherical wavefront of radius R. The comparison of the phase term $i\beta_i r^2$ in (8.12) with the phase term $ikr^2/2R$ in (8.14) then shows that β_i is related to the radius of curvature of the wavefront by

$$\beta_i = k/2R \qquad (8.15)$$

We are now ready to substitute (8.12) into the wave equation (8.10) and use the boundary condition (8.11). The substitution gives

$$r^2\left(ik\frac{d\beta}{dz} + 2\beta^2\right) - \left(ik\frac{d\alpha}{dz} + 2\beta\right) = 0 \qquad (8.16)$$

Since this expression must be zero for any r, each of the two terms in brackets must be zero, i.e.,

$$ik\frac{d\beta}{dz} + 2\beta^2 = 0 \qquad (8.17a)$$

$$ik\frac{d\alpha}{dz} + 2\beta = 0 \qquad (8.17b)$$

The solution of (8.17a) with the boundary condition (8.11) gives

$$\beta = \frac{ik}{2\left(z + i\dfrac{\pi w_0^2}{\lambda}\right)} \qquad (8.18)$$

With the help of (8.18) and again using the boundary condition (8.11), we obtain from (8.17b)

$$\alpha = -\ln\left(1 - \frac{iz\lambda}{\pi w_0^2}\right) \qquad (8.19)$$

Calculating the real and imaginary parts of β from (8.18) and using the relations (8.13) and (8.15) then yields equations (8.1a) and (8.1b) respectively. Equation (8.19), with the help of (8.1a), can be expressed in the form

$$\exp\alpha = \frac{w_0}{w} \exp i\phi \qquad (8.20)$$

where

$$\phi = \tan^{-1}\left(\frac{\lambda z}{\pi w_0^2} \right) \tag{8.21}$$

represents the additional phase shift to be added to the usual plane wave phase shift. From (8.9), (8.12), (8.13), (8.15), and (8.20) we finally get the complete expression for the field amplitude as [compare with (4.33)]

$$E(r,z) = \frac{w_0}{w} \exp\left[-i(kz - \phi) - r^2\left(\frac{1}{w^2} + \frac{ik}{2R} \right) \right] \tag{8.22}$$

Note that the field amplitude at $r = 0$ (i.e., on the beam axis) scales as w_0/w. Note also that, strictly speaking, the equiphase surfaces are really paraboloids rather than spheres. The equiphase surface which cuts the z axis at $z = z_0$ must have a phase which is at all points the same as the phase given by (8.22) for the particular point $r = 0$ and $z = z_0$, i.e., $\phi(z) - k[z + (r^2/2R)] = \phi(z_0) - kz_0$. Neglecting the small term ϕ, we see that the previous equation corresponds to a paraboloid cutting the z axis at $z = z_0$. Thus the radius of curvature R is in fact the radius of curvature of this paraboloid on the beam axis (i.e., for $r = 0$).

8.3 TRANSFORMATION IN AMPLITUDE: LASER AMPLIFICATION [4-6]

In this section we consider the rate-equation treatment of a laser amplifier. We assume that a plane wave of uniform intensity I enters (at $z = 0$) a laser amplifier extending for a length l along the z direction. We limit our considerations to a situation where the incident radiation is in the form of a pulse of duration τ_p such that $\tau_1 \ll \tau_p \ll (\tau, W_p^{-1})$, where τ_1 and τ are respectively the lifetime of the lower and upper levels of the amplifier medium and where W_p is the amplifier pump rate. This is perhaps the most relevant set of conditions for laser amplification and applies, for instance, when a Q-switched laser pulse from a Nd:YAG laser needs to be amplified. The case of cw amplification (steady-state amplification) is therefore not considered here and we refer the reader elsewhere for a discussion of this topic.[5-6]

Given these assumptions, the population of the lower level of the amplifier can be set equal to zero, and pumping and decay of the upper level of the amplifier during the passage of the pulse can be neglected. The rate of change of population inversion $N(t,z)$ at a point z within the amplifier can then be written with the help of (2.60) [in which we put

$F = I/h\nu]$ as

$$\frac{\partial N}{\partial t} = -WN = -\frac{NI}{\Gamma_s} \qquad (8.23)$$

where

$$\Gamma_s = \frac{h\nu}{\sigma} \qquad (8.24)$$

is a parameter which depends only on the laser material. Note that a partial derivative is required in (8.23) since we expect N to be a function of both z and t, i.e., $N = N(t,z)$, on account of the fact that $I = I(t,z)$. Next we derive a differential equation describing the temporal and spatial variation of intensity I. To do this we first consider the rate of change of energy density ρ of the light wave (where $\rho = I/c$, hence $\partial I/c\partial t = \partial\rho/\partial t$). By considering the net rate of change of photon energy within a small volume of the amplifier (see Fig. 8.4) we can write

$$\frac{1}{c}\frac{\partial I}{\partial t} = \frac{\partial\rho}{\partial t} = \left(\frac{\partial\rho}{\partial t}\right)_1 + \left(\frac{\partial\rho}{\partial t}\right)_2 + \left(\frac{\partial\rho}{\partial t}\right)_3 \qquad (8.25)$$

where $(\partial\rho/\partial t)_1$ accounts for stimulated emission and absorption in the amplifier, $(\partial\rho/\partial t)_2$ for the amplifier loss (e.g., scattering losses), and $(\partial\rho/\partial t)_3$ for the net photon flux which flows into the volume. With the help of (2.60) $[F = I/h\nu]$ we obtain

$$\left(\frac{\partial\rho}{\partial t}\right)_1 = WNh\nu = \sigma NI \qquad (8.26)$$

and from (2.60) and (2.64) we obtain

$$\left(\frac{\partial\rho}{\partial t}\right)_2 = -W_a N_a h\nu = -\alpha I \qquad (8.27)$$

where N_a is the density, W_a the absorption rate, and α the absorption coefficient of the loss centers. To calculate $(\partial\rho/\partial t)_3$, we refer to Fig. 8.4 where an elemental volume of the amplifier material of length dz and unit cross section is indicated by the shaded region. The quantity $(\partial\rho/\partial t)_3\,dz$ is the rate of change of photon energy in this volume due to the difference

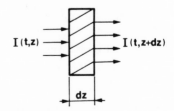

FIG. 8.4. Rate of change of the photon energy contained in an elemental volume dz (unit cross section) of the laser amplifier.

between the incoming and outgoing laser intensity, i.e.,

$$\left(\frac{\partial\rho}{\partial t}\right)_3 dz = I(t,z) - I(t,z+dz) = -\frac{\partial I}{\partial z}\, dz \qquad (8.28)$$

From (8.25) to (8.28) we then obtain the equation

$$\frac{1}{c}\frac{\partial I}{\partial t} + \frac{\partial I}{\partial z} = \sigma NI - \alpha I \qquad (8.29)$$

which together with (8.23) completely describes the amplification behavior. Note that (8.29) has the usual form of a time-dependent transport equation. Note also that, in the steady state and for $\alpha = 0$, it reduces to (1.7).

Equations (8.23) and (8.29) must now be solved with the appropriate boundary and initial conditions. As the initial condition we take $N(0,z)$ $= N_0 = $ const, where N_0 is established by pumping of the amplifier before the arrival of the laser pulse. The boundary condition is obviously established by the intensity $I_0(t)$ of the light pulse which is injected into the amplifier, i.e., $I(t,0) = I_0(t)$. For negligible amplifier losses (i.e., neglecting the term $-\alpha I$), the solution to (8.29) can be written as

$$I(z,\tau) = I_0(\tau)\left\{ 1 - \left[1 - \exp\left(-g_0 z\right)\right]\exp\left[-\int_{-\infty}^{\tau} I_0(\tau')\,d\tau'/\Gamma_s\right]\right\}^{-1}$$

$$(8.30)$$

where $\tau = t - (z/c)$ and where $g_0 = \sigma N_0$ is the unsaturated gain coefficient of the amplifier. From (8.29), one can also readily obtain an equation for the total laser energy fluence

$$\Gamma(z) = \int_{-\infty}^{+\infty} I(z,t)\,dt \qquad (8.31)$$

Integrating both sides of (8.29) with respect to time, from $t = -\infty$ to $t = +\infty$, and using (8.23), we get

$$\frac{d\Gamma}{dz} = g_0\Gamma_s\left[1 - \exp\left(-\Gamma/\Gamma_s\right)\right] - \alpha\Gamma \qquad (8.32)$$

Again neglecting amplifier losses, (8.32) gives

$$\Gamma(l) = \Gamma_s \ln\left\{ 1 + \left[\exp\left(\frac{\Gamma_{in}}{\Gamma_s}\right) - 1\right]G_0\right\} \qquad (8.33)$$

where $G_0 = \exp g_0 l$ is the unsaturated amplifier gain and Γ_{in} the energy fluence of the input beam. As a representative example the ratio Γ/Γ_s is plotted in Fig. 8.5 versus Γ_{in}/Γ_s for $G_0 = 3$. Note that, for $\Gamma_{in} \ll \Gamma_s$, (8.33) can be approximated as

$$\Gamma(l) = G_0\Gamma_{in} \qquad (8.34)$$

and the output fluence increases linearly with the input fluence (linear

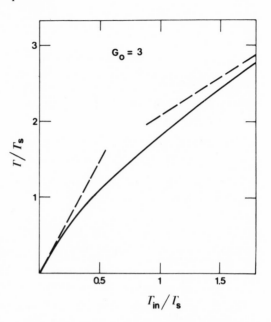

FIG. 8.5. Output laser energy fluence Γ versus input fluence Γ_{in} for a laser amplifier with a small signal gain $G_0 = 3$. The energy fluence is normalized to the laser saturation fluence $\Gamma_s = h\nu/\sigma$.

amplification regime). Equation (8.34) is also plotted in Fig. 8.5 as a dashed straight line starting from the origin. At higher input fluences, however, Γ increases with Γ_{in} at a lower rate than that predicted by (8.34) (see Fig. 8.5), i.e., amplifier saturation occurs. Thus Γ_s may be called the saturation energy fluence of the amplifier. For $\Gamma_{in} \gg \Gamma_s$ (saturation regime) we get

$$\Gamma(l) = \Gamma_{in} + \Gamma_s g_0 l \qquad (8.35)$$

Equation (8.35) has also been plotted in Fig. 8.5 as a dashed straight line. Note that (8.35) shows that, for high input fluences, the output fluence is linearly dependent on the length l of the amplifier. Since $\Gamma_s g_0 l = N_0 l h\nu$, we see that every excited atom undergoes stimulated emission and thus contributes its energy to the beam. Such a condition obviously represents the most efficient conversion of stored energy to beam energy, and for this reason amplifier designs operating in the saturation regime are used wherever practical.

If amplifier losses are present, the above picture is somewhat modified. In particular the output fluence $\Gamma(l)$ does not continue to increase with input fluence (as in Fig. 8.5) but reaches a maximum and then decreases. This can be understood by noting that the output as a function of amplifier length tends to grow linearly due to amplification [at least for high input fluences, see (8.35)] and to decrease exponentially due to loss [on account of the term $-\alpha\Gamma$ in (8.32)]. The competition of these two terms then gives a maximum for the output fluence Γ. For $\alpha \ll g_0$ this maximum value of the output fluence is

$$\Gamma \simeq g_0 \Gamma_s / \alpha \qquad (8.36)$$

Since, however, amplifier losses are typically quite small, other phenomena usually limit the maximum energy fluence that can be extracted from an amplifier. In fact, the limit is usually set by the amplifier damage fluence Γ_d (of the order of 10 J/cm^2 in some practical cases). From (8.35) we then get

$$\Gamma \simeq \Gamma_s g_0 l < \Gamma_d \qquad (8.37)$$

On the other hand, the unsaturated gain $G_0 = \exp(g_0 l)$ must not be made too high, otherwise two undesirable effects can occur in the amplifier: (i) parasitic oscillations, (ii) amplified spontaneous emission (ASE). Parasitic oscillation occurs when the amplifier starts lasing by virtue of some internal feedback which will always be present to some degree (e.g., due to the amplifier end faces). The phenomenon of ASE has already been discussed in Section 2.3.4. Both these phenomena tend to depopulate the available inversion and hence decrease the laser gain. To minimize parasitic oscillations one should avoid elongated amplifiers and in fact ideally use amplifiers with approximately equal dimensions in all directions. Even in this case, however, parasitic oscillations set an upper limit $(g_0 l)_{max}$ to the available gain coefficient g_0 times the amplifier length, l, i.e.,

$$g_0 l < (g_0 l)_{max} \qquad (8.38)$$

and $(g_0 l)_{max}$ may be of the order of 3 to 5 in practical cases. The threshold for ASE has already been given in Section 2.3.4 [equation (2.91b)]. For an amplifier in the form of a cube (i.e., for $\Omega \simeq 1$) we get $G \simeq 5.1$ [i.e., $g_0 l \simeq 1.6$] which is of the same order as that established by parasitic oscillations. For smaller values of solid angle Ω (as is more common), the value of G for the onset of ASE increases [equation (2.91b)]. Hence parasitic oscillations, rather than ASE, usually determine the maximum gain that can be achieved. Taking into account both the limit due to damage, (8.37), and the limit due to parasitic oscillations, (8.38), we can readily obtain an expression for the maximum energy E_m which can be extracted from an amplifier as

$$E_m = \Gamma_d l_m^2 = \Gamma_d (g_0 l)_m^2 / g_0^2 \qquad (8.39)$$

where l_m is the maximum amplifier dimension (for a cubic amplifier) implied by (8.38). Equation (8.39) shows that E_m is increased by decreasing the amplifier gain coefficient g_0. A limit to this reduction of g_0 would ultimately be established by the amplifier losses α. Taking, as an example, $g_0 \simeq 10^{-2}$ cm^{-1} and $\Gamma_d = 10$ J/cm^2, we get from (8.39) $E_m \simeq 1$ MJ. This, however, would require an amplifier dimension of the order of l_m $\simeq (g_0 l)_m / g_0 \simeq 4$ m, which is somewhat impracticable.

So far we have concerned ourselves mostly with the change of a laser pulse energy as it passes through an amplifier. In the saturation regime, however, important changes in both the temporal and spatial shape of the input beam also occur. The spatial distortions can be readily understood with the help of Fig. 8.5. For an input beam with a bell-shaped transverse intensity profile (e.g., a Gaussian beam), the beam center, as a result of saturation, will experience less gain than the periphery of the beam. Thus, the width of the beam's spatial profile is enlarged as the beam passes through the amplifier. The reason for temporal distortions can also be seen quite readily. Stimulated emission caused by the leading edge of the pulse implies that some of the stored energy has already been extracted from the amplifier by the time the trailing edge of the pulse arrives, which will therefore see a smaller population inversion and thus experience a reduced gain. As a result, less energy is added to the trailing edge than to the leading edge of the pulse, and this leads to considerable pulse reshaping. The output pulse shape can be calculated from (8.30), and it is found that the pulse may either get broader or narrower (or even remain unchanged) due to this phenomenon, the outcome depending upon the shape of the input pulse.[4]

To conclude this section we will briefly examine two further examples of laser amplification, involving conditions different from those considered above. In the first case the duration τ_p of the pulse to be amplified is assumed to be much shorter than the lifetime of the lower laser level.† This is, for instance, the situation in a ruby amplifier in which the lower laser level is coincident with the ground level. This is also the situation in a neodymium amplifier when $\tau_p < 1$ ns. In both previous cases the amplifier behaves like a three-level system, and it can be readily shown that the previous equations still apply provided Γ_s is now given by

$$\Gamma_s = h\nu / 2\sigma \qquad (8.39a)$$

The second case we briefly consider is that of an amplifier in which both the upper and lower levels are made up of many sublevels which are strongly coupled. This applies, for instance, to CO_2 or HF amplifiers whose upper and lower (vibrational) states consist of many rotational levels (see, for example, Fig. 6.13). If the pulse duration is much longer than the time for relaxation between rotational levels, then the thermal equilibrium

---

†We will, however, assume $\tau_p \gg T_2$, where $T_2 = 1/\pi\Delta\nu_0$, this being a necessary condition for the validity of the rate-equation approximation (see Section 5.5).

population distribution will be maintained among these levels. The population N_J of a rotational level belonging to a given vibrational state can then be written as a fraction z of the total population N of the vibrational state (see Section 2.7), where z (the partition function) can be calculated according to Boltzmann statistics. We further assume that: (i) The pulse duration τ_p is much shorter than the relaxation time of the lower laser level (so that the system effectively behaves like a three-level system). (ii) The wavelength of the incoming light pulse corresponds to just one rotational–vibrational line. In this case all the previous results will again apply provided we take[6]

$$\Gamma_s = h\nu/2\sigma z \qquad (8.39b)$$

where σ is the stimulated emission cross section for the rotational–vibrational transition involved in the amplification process. A comparison of (8.39b) with (8.39a) then shows that we can define an effective cross section as $z\sigma$ [see also equations (2.142m) and (2.142n)]. When the pulse duration becomes comparable with the rotational relaxation time, the picture becomes much more involved, and calculations using the resulting equations generally require the use of computers.[6]

8.4 TRANSFORMATION IN FREQUENCY: SECOND-HARMONIC GENERATION AND PARAMETRIC OSCILLATION [7–9]

In classical linear optics one assumes that the induced dielectric polarization of a medium is linearly related to the applied electric field, i.e.,

$$\mathbf{P} = \varepsilon_0 \chi \mathbf{E} \qquad (8.40)$$

where χ is the dielectric susceptibility. With the high electric fields involved in laser beams the above linear relation is no longer a good approximation and further terms in which $\mathbf{P}$ is related to higher-order powers of $\mathbf{E}$ must also be considered. This nonlinear response can lead to an exchange of energy between e.m. waves at different frequencies.

In this section we will consider some of the effects produced by a nonlinear polarization term which is proportional to the square of the electric field. The two effects that we will consider are: (i) Second-Harmonic Generation (SHG) in which a laser beam at frequency ω is partially converted, in the nonlinear material, to a coherent beam at frequency 2ω (as first shown by Franken et al.[10]); (ii) Optical Parameter Oscillation (OPO) in which a laser beam at frequency ω_3 causes the simultaneous generation, in the nonlinear material, of two coherent beams at frequency ω_1 and ω_2 such that $\omega_1 + \omega_2 = \omega_3$ (as first shown by

Giordmaine and Miller[11]). With the high electric fields available in laser beams the conversion efficiency of both these processes can be very high (approaching 100% in SHG). These techniques are therefore currently being used to generate new coherent waves at different frequencies from that of the incoming wave.

8.4.1 Physical Picture

We will first introduce some ideas using the simplifying assumption that the induced nonlinear polarization P^{NL} is related to the electric field E of the e.m. wave by a scalar equation, i.e.,

$$P^{NL} = 2\varepsilon_0 dE^2 \qquad (8.41)$$

where d is a coefficient whose dimension is the inverse of an electric field.[†] The physical origin of (8.41) is due to the nonlinear deformation of the outer, loosely bound, electrons of an atom or atomic system when subjected to high electric fields. This is analogous to a breakdown of Hooke's law for an extended spring, i.e., the restoring force is no longer linearly dependent on the displacement from equilibrium. A comparison of (8.41) and (8.40) shows that the nonlinear polarization term becomes comparable to the linear one for an electric field $E \simeq \chi/d$. Since $\chi \simeq 1$, we see that $(1/d)$ must be that field strength at which the linear and nonlinear terms become comparable, i.e., at which a sizable nonlinear deformation of the outer electrons will occur. Thus $1/d$ is expected to be of the order of the electric field which an electronic charge produces at a distance corresponding to a typical atomic dimension a, i.e., $(1/d) \simeq e/4\pi\varepsilon_0 a^2$ [thus $(1/d) \sim 10^{11}$ V/m for $a \simeq 1$ Å]. We note that, for symmetry reasons, d must be zero for a centrosymmetric material (such as for a centrosymmetric crystal and usually for liquids and gases). For symmetry reasons, in fact, if we reverse the sign of E, the sign of the total polarization $P_T = P + P^{NL}$ must also reverse. Since, however, $P^{NL} \propto E^2$, this can only occur if $d = 0$. From now on we will therefore confine ourselves to a consideration of noncentrosymmetric materials. We will see that the simple equation (8.41) is in this case able to account for both SHG and OPO.

8.4.1.1 Second-Harmonic Generation

We consider a monochromatic plane wave of frequency ω propagating in the z direction through a nonlinear crystal. For a plane wave of uniform

---

[†]We use $2\varepsilon_0 dE^2$ rather than dE^2 (as often used in other textbooks) to make d conform to increasingly accepted practice.

intensity we can write the following expression for the electric field $E_\omega(z,t)$ of the wave:

$$E_\omega(z,t) = \tfrac{1}{2} \left\{ E(z,\omega) \exp\left[i(\omega t - k_\omega z) \right] + \text{c.c.} \right\} \qquad (8.42)$$

In the above expression c.c. means the complex conjugate of the other term appearing in the braces and

$$k_\omega = \frac{\omega}{c_\omega} = \frac{n_\omega \omega}{c_0} \qquad (8.43)$$

where c_ω is the light velocity in the crystal, n_ω is the refractive index at frequency ω, and c_0 is the velocity of light *in vacuo*. Substitution of (8.42) into (8.41) shows that P^{NL} contains a term[†] oscillating at frequency 2ω, namely

$$P_{2\omega}^{\text{NL}} = \frac{\varepsilon_0 d}{2} \left\{ E^2(z,\omega) \exp\left[i(2\omega t - 2k_\omega z) \right] + \text{c.c.} \right\} \qquad (8.44)$$

Equation (8.44) describes a polarization oscillating at frequency 2ω and whose spatial variation is in the form of a wave. This polarization wave will radiate at frequency 2ω. Thus it generates an e.m. wave at the second harmonic frequency 2ω [the analytical treatment, given later, involves substituting this polarization in the wave equation (8.65)], and this e.m. wave has the form

$$E_{2\omega}(z,t) = \tfrac{1}{2} \left\{ E(z,2\omega) \exp\left[i(2\omega t - k_{2\omega} z) \right] + \text{c.c.} \right\} \qquad (8.45)$$

where

$$k_{2\omega} = \frac{2\omega}{c_{2\omega}} = \frac{2n_{2\omega}\omega}{c_0} \qquad (8.46)$$

is the wavevector (magnitude) at frequency 2ω. The physical origin of SHG can thus be traced back to the fact that, as a result of the nonlinear relation (8.41), the e.m. wave at the fundamental frequency ω will beat with itself to produce a polarization at 2ω. A comparison of (8.44) with (8.45) reveals a very important condition which must be satisfied if this process is to occur efficiently, viz., that the phase velocity of the polarization wave ($v_P = 2\omega/2k_\omega$) be equal to that of the generated e.m. wave ($v_E = 2\omega/k_{2\omega}$). The condition can thus be written

$$k_{2\omega} = 2k_\omega \qquad (8.47)$$

In fact, if this condition is not satisfied, the phase of the polarization wave at some point a distance l into the crystal (i.e., where the phase is $2k_\omega l$) will

---

[†]The quantity P^{NL} also contains a term at frequency $\omega = 0$ which leads to development of a dc voltage across the crystal (optical rectification).

be different from that of the generated wave (phase $k_{2\omega}l$). This increasing phase difference with distance l means that the generated wave will not grow cumulatively with distance l since it is not being driven by a polarization with the appropriate phase. Condition (8.47) is therefore referred to as the *phase-matching* condition. Note that, according to (8.43) and (8.46), equation (8.47) reduces to

$$n_{2\omega} = n_\omega \qquad (8.48)$$

If the directions of E_ω and P^{NL} (and hence of $E_{2\omega}$) were indeed the same [as implied by (8.41)] it would not be possible to satisfy the condition (8.48) owing to the dispersion ($\Delta n = n_{2\omega} - n_\omega$) of the crystal. This would then set a severe limit to the crystal length l_c over which P^{NL} can give contributions which keep adding cumulatively to form the second harmonic wave. This length l_c (the coherence length) must in fact correspond to the distance over which the P wave and the $E_{2\omega}$ wave get out of phase with each other by an amount π, i.e., $k_{2\omega}l_c - 2k_\omega l_c = \pi$. From this, with the help of (8.43) and (8.46), we get

$$l_c = \frac{\lambda}{4\Delta n} \qquad (8.49)$$

where $\lambda = 2\pi c_0/\omega$ is the wavelength *in vacuo* of the fundamental wave. Taking, as an example, $\lambda \simeq 1$ μm and $\Delta n = 10^{-2}$, we get $l_c = 25$ μm. Note that, at this distance into the crystal, the contribution of the P wave to the $E_{2\omega}$ wave is 180° out of phase and the $E_{2\omega}$ wave thus begins to decrease rather than continuing to increase. In this case, with l_c having such a small value, only a very small fraction of the incident power can then be transformed into the second harmonic wave.

At this point it is worth pointing out another useful way of visualizing the SHG process, in terms of photons rather than fields. First we write the relation between the frequency of the fundamental (ω) and second-harmonic (ω_{SH}) wave, viz.,

$$\omega_{SH} = 2\omega \qquad (8.50)$$

If we now multiply both sides of (8.47) and (8.50) by $\hbar$, we get

$$\hbar\omega_{SH} = 2\hbar\omega \qquad (8.51a)$$

$$\hbar k_{2\omega} = 2\hbar k_\omega \qquad (8.51b)$$

For energy to be conserved in the SHG process, we must have $dI_{2\omega}/dz = -dI_\omega/dz$, where $I_{2\omega}$ and I_ω are the wave intensities at the two frequencies. With the help of (8.51a) we get $dF_{2\omega}/dz = -2dF_\omega/dz$, where $F_{2\omega}$ and F_ω are the photon fluxes of the two waves. From this last equation we can then say that, in the SHG process, whenever two photons at frequency ω

disappear, one photon at frequency 2ω is produced. Thus the relation (8.51a) can be regarded as a statement of conservation of photon energy. Recalling that a photon has a momentum $\hbar k$, then equation (8.51b) is seen to correspond to the requirement that photon momentum also be conserved in the process.

We now reconsider the phase-matching condition (8.48) to see how it can be satisfied in a suitable, optically anisotropic crystal.[12,13] To understand this we will first need to make a small digression to explain the propagation behavior of waves in an anisotropic crystal, and also how the simple nonlinear relation (8.41) should be generalized for anisotropic media.

In an anisotropic crystal it can be shown that, for a given direction of propagation, there are two different, linearly polarized plane waves which can propagate. Corresponding to these two different polarizations there are two different refractive indices. The difference of refraction index is referred to as birefringence. This behavior is usually described in terms of the so-called index ellipsoid which, for a uniaxial crystal, is an ellipsoid of revolution around the optic axis (the z axis of Fig. 8.6). The two allowed directions of polarization and their corresponding refractive indices are found as follows: Through the center of the ellipsoid we draw a line in the direction of beam propagation (line OP of Fig. 8.6) and a plane perpendicular to this line. The intersection of this plane with the ellipsoid is an ellipse. The two axes of this ellipse are parallel to the two directions of polarization, and the length of each semiaxis is equal to the refractive index for that direction of polarization. One of these directions is necessarily perpendicular to the optic axis, and the wave having this polarization

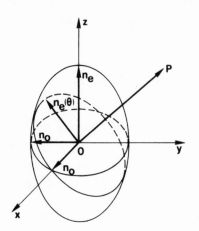

FIG. 8.6. Index ellipsoid for a positive uniaxial crystal.

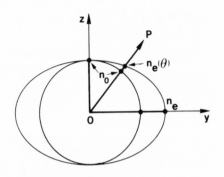

FIG. 8.7. Normal (index) surface for both the ordinary and extraordinary waves (for a positive uniaxial crystal).

direction is called the ordinary wave. Its refractive index n_0 can be seen from the figure to be independent of the direction of propagation. The wave with the other direction of polarization is called the extraordinary wave and the corresponding index $n_e(\theta)$ ranges in value from that of the ordinary wave n_0 (when OP is parallel to z) to a value n_e, called the extraordinary index (when OP is perpendicular to z). A positive uniaxial crystal corresponds to the case $n_e > n_o$, and a negative uniaxial crystal, to the case $n_e < n_0$. An equivalent way to describe wave propagation is through the so-called normal (index) surfaces for the ordinary and extraordinary waves (Fig. 8.7). In this case, for a given direction of propagation OP and for either ordinary or extraordinary waves the length of the ray OP (P being the point of interception with the surface) gives the refractive index of the wave. The normal surface for the ordinary wave is thus a sphere, while the normal surface for the extraordinary wave is an ellipsoid of revolution around the z axis. In Fig. 8.7 the intersections of these two normal surfaces with the y–z plane are indicated for the case of a positive uniaxial crystal.

After this brief discussion of wave propagation in anisotropic crystals, we now return to the problem of the induced nonlinear polarization. In general, in an anisotropic medium, the scalar relation (8.41) does not hold and a tensor relation needs to be introduced. First, we write the electric field $\mathbf{E}^\omega(r, t)$ of the e.m. wave at frequency ω and at a given point $\mathbf{r}$ and the nonlinear polarization vector at frequency 2ω, $\mathbf{P}^{2\omega}_{\mathrm{NL}}(r, t)$, in the form

$$\mathbf{E}^\omega(\mathbf{r}, t) = \tfrac{1}{2}\left[\mathbf{E}^\omega(\mathbf{r}, \omega)\exp\left(i\omega t\right) + \text{c.c.}\right] \tag{8.52a}$$

$$\mathbf{P}^{2\omega}_{\mathrm{NL}}(\mathbf{r}, t) = \tfrac{1}{2}\left[\mathbf{P}^{2\omega}(\mathbf{r}, 2\omega)\exp\left(2i\omega t\right) + \text{c.c.}\right] \tag{8.52b}$$

A tensor relation can then be established between $\mathbf{P}^{2\omega}(\mathbf{r}, 2\omega)$ and $\mathbf{E}^\omega(\mathbf{r}, \omega)$. The second harmonic polarization component along, say, the i direction of the crystal can be written as

$$P_i^{2\omega} = \sum_{j,\,k\,=\,1,\,2,\,3} \varepsilon_0 d_{ijk}^{2\omega} E_j^\omega E_k^\omega \tag{8.53}$$

Note that (8.53) is often written in condensed notation as

$$P_i^{2\omega} = \sum_1^6 {}_m \varepsilon_0 d_{im}^{2\omega} (EE)_m \tag{8.54}$$

where m runs from 1 to 6. The abbreviated field notation is that $(EE)_1$ $\equiv E_1^2 \equiv E_x^2$, $(EE)_2 \equiv E_2^2 \equiv E_y^2$, $(EE)_3 \equiv E_3^2 \equiv E_z^2$, $(EE)_4 \equiv 2E_2 E_3$ $\equiv 2E_y E_z$, $(EE)_5 \equiv 2E_1 E_3 \equiv 2E_x E_z$, and $(EE)_6 \equiv 2E_1 E_2 \equiv 2E_x E_y$, where both the 1, 2, 3 and the x, y, z notation for axes have been indicated. Note that, expressed in matrix form, d_{im} is a 3×6 matrix which operates on the column vector $(EE)_m$. Depending on the crystal symmetry, some of the values of the d_{im} matrix may be equal and some may be zero. For the $\overline{4}2m$ point group symmetry, which includes the important nonlinear crystals of the KDP type and the chalcopyrite semiconductors, only d_{14}, d_{25}, and d_{36} are nonzero and these three d coefficients are themselves equal. Thus only one coefficient, for example, d_{36}, needs to be specified, and we can write

$$P_x = 2\varepsilon_0 d_{36} E_y E_z \tag{8.55a}$$

$$P_y = 2\varepsilon_0 d_{36} E_z E_x \tag{8.55b}$$

$$P_z = 2\varepsilon_0 d_{36} E_x E_y \tag{8.55c}$$

where the z axis is again taken along the optic axis of the uniaxial crystal. The nonlinear optical coefficients, the symmetry class, and the transparency range of some selected nonlinear materials are indicated in Table 8.1.

Following this digression on the properties of anisotropic media, we can now go on to show how phase matching can be achieved for the particular case of a crystal of $\overline{4}2m$ point group symmetry. From (8.55) we note that, if $E_z = 0$, only P_z will be nonvanishing and will thus tend to generate a second-harmonic wave with a nonzero z component. We recall (see Fig. 8.6) that a wave with $E_z = 0$ is an ordinary wave while a wave with $E_z \neq 0$ is an extraordinary wave. Thus an ordinary wave at the fundamental frequency ω tends, in this case, to generate an extraordinary wave at 2ω. To satisfy the phase-matching condition one can then propagate the fundamental wave at an angle θ_m to the optic axis, in such a way that

$$n_e(2\omega, \theta_m) = n_0(\omega) \tag{8.56}$$

This can be better understood with the help of Fig. 8.8 which shows the intercepts of the normal surfaces $n_o(\omega)$ and $n_e(2\omega, \theta)$ with the plane containing the z axis and the propagation direction. Note that, due to dispersion

TABLE 8.1. Nonlinear Optical Coefficients for Selected Materials

Material	Symbol	Formula	Nonlinear d coefficient (relative to KDP)	Symmetry class	Transparence range (μm)
Potassium dihydrogen phosphate	KDP	KH_2PO_4	$d_{36} = d_{14} = 1$	$\bar{4}2m$	0.22–1.1
Potassium dideuterium phosphate	KD*P	KD_2PO_4	$d_{36} = d_{14} = 1.06$	$\bar{4}2m$	0.22–1.1
Ammonium dihydrogen phosphate	ADP	$NH_4H_2PO_4$	$d_{36} = d_{14} = 1.2$	$\bar{4}2m$	0.2–1.1
Cesium dihydrogen arsenate	CDA	CsH_2AsO_4	$d_{36} = d_{14} = 0.92$	$\bar{4}2m$	0.26–1.6
Lithium iodate	—	$LiIO_3$	$d_{31} = d_{32} = d_{24} = d_{15} = 14$	6	0.31–5.5
Cadmium germanium arsenide	—	$CdGeAs_3$	$d_{36} = d_{14} = 472$	$\bar{4}2m$	2–20
Lithium niobate		$LiNbO_3$	$d_{31} = 10.6$ $d_{22} = 5.1$	$3m$	0.35–4.5
Proustite		Ag_3AsS_3	$d_{31} = 30$ $d_{22} = 50$	$3m$	0.6–13

(normal dispersion), we have $n_o(\omega) < n_o(2\omega) = n_e(2\omega, 0)$. Thus the ordinary circle (for frequency ω) intersects the extraordinary ellipse (for frequency 2ω) at some angle θ_m.[†] For light propagating at this angle θ_m to the optic axis (i.e., for all ray directions lying in a cone around the z axis, with cone angle θ_m), equation (8.56) is satisfied and hence the phase-matching condition is satisfied. Note, however, that if $\theta \neq 90°$, the phenomenon of double refraction will occur, i.e., the direction of the energy flow for the extraordinary (SH) beam will be at an angle slightly different from θ_m. Thus the fundamental and SH beams will travel in slightly different directions (although satisfying the phase-matching condition). For a fundamental beam of finite transverse dimensions this will put an upper limit on the interaction length in the crystal. This limitation can be overcome if it is

[†]It should be noted that for this intersection to occur at all it is necessary for $n_e(2\omega, 90°)$ to be less than $n_o(\omega)$, otherwise the ellipse for $n_e(2\omega)$ (see Fig. 8.8) will lie wholly outside the circle for $n_o(\omega)$. Thus $n_e(2\omega, 90°) = n_e(2\omega) < n_o(\omega) < n_o(2\omega)$ which shows that crystal birefringence $n_o(2\omega) - n_e(2\omega)$ must be larger than crystal dispersion $n_o(2\omega) - n_o(\omega)$.

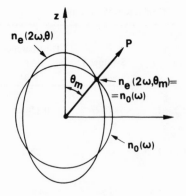

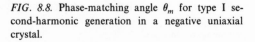

FIG. 8.8. Phase-matching angle θ_m for type I second-harmonic generation in a negative uniaxial
crystal.

possible to operate with $\theta_m = 90°$, i.e., $n_e(2\omega, 90°) = n_o(\omega)$. This kind of
phase matching is called 90° phase matching and in some cases can be
achieved by changing the crystal temperature, since n_e and n_o in general
undergo different changes with the temperature. To summarize the above
discussion, we can say that phase matching can be achieved in a
(sufficiently birefringent) negative uniaxial crystal when an ordinary ray at
ω [E_x beam of (8.55c)] combines with an ordinary ray at ω [E_y beam of
(8.55c)] to give an extraordinary ray at 2ω, or, in symbols, $o_\omega + o_\omega \rightarrow e_{2\omega}$.
This is called type I second-harmonic generation. In a negative uniaxial
crystal another scheme for phase-matched SHG, called type II, is also
possible. In this case an ordinary wave at ω can combine with an extraordinary wave at ω to give an extraordinary wave at 2ω, or, in symbols,
$o_\omega + e_\omega \rightarrow e_{2\omega}$.[†]

Second-harmonic generation is currently used to provide coherent
sources at new wavelengths. The nonlinear crystal may be placed either
outside or inside the cavity of the laser producing the fundamental beam.
In the latter case one takes advantage of the greater e.m. field strength
inside the resonator to increase the conversion efficiency. Very high conversion efficiencies (approaching 100%) have been obtained with both arrange-
ments. Among the most frequent applications of SHG are frequency
doubling the output of a Nd:YAG laser (thus producing a green beam,
$\lambda = 532$ nm, from an infrared one, $\lambda = 1.06$ μm) and generation of tunable
UV radiation (down to $\lambda = 210$ nm) by frequency doubling a tunable dye
laser. In both of these cases either cw or pulsed laser sources are used.

[†]More generally, interactions in which the polarizations of the two fundamental waves are the
same are termed type I (e.g., also $e_\omega + e_\omega \rightarrow o_{2\omega}$), and interactions in which the polarization of
the fundamental waves are orthogonal are termed type II.

The nonlinear crystals most commonly used for SHG belong to the $\overline{4}2m$ point group symmetry, in particular the materials KDP, KD*P, and CDA. For intracavity SHG, Lithium Iodate ($LiIO_3$) is also often used. Efficient frequency conversion of infrared radiation from CO_2 or CO lasers in chalcopyrite semiconductors (e.g., $CdGeAs_2$) is another interesting example.

8.4.1.2 Parametric Oscillation

We now go on to discuss the process of parametric oscillation. We begin by noticing that the previous ideas introduced in the context of SHG can be readily extended to the case of two incoming waves at frequencies ω_1 and ω_2 combining to give a wave at frequency $\omega_3 = \omega_1 + \omega_2$ (sum-frequency generation). Harmonic generation can in fact be thought of as a limiting case of sum-frequency generation with $\omega_1 = \omega_2 = \omega$ and $\omega_3 = 2\omega$. The physical picture is again very similar to the SHG case: By virtue of the nonlinear relation (8.41) between P^{NL} and the total field E [$E = E_{\omega_1}(z, t) + E_{\omega_2}(z, t)$], the wave at ω_1 will beat with that at ω_2 to give a polarization component at $\omega_3 = \omega_1 + \omega_2$. This will then radiate an e.m. wave at ω_3. Thus for sum-frequency generation we can write

$$\hbar\omega_1 + \hbar\omega_2 = \hbar\omega_3 \qquad (8.57a)$$

which, according to a description in terms of photons rather than fields, implies that one photon at ω_1 and one photon at ω_2 disappear while a photon at ω_3 is created. We therefore expect the photon momentum to be also conserved in the process, i.e.,

$$\hbar\mathbf{k}_1 + \hbar\mathbf{k}_2 = \hbar\mathbf{k}_3 \qquad (8.57b)$$

where the relationship is put in its general form, with the k denoted by vectors. Equation (8.57b), which expresses the phase-matching condition for sum-frequency generation, can be seen to be a straightforward generalization of that for SHG [compare with (8.51b)].

Optical parametric generation is in fact just the reverse of sum-frequency generation. Here a wave at frequency ω_3 (the pump frequency) generates two waves (called the idler and signal waves) at frequencies ω_1 and ω_2, in such a way that the total photon energy and momentum is conserved, i.e.,

$$\hbar\omega_3 = \hbar\omega_1 + \hbar\omega_2 \qquad (8.58a)$$

$$\hbar\mathbf{k}_3 = \hbar\mathbf{k}_1 + \hbar\mathbf{k}_2 \qquad (8.58b)$$

The physical process occurring in this case can be visualized in the following way. Imagine first that a strong wave at ω_3 and a weak wave at ω_1 are both present in the nonlinear crystal. As a result of the nonlinear relation (8.41), the wave at ω_3 will beat with the wave at ω_1 to give a polarization component at $\omega_3 - \omega_1 = \omega_2$. If the phase-matching condition (8.58b) is satisfied, a wave at ω_2 will thus build up as it travels through the crystal. Then the total E field will in fact be the sum of three fields $[E = E_{\omega_1}(z,t) + E_{\omega_2}(z,t) + E_{\omega_3}(z,t)]$ and the wave at ω_2 will in turn beat with the wave at ω_3 to give a polarization component at $\omega_3 - \omega_2 = \omega_1$. This polarization will cause the ω_1 wave to grow also. Thus power will be transferred from the beam at ω_3 to those at ω_1 and ω_2, and the weak wave at ω_1 which was assumed to be initially present will be amplified. From this picture we see a fundamental difference between parametric generation and SHG. In the latter case only a strong beam at the fundamental frequency is needed for the SHG process to occur. In the former case, however, a weak beam at ω_1 is also needed and the system behaves like an amplifier at frequency ω_1 (and ω_2). In practice, however, the weak beam need not be supplied by an external source (such as another laser) since it is generated internally to the crystal as a form of noise (so-called parametric noise). One can then generate coherent beams from this noise in a way analogous to that used in a laser oscillator. Thus, the nonlinear crystal, which is pumped by an appropriately focused pump beam, is placed in an optical resonator (Fig. 8.9). The two mirrors (1 and 2) of this parametric oscillator have high reflectivity (e.g., $R_1 = 1$ and $R_2 \simeq 1$) either at ω_1 only (singly resonant oscillator, SRO) or at both ω_1 and ω_2 (doubly resonant oscillator, DRO). The mirrors are ideally transparent to the pump beam. Oscillation will start when the gain arising from the parametric effect just exceeds the losses of the optical resonator. Some threshold power of the input pump beam is therefore required before oscillation will begin. When this threshold is reached, oscillation occurs at both ω_1 and ω_2, and the particular pair of values of ω_1 and ω_2 is determined by the two equations (8.58). For instance, with type I phase matching involving an extraordinary wave at ω_3 and

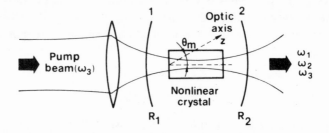

FIG. 8.9. Schematic diagram of an optical parametric oscillator.

ordinary waves at ω_1 and ω_2 (i.e., $e_{\omega_3} \rightarrow o_{\omega_1} + o_{\omega_2}$), (8.58b) would give

$$\omega_3 n_e(\omega_3, \theta) = \omega_1 n_o(\omega_1) + \omega_2 n_o(\omega_2) \tag{8.59}$$

For a given θ, i.e., for a given inclination of the nonlinear crystal with respect to the cavity axis, (8.59) provides a relation between ω_1 and ω_2 which, together with the relation (8.58a), determines the values of both ω_1 and ω_2. Phase-matching schemes of both type I and type II (e.g., $e_{\omega_3} \rightarrow o_{\omega_1} + e_{\omega_2}$ for a negative uniaxial crystal) are possible and tuning can be achieved by either changing the crystal inclination (angle tuning) or its temperature (temperature tuning). As a final comment, we note that, if the gain from the parametric effect is large enough, one can dispense with the mirrors altogether, and an intense emission at ω_1 and ω_2 grows from parametric noise in a single pass through the crystal. This behavior is superficially rather similar to the phenomena of superfluorescence and amplified spontaneous emission discussed in Section 2.3.4 and is sometimes (rather inappropriately) called superfluorescent parametric emission.

Singly resonant and doubly resonant optical parametric oscillators have both been used. Doubly resonant parametric oscillation has been achieved with both cw and pulsed pump lasers. For cw excitation, threshold powers as low as a few milliwatts have been demonstrated. However, the doubly resonant character of the resonator causes the output to be somewhat unstable both in amplitude and frequency. Singly resonant parametric oscillation has only been achieved using pulsed pump lasers since the threshold pump power for the singly resonant case is much higher (as much as about two orders of magnitude) than that of the doubly resonant case. However, singly resonant oscillators produce a much more stable output and impose less stringent demands on the mirror coating design. For these reasons the singly resonant configuration is the one most frequently used. Optical parametric oscillators producing coherent radiation from the visible to the near infrared (0.5–3.5 μm) are now well developed, with the most successful device based on a lithium niobate (LiNbO$_3$) crystal pumped by a Nd:YAG laser. They face competition, however, from color-center lasers, which operate in a similar range in the infrared. Optical parametric oscillators can also generate coherent radiation at longer infrared wavelengths (to $\sim$14 μm) using crystals such as proustite (Ag$_3$AsS$_3$) and cadmium selenide (CdSe).[†] The efficiency of an OPO can also be very high (approaching the theoretical 100% photon efficiency).

[†]Materials such as the chalcopyrite semiconductors have shown much promise, but unfortunately suffer from high loss and hence have not been operated as parametric oscillators. Nevertheless, these materials and many others have been widely used for difference frequency generation, i.e., where two beams, ω_3 and ω_1, are used to generate radiation at their difference frequency $\omega_2 = \omega_3 - \omega_1$.

8.4.2 Analytical Treatment

To arrive at an analytical description of both SHG and parametric processes, we need to see how the nonlinear polarization [e.g., (8.41)] which acts as the source term to drive the generated waves is introduced into the wave equation. The fields within the material obey Maxwell's equations:

$$\nabla \times \mathbf{E} = - \frac{\partial \mathbf{B}}{\partial t} \tag{8.60a}$$

$$\nabla \times \mathbf{H} = \mathbf{J} + \frac{\partial \mathbf{D}}{\partial t} \tag{8.60b}$$

$$\nabla \cdot \mathbf{D} = \rho \tag{8.60c}$$

$$\nabla \cdot \mathbf{B} = 0 \tag{8.60d}$$

where ρ is the free-charge density. For the media of interest here we can assume the magnetization $\mathbf{M}$ to be zero; thus

$$\mathbf{B} = \mu_0 \mathbf{H} + \mu_0 \mathbf{M} = \mu_0 \mathbf{H} \tag{8.61}$$

Losses within the material (e.g., scattering losses) can be simulated by the introduction of a fictitious conductivity σ_s such that

$$\mathbf{J} = \sigma_s \mathbf{E} \tag{8.62}$$

Finally we can write

$$\mathbf{D} = \varepsilon_0 \mathbf{E} + \mathbf{P}^L + \mathbf{P}^{NL} = \varepsilon \mathbf{E} + \mathbf{P}^{NL} \tag{8.63}$$

where $\mathbf{P}^L$ is the linear polarization of the medium and is taken account of, in the usual way, by introducing the dielectric constant ε. As we shall now see, when $\mathbf{D}$ given by (8.63) is substituted in Maxwell's equations, the nonlinear polarization term $\mathbf{P}^{NL}$ is introduced into the wave equation. Applying the $\nabla \times$ operator to both sides of (8.60a) [interchanging the order of $\nabla \times$ and $\partial / \partial t$ operators on the right-hand side of (8.60a)] and making use of (8.61), (8.60b), (8.62), and (8.63), we first obtain

$$\nabla \times \nabla \times \mathbf{E} = - \mu_0 \left(\sigma_s \frac{\partial \mathbf{E}}{\partial t} + \varepsilon \frac{\partial^2 \mathbf{E}}{\partial t^2} + \frac{\partial^2 \mathbf{P}^{NL}}{\partial t^2} \right) \tag{8.64}$$

Using the identity $\nabla \times \nabla \times \mathbf{E} = (\nabla \cdot \mathbf{E}) - \nabla^2 \mathbf{E}$ and making the assumption that $\nabla \cdot \mathbf{E} \simeq 0$, we find from (8.64) that

$$\nabla^2 \mathbf{E} - \frac{\sigma_s}{\varepsilon c^2} \frac{\partial \mathbf{E}}{\partial t} - \frac{1}{c^2} \frac{\partial^2 \mathbf{E}}{\partial t^2} = \frac{1}{\varepsilon c^2} \frac{\partial^2 \mathbf{P}^{NL}}{\partial t^2} \tag{8.65}$$

where $c = (\varepsilon \mu_0)^{-1/2}$ is the phase velocity in the material. Equation (8.65) is the wave equation with the nonlinear polarization term included. Note that the linear part of the medium polarization has been transferred to the left-hand side of (8.65) and is contained in the dielectric constant ε. The nonlinear part P^{NL} has been kept on the right-hand side, and it will be shown to act as a source term for the waves being generated at new frequencies as well as a loss term for the incoming wave. Confining ourselves to the simple scalar case of plane waves propagating along the z direction, we see that (8.65) reduces to

$$\frac{\partial^2 E}{\partial z^2} - \frac{\sigma_s}{\epsilon c^2} \frac{\partial E}{\partial t} - \frac{1}{c^2} \frac{\partial^2 E}{\partial t^2} = \frac{1}{\epsilon c^2} \frac{\partial^2 P^{NL}}{\partial t^2} \qquad (8.65a)$$

The field amplitude at frequency ω_j will be written as

$$E^{\omega_j}(z,t) = \tfrac{1}{2} \left\{ E_j(z) \exp \left[i(\omega_j t - k_j z) \right] + \text{c.c.} \right\} \qquad (8.66a)$$

where E_j is taken to be complex in general. Likewise, the amplitude of the nonlinear polarization at frequency ω_j will be written as

$$P^{NL}_{\omega_j} = \tfrac{1}{2} \left\{ P^{NL}_j(z) \exp \left[i(\omega_j t - k_j z) \right] + \text{c.c.} \right\} \qquad (8.66b)$$

Since (8.65a) must hold separately for each frequency corresponding to waves which are present in the crystal, equations (8.66a) and (8.66b) can be substituted into the left- and right-hand sides of (8.65a) respectively. Within the slowly varying amplitude approximation, we can neglect the second derivative of $E_j(z)$ (i.e., assume that $d^2E_j/dz^2 \ll k_j dE_j/dz$), and (8.65a) then yields

$$2 \frac{dE_j}{dz} + \frac{\sigma_j}{n_j \epsilon_0 c_0} E_j = -i \left(\frac{\omega_j}{n_j \epsilon_0 c_0} \right) P^{NL}_j \qquad (8.67)$$

where the relations $k_j = n_j \omega_j / c_0$ and $\epsilon_j = n_j^2 \epsilon_0$ have been used (c_0 is the light velocity *in vacuo* and n_j is the refractive index at ω_j).

Equation (8.67) is the basic equation that will be used in the next sections. Note that it has been obtained subject to the assumption of a scalar relation between P^{NL} and E [see (8.41)]. This assumption is not correct, and actually a tensor relation should be used [see (8.54)]. However, it can be shown that one can still use this scalar equation provided that E_j now refers to the field component along an appropriate axis and an effective coefficient, d_{eff}, is substituted for d in (8.41). In general, d_{eff} is a combination of one or several of the d_{im} coefficients appearing in (8.54) and of the angles θ and ϕ which define the direction of wave propagation in the crystal[14] (θ is the angle to the z axis and ϕ is the angle that the projection of the propagation vector in the $x-y$ plane makes with the x axis of the crystal). For example, for a crystal of $\overline{4}2m$ point group symmetry and for type I phase matching, one obtains $d_{eff} = d_{36} \sin 2\phi \sin \theta$. As a short-hand notation, however, we will still retain the symbol d in (8.41) while bearing in mind that it means the effective value of the d coefficient, d_{eff}.

8.4.2.1 Parametric Oscillation

We now consider three waves at frequencies ω_1, ω_2, and ω_3 [where $\omega_3 = \omega_1 + \omega_2$] interacting in the crystal. We thus write the total field $E(z,t)$ as

$$E(z,t) = E^{\omega_1}(z,t) + E^{\omega_2}(z,t) + E^{\omega_3}(z,t) \qquad (8.68)$$

where each of the fields can be written in the form of (8.66a). Upon substituting (8.68) into (8.41) and using (8.66a) we obtain an expression for the components $P_j^{\mathrm{NL}}(z)$ [as defined by (8.66b)] of the nonlinear polarization at the various frequencies ω_j. After some lengthy but straightforward algebra we find that, for instance, the component P_1^{NL} at frequency ω_1 is given by

$$P_1^{\mathrm{NL}} = 2\varepsilon_0 dE_3(z)E_2^*(z)\exp\left[i(k_1 + k_2 - k_3)z \right] \qquad (8.69)$$

The components of P^{NL} at ω_2 and ω_3 are obtained in a similar way. The field equations for each of the three frequencies are then obtained by substituting the appropriate P^{NL} into (8.67). We thus arrive at the following three equations:

$$\frac{dE_1}{dz} = -\left(\frac{\sigma_1}{2n_1\varepsilon_0 c_0} \right)E_1 - i\left(\frac{\omega_1}{n_1 c_0} \right)dE_3 E_2^* \exp\left[-i(k_3 - k_2 - k_1)z \right]$$

$$(8.70a)$$

$$\frac{dE_2}{dz} = -\left(\frac{\sigma_2}{2n_2\varepsilon_0 c_0} \right)E_2 - i\left(\frac{\omega_2}{n_2 c_0} \right)dE_3 E_1^* \exp\left[-i(k_3 - k_1 - k_2)z \right]$$

$$(8.70b)$$

$$\frac{dE_3}{dz} = -\left(\frac{\sigma_3}{2n_3\varepsilon_0 c_0} \right)E_3 - i\left(\frac{\omega_3}{n_3 c_0} \right)dE_1 E_2 \exp\left[-i(k_1 + k_2 - k_3)z \right]$$

$$(8.70c)$$

These are the basic equations describing the nonlinear parametric interaction. We note that they are coupled to each other via the nonlinear coefficient d.

It is convenient at this point to define new field variables A_j as

$$A_j = \left(n_j/\omega_j \right)^{1/2} E_j \qquad (8.71)$$

Since the intensity of the wave is $I_j = n_j\varepsilon_0 c_0 |E_j|^2/2$, the corresponding photon flux F_j is $F_j = I_j/\hbar\omega_j = (\varepsilon_0 c_0/2\hbar)|A_j|^2$. Thus $|A_j|^2$ is proportional to the photon flux at ω_j with the proportionality constant being independent of n_j and ω_j. When re-expressed in terms of these new field variables, equations (8.70) transform to

$$\frac{dA_1}{dz} = -\frac{\alpha_1 A_1}{2} - i\lambda A_3 A_2^* \exp\left[-i(\Delta kz)\right] \tag{8.72a}$$

$$\frac{dA_2}{dz} = -\frac{\alpha_2 A_2}{2} - i\lambda A_3 A_1^* \exp\left[-i(\Delta kz)\right] \tag{8.72b}$$

$$\frac{dA_3}{dz} = -\frac{\alpha_3 A_3}{2} - i\lambda A_1 A_2 \exp\left[i(\Delta kz)\right] \tag{8.72c}$$

where we have put $\alpha_j = \sigma_j/n_j\varepsilon_0 c_0$, $\Delta k = k_3 - k_2 - k_1$, and

$$\lambda = \frac{d}{c_0}\left[\frac{\omega_1\omega_2\omega_3}{n_1 n_2 n_3}\right]^{1/2} \tag{8.73}$$

The advantage of using A_j instead of E_j is now apparent since, unlike (8.70), relations (8.72) now involve a single coupling parameter λ.

Neglecting the losses (i.e., putting $\alpha_j = 0$), multiplying both sides of (8.72a) by A_1^* and both sides of (8.72b) by A_2^*, and comparing the resulting expressions, we arrive at the following relation: $d|A_1|^2/dz = d|A_2|^2/dz$. Similarly from (8.72b) and (8.72c) we get $d|A_2|^2/dz = -d|A_3|^2/dz$. We can therefore write

$$\frac{d|A_1|^2}{dz} = \frac{d|A_2|^2}{dz} = -\frac{d|A_3|^2}{dz} \tag{8.74}$$

which are known as the Manley–Rowe relations. Since $|A_j|^2$ is proportional to the corresponding photon flux, (8.74) implies that whenever a photon at ω_3 is destroyed, a photon at ω_1 and a photon at ω_2 are created. This is consistent with the photon model for the parametric process, as discussed in Section 8.4.1.2. Note that (8.74) means, for instance, that $(dP_1/dz) = -(\omega_1/\omega_3)(dP_3/dz)$, where P_1 and P_3 are the powers of the two waves. Thus only the fraction (ω_1/ω_3) of the power at frequency ω_3 can be converted into that at frequency ω_1.

Strictly speaking, equations (8.72) apply to a traveling wave situation in which an arbitrarily long crystal is being traversed by the three waves at ω_1, ω_2, ω_3. We now want to see how these equations might be applied to the case of an optical parametric oscillator as in Fig. 8.9. Here we will first consider the DRO scheme. The waves at ω_1 and ω_2 will therefore travel back and forth within the cavity, and the parametric process will only occur when their propagation direction is the same as that of the pump wave (since it is only under these circumstances that phase matching can be satisfied). If we unfold the optical path, it will look like that of Fig. 8.10a, and it can be seen that loss occurs on every pass while parametric gain

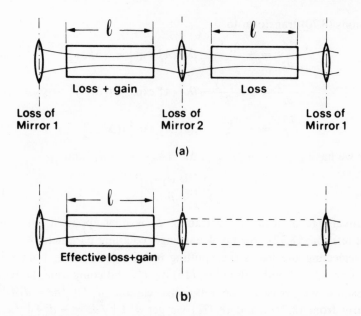

FIG. 8.10. (a) Unfolded path of an optical parametric oscillator; (b) Reduction to a single-pass scheme with mirror losses incorporated into the distributed losses of the crystal.

occurs only once in every two passes. This situation can be reduced to that of Fig. 8.10b if we choose an appropriate definition of the effective loss coefficient α_j ($j = 1, 2$). The loss due to a crystal of length l in Fig. 8.10b must in fact equal the losses incurred in a double pass in Fig. 8.10a. The latter losses must account for the actual losses in the crystal, as well as the mirror and diffraction losses. Thus the coefficients α_1 and α_2 in (8.72) must be appropriately defined so as to incorporate these various losses. From (8.72), neglecting the parametric interaction [i.e., setting $\lambda = 0$], we see that after traversing the length l of the crystal, the power at ω_j ($j = 1, 2$) is reduced to a fraction $\exp(-\alpha_j l)$ of its power at the entrance face of the crystal. This reduction must account for the round-trip cavity losses, which requires that

$$\exp(-\alpha_j l) = R_{1j} R_{2j} (1 - T)^2 \qquad (8.74a)$$

where R_{1j} and R_{2j} are the two mirror reflectivities and T is the crystal loss (plus diffraction loss) per pass at ω_j. If we now define [compare with (5.4)] $\gamma_{1j} = -\ln R_{1j}$, $\gamma_{2j} = -\ln R_{2j}$, $\gamma'_j = -\ln(1 - T)$, and $\gamma_j = [(\gamma_{1j} + \gamma_{2j})/2] + \gamma'_j$, we can rewrite (8.74a) as

$$\alpha_j l = 2\gamma_j \qquad (8.75)$$

where γ_j is the overall cavity loss per pass. Note that this amounts to simulating the mirror losses by losses distributed through the crystal and then including them in the effective crystal absorption coefficient α_j ($j = 1$, 2). The loss α_3, on the other hand, only involves crystal losses and can in general be neglected. Thus at this point we can say that, for a DRO, equations (8.72) will still apply provided that α_1 and α_2 are given by (8.75). To obtain the threshold condition of a DRO, equations (8.72) can be further simplified if we neglect depletion of the pump wave by the parametric process. This assumption together with the assumption $\alpha_3 = 0$ means that we can take $A_3(z) \simeq A_3(0)$, where $A_3(0)$, the field amplitude of the incoming pump wave, is taken to be real. With the further assumption of $\Delta k = 0$ (perfect phase matching), (8.72) is considerably simplified and becomes

$$\frac{dA_1}{dz} = -\frac{\alpha_1 A_1}{2} - i\frac{g}{2} A_2^* \tag{8.76a}$$

$$\frac{dA_2}{dz} = -\frac{\alpha_2 A_2}{2} - i\frac{g}{2} A_1^* \tag{8.76b}$$

where

$$g = 2\lambda A_3(0) = 2d\frac{E_3(0)}{c_0}\left(\frac{\omega_1 \omega_2}{n_1 n_2}\right)^{1/2} \tag{8.77}$$

The threshold condition for a DRO is then readily obtained from (8.76) by putting $dA_1/dz = dA_2/dz = 0$. This leads to

$$\alpha_1 A_1 + igA_2^* = 0 \tag{8.78a}$$

$$igA_1 - \alpha_2 A_2^* = 0 \tag{8.78b}$$

where the complex conjugate of (8.76b) has been taken. The solution of this homogeneous system of equations will yield nonzero values for A_1 and A_2 only if

$$g^2 = \alpha_1 \alpha_2 = 4\frac{\gamma_1 \gamma_2}{l^2} \tag{8.79}$$

where (8.75) has been used. According to (8.77), g^2 is proportional to $E_3^2(0)$, i.e., to the intensity of the pump wave. Thus condition (8.79) means that a certain threshold intensity of the pump wave is needed in order for parametric oscillation to start. This intensity is proportional to the product of the single-pass (power) losses, γ_1 and γ_2, of the two waves at ω_1 and ω_2, and inversely proportional to d^2 and l^2.

     The SRO case is somewhat more involved. If the laser cavity is resonant only at ω_1, then α_1 can again be written as in (8.75). Since the wave at ω_2 is no longer reflected back through the cavity, α_2 will involve only the crystal losses and it can therefore be neglected. Again, neglecting depletion of the pump wave and assuming perfect phase matching, (8.76) will still be applicable provided we now set $\alpha_2 = 0$. For small parametric conversion we can put $A_1^*(z) \simeq A_1^*(0)$ on the right-hand side of (8.76*b*). We thus get

$$A_2(z) = -igA_1^*(0)z/2 \tag{8.80}$$

where the condition $A_2(0) = 0$ has been assumed (i.e., no field at ω_2 is fed back into the crystal by the resonator). If we substitute (8.80) in (8.76*a*) and put $A_1(z) \simeq A_1(0)$ in the right side of (8.76*a*), we get

$$\frac{dA_1}{dz} = \left[-\frac{\alpha_1}{2} + \frac{g^2 z}{4} \right] A_1(0) \tag{8.81}$$

Integration of (8.81) gives

$$A_1(l) = A_1(0) \left[1 - \frac{\alpha_1 l}{2} + \frac{g^2 l^2}{8} \right] \tag{8.82}$$

for the field at ω_1 after traversing the length l of the crystal. The threshold condition is reached when $A_1(l) = A_1(0)$, i.e., when

$$g^2 = \frac{4\alpha_1}{l} = \frac{8\gamma_1}{l^2} \tag{8.83}$$

Since g^2 is proportional to the intensity I of the pump wave, a comparison of (8.83) with (8.79) gives the ratio of threshold pump intensities as

$$\frac{I_{\text{SRO}}}{I_{\text{DRO}}} = \frac{2}{\gamma_2} \tag{8.84}$$

If, for example, we take a loss per pass of $\gamma_2 = 2\%$, we find from (8.84) that the threshold power for SRO is 100 times larger than that for DRO.

8.4.2.2 Second-Harmonic Generation

In the case of SHG we take

$$E(z, t) = \tfrac{1}{2} \left\{ E_\omega \exp \left[i(\omega t - k_\omega z) \right] + E_{2\omega} \exp \left[i(2\omega t - k_{2\omega} z) \right] + \text{c.c.} \right\} \tag{8.85}$$

$$P^{\text{NL}}(z, t) = \tfrac{1}{2} \left\{ P_\omega^{\text{NL}} \exp \left[i(\omega t - k_\omega z) \right] + P_{2\omega}^{\text{NL}} \exp \left[i(2\omega t - k_{2\omega} z) \right] + \text{c.c.} \right\} \tag{8.86}$$

Substituting (8.85) and (8.86) into (8.41) gives

$$P_{2\omega}^{NL} = \varepsilon_0 dE_\omega^2 \exp\left[-i(2k_\omega - k_{2\omega})z\right] \tag{8.87a}$$

$$P_\omega^{NL} = 2\varepsilon_0 dE_{2\omega} E_\omega^* \exp\left[-i(k_{2\omega} - 2k_\omega)z\right] \tag{8.87b}$$

Then, substituting (8.87) into (8.67) and neglecting crystal loses (i.e., putting $\sigma_j = 0$), we get

$$\frac{dE_{2\omega}}{dz} = -i\frac{\omega}{n_{2\omega}c_0} dE_\omega^2 \exp(i\Delta kz) \tag{8.88a}$$

$$\frac{dE_\omega}{dz} = -i\frac{\omega}{n_\omega c_0} dE_{2\omega} E_\omega^* \exp(-i\Delta kz) \tag{8.88b}$$

where $\Delta k = k_{2\omega} - 2k_\omega$. These are the basic equations describing SHG. To solve them, it is first convenient to define new field variables E_ω' and $E_{2\omega}'$ such that

$$E_\omega' = (n_\omega)^{1/2} E_\omega \tag{8.89a}$$

$$E_{2\omega}' = (n_{2\omega})^{1/2} E_{2\omega} \tag{8.89b}$$

Since the intensity I_ω of the wave at ω is proportional to $n_\omega |E_\omega|^2$, the quantity $|E_\omega'|^2$ is also proportional to I_ω but with the proportionality constant independent of refractive index. Substituting (8.89) into (8.88) gives

$$\frac{dE_{2\omega}'}{dz} = -\frac{i}{l_{SH}} \frac{E_\omega'^2}{E_\omega'(0)} \exp\left[i(\Delta kz)\right] \tag{8.90a}$$

$$\frac{dE_\omega'}{dz} = -\frac{i}{l_{SH}} \frac{E_{2\omega}' E_\omega'^*}{E_\omega'(0)} \exp\left[-i(\Delta kz)\right] \tag{8.90b}$$

where $E_\omega'(0)$ is the value of E_ω' at $z = 0$ and l_{SH} is a characteristic length for the second-harmonic interaction, given by

$$l_{SH} = \frac{\lambda_0 (n_\omega n_{2\omega})^{1/2}}{2\pi dE_\omega(0)} \tag{8.91}$$

where λ_0 is the wavelength and $E_\omega(0)$ the incident field amplitude of the fundamental wave at frequency ω. Note again that the advantage of using the new field variables E_ω' and $E_{2\omega}'$ is apparent from equations (8.90) since they involve a single coupling parameter l_{SH}. Note also that $E_\omega(0)$ and hence $E_\omega'(0)$ have been taken to be real. From (8.90) we find

$$\frac{d|E_{2\omega}'|^2}{dz} = -\frac{d|E_\omega'|^2}{dz} \tag{8.92}$$

According to this relation (Manley–Rowe relation), it is possible in this case to have 100% conversion of fundamental-beam power into second-harmonic power.

As a first example of the solution of (8.90), we consider the case where there is an appreciable phase mismatch (by which we mean $l_{SH}\Delta k \gg 1$) so that little conversion of fundamental into SH is expected to occur. We therefore put $E'_\omega(z) = E'_\omega(0)$ on the right-hand side of (8.90a). The resulting equation can then be readily integrated [with the boundary condition $E'_{2\omega}(0) = 0$] to give

$$E'_{2\omega}(l) = -\frac{E'_\omega(0)}{l_{SH}} \left\{ \frac{\exp(-i\Delta kl) - 1}{\Delta k} \right\} \qquad (8.93)$$

from which we get

$$\left| \frac{E'_{2\omega}(l)}{E'_\omega(0)} \right|^2 = \frac{\sin^2(\Delta kl/2)}{(\Delta kl_{SH}/2)^2} \qquad (8.94)$$

Since $|E'_{2\omega}|^2$ is proportional to the SH intensity $I_{2\omega}$, the variation of this intensity with crystal length is readily obtained from (8.94). According to (8.92) the behavior of I_ω must then be such that $I_\omega + I_{2\omega} = I_\omega(0)$. The dependence of $[I_\omega/I_\omega(0)]$ and $[I_{2\omega}/I_\omega(0)]$ on (l/l_{SH}) for $l_{SH}\Delta k = 10$ have both been plotted as dashed curves in Fig. 8.11. Note that, due to the large phase mismatch, only little conversion to second harmonic occurs. It can readily be shown from (8.94) that the first maximum of $[I_{2\omega}/I_\omega(0)]$ occurs at $l = l_c$, where l_c (the coherence length) is given by (8.49).

As a second example of the solution to (8.90), we consider the case of perfect phase matching ($\Delta k = 0$). In this case appreciable conversion to second harmonic may occur and the depletion of the fundamental beam must therefore be considered. This means that (8.90) must now be solved without the approximation that $E'_\omega(z) = E'_\omega(0)$. If $\Delta k = 0$, however, it can be shown from (8.90) that $E'_{2\omega}$ and E'_ω are imaginary and real respectively. We can therefore write

$$E'_\omega = |E'_\omega| \qquad (8.95a)$$

$$E'_{2\omega} = -i|E'_{2\omega}| \qquad (8.95b)$$

and (8.90) then gives

$$\frac{d|E'_\omega|}{dz} = -\frac{1}{l_{SH}} \frac{|E'_{2\omega}||E'_\omega|}{E'_\omega(0)} \qquad (8.96a)$$

$$\frac{d|E'_{2\omega}|}{dz} = \frac{1}{l_{SH}} \frac{|E'_\omega|^2}{E'_\omega(0)} \qquad (8.96b)$$

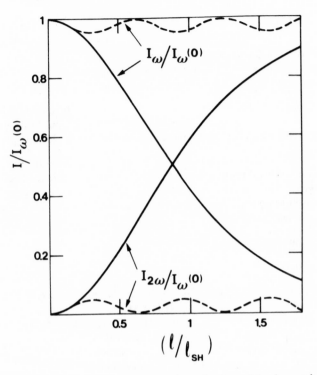

FIG. 8.11. Normalized plots of second-harmonic intensity $I_{2\omega}$ and fundamental intensity I_ω versus crystal length l for perfect phase-matching (continuous curves) and for a finite phase mismatch (dashed curves).

The solution of (8.96) with the boundary conditions $E'_\omega(l = 0) = E'_\omega(0)$ and $E'_{2\omega}(0) = 0$ is

$$|E'_{2\omega}| = E'_\omega(0) \tanh(z/l_{SH}) \tag{8.97a}$$

$$|E'_\omega| = E'_\omega(0) \operatorname{sech}(z/l_{SH}) \tag{8.97b}$$

Since the intensity of the wave is proportional to $|E'|^2$, we have $I_{2\omega}/I_\omega(0)$ $= |E'_{2\omega}|^2/E'^2_\omega(0)$ and $I_\omega/I_\omega(0) = |E'_\omega|^2/E'^2_\omega(0)$. The dependence of $I_{2\omega}/I_\omega(0)$ and $I_\omega/I_\omega(0)$ on crystal length as predicted by (8.97) is shown by the solid curves in Fig. 8.11. Note that, that for $l = l_{SH}$, an appreciable fraction ($\sim 59\%$) of the incident wave has been converted into SH. This illustrates the role of l_{SH} as a characteristic length for the second-harmonic interaction, with a value which is inversely proportional to the square root of the fundamental-beam intensity [see (8.91)]. Note also that for $l \gg l_{SH}$ the fundamental radiation can be completely converted into second-harmonic radiation, in agreement with the Manley–Rowe relation (8.92).

PROBLEMS

8.1. A Gaussian beam emitted by a visible He–Ne laser has a spot size (at the beam waist) of $w_0 = 0.5$ mm. Calculate the beam spot size and the radius of curvature of the equiphase surface at a distance of 10 m from the beam waist.

8.2. The Gaussian beam of the previous problem is to be focused to a beam waist of spot size 50 μm at a distance of 1 m from the original beam waist. What focal length should the lens have and where should the lens be placed?

8.3. A laser has a hemifocal resonator of length 50 cm. We want to place a lens after the spherical (output) mirror of the resonator to reduce the output beam divergence. If we require the spot size at the beam waist formed after the lens to be 0.95 times the beam spot size on the spherical mirror, what focal length must the lens have?

8.4. Prove equations (8.4).

8.5. Prove equation (8.10).

8.6. The output of a Q-switched Nd:YAG laser ($E = 100$ mJ, $\tau_p = 20$ ns) is to be amplified by a 6.3-mm-diameter Nd/YAG amplifier with a small signal gain $G_0 = 100$. Assuming a peak cross section of the laser transition of $\sigma \simeq 3.5 \times 10^{-19}$ cm^2, calculate the energy of the beam after the amplifier and hence the energy amplification. Also calculate the fraction of the stored energy in the amplifier which is extracted by the incident pulse.

8.7. A large Nd:glass amplifier to be used for fusion experiments uses a rod of 9 cm diameter and 15 cm length. The small signal gain of such an amplifier is 4. Taking the peak cross section of Nd:glass as $\sigma = 3 \times 10^{-20}$ cm^2, calculate the required input pulse energy (1-ns pulse) to generate an output of 450 J. What is the total energy stored in the amplifier?

8.8. A large CO_2 TEA amplifier (with a gas mixture $CO_2:N_2:He$ in the proportion $3:1.4:1$) has dimensions of $10 \times 10 \times 100$ cm. The small signal gain coefficient for the $P(20)$ transition is $g_0 = 4 \times 10^{-2}$ cm^{-1}. The duration of the input light pulse is 200 ns, which can therefore be taken to be much longer than the thermalization time of the rotational levels and much shorter than the decay time of the lower laser level. The peak cross section for the $P(20)$ transition under these conditions can be taken to be $\sigma = 1.54 \times 10^{-18}$ cm^2 and the partition function is $z = 0.07$ ($T = 300°$K). Calculate the output energy and gain available from this amplifier for an input energy of 17 J. Also calculate the energy per unit volume stored in the amplifier.

8.9. Prove that equation (8.39a) holds for a three-level system.

8.10. Prove equation (8.32).

8.11. Show that (8.56) gives $\sin^2 \theta_m = [(n_2^o/n_1^o)^2 - 1]/[(n_2^o/n_2^e)^2 - 1]$, where n_2^o and n_2^e are the ordinary and extraordinary refractive indices at 2ω and where n_1^o is the ordinary refractive index at ω.

8.12. The frequency of a Nd:YAG laser output ($\lambda = 1.06$ μm) is to be doubled in a KDP crystal. Knowing that, for KDP, $n_o(\lambda = 1.06$ μm) $\equiv n_1^o = 1.507$, $n_o(\lambda = 0.532$ μm) $\equiv n_2^o = 1.5283$, and $n_e(\lambda = 0.532$ μm) $\equiv n_2^e = 1.48222$, calculate the phase-matching angle θ_m.

8.13. Prove (8.69).

8.14. From (8.77) and (8.79) show that the threshold intensity of the pump wave for a DRO is $I = (n_3/2Zd^2)[n_1 n_2 \lambda_1 \lambda_2 / (2\pi l)^2]\gamma_1\gamma_2$, where $Z = 1/\varepsilon_0 c_0 = 377$ Ω is the free-space impedance and λ_1 and λ_2 are the wavelengths of the signal and idler waves.

8.15. Using the result of the previous problem calculate the threshold pump intensity for parametric oscillation at $\lambda_1 \simeq \lambda_2 = 1$ μm in a 5-cm-long LiNbO$_3$ crystal pumped at $\lambda_3 \simeq 0.5$ μm [$n_1 = n_2 = 2.16, n_3 = 2.24, d \simeq 6 \times 10^{-12}$ m/V, $\gamma_1 = \gamma_2 = 2 \times 10^{-2}$]. If the beam is focused in the crystal to a spot of $\sim$100 μm diameter, calculate the resulting threshold pump power.

8.16. Calculate the second-harmonic conversion efficiency for type I harmonic generation in a perfectly phase-matched 2.5-cm-long KDP crystal with an incident beam at $\lambda = 1.06$ μm having an intensity of 100 MW/cm^2 [for KDP $n \simeq 1.5$, $d_{eff} = d_{36} \sin\theta_m = 0.28 \times 10^{-12}$ m/V, where $\theta_m \simeq 50°$ is the phase-matching angle].

REFERENCES

1. (a) N. Bloembergen, *Nonlinear Optics* (Benjamin, New York, 1965); (b) S. A. Akhmanov and R. V. Khokhlov, *Problems of Nonlinear Optics* (Gordon and Breach, New York, 1972).
2. O. Svelto, in *Progress in Optics*, Vol. XII, ed. by Emil Wolf (North-Holland, Amsterdam, 1974), pp. 3–50.
3. M. Born and E. Wolf, *Principles of Optics*, 4th ed. (Pergamon Press, London, 1970) Chapter X.
4. P. G. Kriukov and V. S. Letokhov, in *Laser Handbook*, ed. by F. T. Arecchi and E. O. Schulz-Dubois (North-Holland, Amsterdam, 1972), Vol. 1, pp. 561–595.
5. W. Koechner, *Solid-State Laser Engineering* (Springer-Verlag, New York, 1976), Chapter 4.
6. O. Judd, in *High-Power Gas Lasers*, ed. by E. R. Pike (The Institute of Physics, Bristol and London, 1976), pp. 45–57.
7. A. Yariv, *Introduction to Optical Electronics* (Holt, Rinehart and Winston, Inc., New York, 1971), Chapter 8.
8. S. A. Akhmanov et al. in *Quantum Electronics*, ed. by H. Rabin and C. L. Tang (Academic Press, New York, 1975), Vol. 1, Part B, pp. 476–583.
9. R. L. Byer, in *Quantum Electronics*, ed. by H. Rabin and C. L. Tang (Academic Press, New York, 1975), Vol. 1, Part B, pp. 588–694.
10. P. A. Franken et al., *Phys. Rev. Lett.* **7**, 118 (1961).
11. J. A. Giordmaine and R. C. Miller, *Phys. Rev. Lett.* **14**, 973 (1965).
12. J. A. Giordmaine, *Phys. Rev. Lett.* **8**, 19 (1962).
13. P. D. Maker et al., *Phys. Rev. Lett.* **8**, (1962).
14. F. Zernike and J. E. Midwinter, *Applied Nonlinear Optics* (John Wiley and Sons, New York, 1973), Section 3.7.
15. M. Lax, W. H. Louisell, and W. B. McKnight, *Phys. Rev. A* **11**, 1365 (1975).

9
Applications of Lasers

9.1 INTRODUCTION

The applications of lasers are already very numerous and cover various scientific and technological fields including physics, chemistry, biology, electronics, and medicine. In general, these applications are a direct consequence of those special characteristics of laser light discussed in Chapter 7. In this chapter, we will limit ourselves to a discussion of the principles underlying some of these applications while referring elsewhere for a more detailed description of each particular application. The applications will be classified as follows: (1) applications in physics and chemistry, (2) applications in biology and medicine, (3) materials processing, (4) optical communications, (5) measurement and inspection, (6) thermonuclear fusion, (7) optical information processing and recording, (8) military uses, and (9) holography.

9.2 APPLICATIONS IN PHYSICS AND CHEMISTRY

The invention of the laser and its subsequent development have depended on fundamental knowledge drawn from the fields of physics and (to a lesser extent) chemistry. It was therefore natural that applications of the laser to physics and chemistry should have been among the first to be considered.

In physics the laser has initiated quite new fields of investigation and has dramatically stimulated some already existing fields. It should also be recognized that the study of laser behavior and the interaction of laser beams with matter themselves constitute new areas of study within the field of physics. A particularly interesting example of a new area of investigation is that of nonlinear optics.[1,2] The high intensity of a laser beam makes it possible to observe new phenomena arising from the nonlinear response of matter. We mention in particular the following processes: (i) harmonic generation (already discussed in Chapter 8) whereby suitable materials, when excited by a laser beam at frequency ν, can produce a new coherent beam at frequency 2ν (second harmonic), 3ν (third harmonic), etc.; (ii) stimulated scattering.[3] In this case the incident laser beam at frequency ν interacts with a material excitation at frequency ν_q (e.g., an acoustic wave) to produce a coherent beam at frequency $\nu - \nu_q$ (Stokes scattering). The energy difference between the incident photon, $h\nu$, and the scattered photon, $h(\nu - \nu_q)$, is given to the material excitation. Particularly important examples of stimulated scattering phenomena are stimulated Raman scattering (which, in its most frequently encountered form, involves a material excitation consisting of an internal vibration of each molecule of the material) and stimulated Brillouin scattering (wherein the material excitation consists of an acoustic wave). Both of these processes can occur with high conversion efficiency (often several tens of percent). For this reason both harmonic generation and stimulated scattering (in particular Raman scattering since it can involve a large frequency shift) are used in practice to generate intense coherent beams at new frequencies.

An area in both physics and chemistry where the laser has dramatically improved previously existing possibilities is that of making time-resolved measurements of the behavior of various media after excitation by short light pulses. In fact, while it is possible with conventional light sources to produce light pulses down to ~ 1 ns, lasers are now able to produce pulses down to ~ 0.1 ps. This has opened up possibilities for investigating a wide variety of phenomena, based on the new capability of ultrashort time-resolved measurements. Since many important processes in physics, chemistry, and biology[4] have time scales in the picosecond range, this is an exciting new development.

Yet another field where the laser has not only improved previously existing possibilities but also introduced quite new concepts is that of spectroscopy.[5,6] It is now possible with some lasers to narrow the oscillation bandwidth down to a few tens of kilohertz (both in the visible and infrared), and this allows spectroscopic measurements to be made with a resolving power many orders of magnitude (from 3 to 6) higher than that obtainable by conventional spectroscopy. The laser has also spawned the new field of nonlinear spectroscopy[6] which allows spectroscopic resolution to be extended well beyond the limit normally imposed by Doppler broadening effects. This has led to new and more detailed studies of the structure of matter.

In the field of chemistry, lasers are used both for diagnostic purposes and for producing an irreversible chemical change (laser photochemistry). In the area of diagnostic techniques, particular mention should be made of "resonant Raman scattering" and "coherent antistokes Raman scattering" (CARS).[7,8] With these techniques it is possible to obtain considerable information on the structure and properties of polyatomic molecules (e.g., frequency of Raman active vibrations, rotational constants, frequency anharmonicity). The CARS technique can also be used to measure the concentration (and temperature) of a molecular species in a given limited spatial region. This capability is being exploited in detailed studies of flame combustion processes and (electrical discharge) plasmas.

Perhaps the most interesting chemical application for lasers (at least potentially so) is the field of photochemistry. It should be borne in mind, however, that, owing to the high cost of laser photons, the commercial exploitation of laser photochemistry can only be justified when the value of the end product is very high. Such a case is that of isotope separation[9] (in particular, for uranium and deuterium). The basic idea here is to selectively excite by a laser beam only the wanted isotopic species. This species, once in the excited state, is easily distinguishable and hence separable (perhaps by chemical means) from the unwanted species left in the ground state. In the case of uranium, for example, two approaches have been followed: (i) Photo-ionization of the wanted species (^{235}U) by light of appropriate wavelength once this species has already been radiatively pumped to some excited state. The ionized species is then collected by applying a suitable dc electric field. In this approach the uranium material is in the form of atomic vapor. (ii) Selective dissociation of a molecular compound of uranium (such as uranium hexafluoride) so that the wanted species (in this case $^{235}UF_6$), having first been selectively pumped to an excited (vibrational) level, is subsequently dissociated as a result of further optical pumping. In this case the uranium hexafluoride is used in the form of a molecular jet at low temperatures ($T < 50°K$).

9.3 APPLICATIONS IN BIOLOGY AND MEDICINE

Lasers are being increasingly used in biological and medical applications. Here again the laser can be used either as a diagnostic tool or to produce an irreversible change of the biomolecule, of the cell or of the tissue (laser photobiology and laser surgery).

In biology, the main use of lasers is that of a diagnostic tool. We mention here the following laser techniques[10-12]: (i) fluorescence induced by ultrashort laser pulses in DNA, in dye–DNA complexes, and in dyes involved in photosynthesis; (ii) resonant Raman scattering as a means of studying biomolecules such as hemoglobin or rhodopsin (the latter being responsible for the mechanism of vision); (iii) photon correlation spectroscopy to obtain information about the structure and the degree of aggregation of various biomolecules; (iv) picosecond flash-photolysis techniques to probe the dynamic behavior of biomolecules in the excited state. Particular mention should be made of the so-called flow microfluorometers.[13] Here, mammalian cells in suspension are made to pass through a special flow chamber where they are aligned and then pass, one at a time, through the focused beam of an Ar^+ laser. By suitably placed photodetectors, one can thus measure: (i) the light that is scattered from the cell (which gives information on its size); (ii) the fluorescence from a dye which is bound to a specific cell constituent, e.g., the DNA (this gives information on the amount of that constituent). The advantage of flow microfluorometry is that it offers the possibility of performing measurements on a large number of cells in a limited time (the cell flow rate is typically 5×10^4 cells/min). This implies a good statistical measurement precision.

In the biological field lasers are also used to produce an irreversible change of a given biomolecule or cellular constituent. In particular we mention the so-called microbeam techniques.[14] Here the laser light (e.g., a pulsed Ar^+ laser) is focused by a suitable microscope objective into some region of the cell with a diameter approximately equal to the laser wavelength (~ 0.5 μm). The principal purpose of the technique is the study of cell function following the laser-induced damage in a particular region.

In the field of medicine, the predominant use of lasers is for surgery (laser surgery).[15-17] A few diagnostic applications have also been developed, however (e.g., clinical use of flow microfluorometers, Doppler velocimetry to measure the blood velocity, laser fluorescence bronchoscope to detect lung tumors in their early phase).

For surgery, the focused laser beam (often a CO_2 laser) is used in place of a conventional (or electric) scalpel. The infrared beam from the CO_2 laser is strongly absorbed by the water molecules in the tissue and produces a rapid evaporation of these molecules and a consequent cutting of the tissue. The principal advantages of a laser beam scalpel can be summarized as follows: (i) The incision can be made with high precision particularly when the beam is directed by means of a suitable microscope (laser microsurgery). (ii) Possibility of operating in inaccessible regions. Thus in practice any region of the body which can be observed by means of a suitable optical system (e.g., lenses or mirrors) can be operated on by a laser. (iii) Drastic reduction in hematic losses due to the cauterising action of the laser beam on the blood vessels (up to a vessel diameter of $\sim$0.5 mm). (iv) Limited damage to the adjacent tissue (a few tens of micrometers). These advantages must, however, be balanced against the following disadvantages: (i) considerable cost and complexity of a laser surgical unit; (ii) smaller velocity of the laser scalpel; (iii) reliability and safety problems associated with the laser scalpel.

Having made these general comments on laser surgery, we now go on to provide some more details concerning a few of these applications. In ophthalmology the laser has already been in use for several years to treat the detachment of the retina and to cure diabetic retinopathies.[18] In this case the laser beam (usually Ar^+) is focused on the retina through the lens of the eye. The green beam of the laser is strongly absorbed by the red blood cells and the consequent thermal effect can lead to re-attachment of the retina or coagulation of its vessels. The laser is now finding increasing use in otolaryngology. Use of the laser is particularly attractive in this branch of surgery since it is concerned with organs such as the trachea, pharynx, and the middle ear whose inaccessibility make them difficult to operate on. Often, in this case, the laser is used in conjunction with a microscope. The laser has also been found useful for surgery within the mouth (e.g., for removal of papillomas or tumors). The principal advantages here are hemostasis, the absence of post-operative edema and pain, and the more rapid recovery of the patient. The laser has also proved its usefulness in treating cases of heavily bleeding lesions in the gastrointestinal tract. In this case the laser beam (usually Nd:YAG or Ar^+) is directed to the region to be treated by means of a special optical fiber inserted in a conventional gastroscope. The laser also appears promising in gynecology where it is mostly used in conjunction with a microscope (the colposcope). The considerable reduction of edema and pain is again a noteworthy feature of the laser scalpel. In dermatology lasers are often used for the removal of tattoos and for treating some vascular diseases (e.g., port wine stains). Finally the use of lasers in certain cases of general surgery[19] and tumor surgery[20] appears to be promising.

9.4 *MATERIAL WORKING* [21–23]

The high intensity available in the focal spot of a high-power ($P >$ 100 W) laser beam offers a number of applications in the field of material working and processing, such as welding, cutting, drilling, surface treatment, and alloying. These applications are, at present, the most important industrial applications of lasers. The principal advantages of a laser beam can be summarized as follows: (i) The heating produced by the laser in a given process is usually less than that involved in the corresponding conventional process. It follows that the material distortion is considerably reduced, and also better control can be exercised over the process. (ii) Possibility of working in inaccessible regions. In practice, any region which can be seen (even though it may require some optical system to permit observation) can be processed by a laser. (iii) High velocity of the region being worked and hence high production rates. For instance, welding velocities as high as 10 m/min can be obtained, i.e., about one order of magnitude greater than that obtainable with the best arc welding systems. As another example, the velocity for surface treatment by a laser is, often, greater than that obtainable by induction heating. (iv) Easy automation of the process. For instance, the movement of the laser beam can be achieved by moving the optical focusing system (and this movement can be computer controlled). This also offers the possibility of precision cutting along complicated profiles. (v) Possibilities for producing new metallurgical processes. With the high intensity and easy control of a laser beam, a number of processes which were previously impracticable are now feasible. For instance, due to the high annealing velocity of a laser beam, one can get new types of surface alloys or induce recrystallization of an amorphous semiconductor surface (laser annealing of semiconductors). (vi) Freedom from wear in the case of a laser tool. This is particularly relevant for laser cutting. Against all these advantages, however, one must balance the following disadvantages: (i) high capital and running cost of the laser system; (ii) problems of reliability and reproducibility of the laser beam; (iii) safety problems.

Following this general discussion, we now go on to give a brief description of a few relevant examples. The lasers most widely used for metalworking are either CO_2 lasers (with powers which, depending on the type of laser, can vary from ~ 100 W up to ~ 15 kW) or Nd:YAG lasers (with powers usually ranging from 50 to 500 W). The CO_2 laser is equally suited to working on metallic or nonmetallic (plastic, ceramic, glass) materials. However, the Nd:YAG laser beam is only appreciably absorbed by metallic materials. The more limited power available from the Nd:YAG laser means that it tends to find use in material-working applications requiring high precision and minimum thermal damage (e.g., welding of relays and microelectronic components, drilling of hard materials such as diamond or sapphire). The most interesting applications for a medium-power CO_2 laser (100 W to 1 kW) are perhaps in the area of processing nonmetallic materials. Examples are the cutting and drilling of ceramic material for microelectronics, the cutting of plastic, cloth, wood, leather, etc. Particular mention should also be made of the application of such a laser in the field of electronics, such as for resistor trimming. Some of the most important applications of high-power CO_2 lasers (1–15 kW) are perhaps in the automotive industry, e.g., welding of gears, heat treatment of cylinders, and surface alloying of the base of valves.

9.5 OPTICAL COMMUNICATIONS [24]

The possibility of using a laser beam for communications through the atmosphere at first aroused quite a lot of enthusiasm since lasers could in principle offer two important advantages: (i) The first arises from the availability of a large oscillating bandwidth, since the amount of information that can be transmitted on a given carrier wave is proportional to its bandwidth. In going from the microwave region to the optical region the carrier frequency increases by about 10^4, thus allowing the possibility of a much larger oscillating bandwidth. (ii) The second arises from the short wavelength of the radiation. Since a typical laser wavelength is about 10^4 times smaller than a typical microwave wavelength, it is seen from (1.11) that, for the same aperture size D, the divergence is 10^4 times smaller for optical waves compared to microwaves. To achieve the same divergence, the antenna of an optical system (mirror or lens) could therefore be much smaller. These two important advantages are, however, nullified by the fact that, under conditions of poor visibility, light beams are strongly attenuated in the atmosphere. The use of lasers for free-space (unguided) communications has therefore only been developed for two particular (although important) cases: (i) Space communications between two satellites or between a satellite and a ground station that is located in a region of particularly favorable climatic conditions. The lasers used in this case are either Nd:YAG (with data rates up to 10^9 bit/s) or CO_2 (with data rates up to 3×10^8 bit/s). The CO_2 laser, although having higher efficiency, requires a more complex detection system and has the further disadvantage that its wavelength is ~10 times larger than that of Nd:YAG. (ii) Point-to-point communications over short distances, e.g., data transmission within the same building. In this case semiconductor lasers are the most attractive sources.

The main area of interest in optical communications, however, relies on transmission through optical fibers. Guided propagation of light along fibers is a phenomenon which has been known for many years. However, these early optical fibers were used over very short distances, a typical application being in medical instruments for endoscopy. Thus around the end of the 1960's the attenuation of the best optical glasses was ~1000 dB/km. Since then, technological development of both glass and quartz fibers has dramatically improved this figure down to attenuations smaller than 0.5 dB/km for quartz (the lower limit being determined by Rayleigh scattering in the fiber medium). These very low attenuations have now established an important future for the use of optical fibers in long-distance communications.

A typical optical fiber transmission system consists of a light source, a suitable optical coupler to inject the light into the fiber, and, at the end of the fiber, a receiver (a photodiode) again coupled to the fiber. Repeaters are placed along the transmission path, with a spacing which may range from 2 to 50 km. The repeaters consist of a receiver and a new emitter. The light source is often a double-heterostructure semiconductor laser. The life of these lasers has recently been improved to $\sim 10^6$ h. Although GaAs devices have often been used so far, a more attractive approach is to use heterojunction lasers in which the active layer is composed of the quaternary alloy $In_{1-x}Ga_xAs_yP_{1-y}$. In this case, the p and n sides of the junctions consist of the simple binary compound InP and, by using the composition with $y = 2.2x$, it can be arranged that the quaternary alloy is lattice matched to InP. By choosing x appropriately, the emission wavelength can be adjusted to fall either at 1.3 μm or near 1.6 μm, corresponding to two absorption minima in a quartz fiber. Depending on the diameter d of its central core, the fiber may either be of the single-mode type ($d \simeq 1$ μm) or multimode type ($d \simeq 50$ μm). For the transmission rates employed to date (~50 Mbit/s), multimode fibers are commonly used. For higher transmission rates, single-mode fibers appear to be more suitable. The receiver is usually an avalanche photodiode, although it is also possible to use a PIN diode followed by a suitable solid-state amplifier.

In concluding this section, it is interesting to note that the application of optical fibers to telecommunications is not confined to long-distance (and high-cost) systems. For data transmission over shorter distance (e.g., within the same building or on board an airplane or ship) use is made of an incoherent light-emitting diode coupled to a multimode fiber.

9.6 MEASUREMENT AND INSPECTION [25]

The properties of directionality, brightness, and monochromaticity make lasers very useful for a variety of measurement and inspection techniques in industrial process control of machine tools and in civil engineering. We also include in this section geodetic measurements and pollution monitoring.

One of the most common industrial applications of lasers is for alignment purposes. The laser's directionality makes it ideal for establishing a straight reference line for aligning machinery for aircraft construction, and for civil engineering uses such as constructing buildings, bridges, or tunnels. In this field the laser has very largely replaced the optical instruments used previously, such as collimators or telescopes. One generally uses a visible He–Ne laser of low power (1–5 mW), and alignment is often achieved with the help of solid-state detectors in the form of quadrants (Fig. 9.1). The position of the laser beam on the receiver is determined by the value of photocurrent from each quadrant. Alignment is then dependent on simple electrical measurements, thus avoiding the reliance on subjective judgment by the operator (which has often been a source of error in the earlier optical systems). The alignment precision obtainable in practice in a workshop ranges from ∼5 µm (at a distance of ∼5 m) to ∼25 µm (at a distance of ∼15 m).

Lasers are also used for distance measurements. The technique used depends on the magnitude of the length to be measured. For short distances (up to ∼50 m) interferometric techniques are used with a frequency-stabilized He–Ne laser as the light source. For intermediate distances (up to ∼1 km), telemetry techniques involving amplitude modulation of He–Ne or GaAs lasers are used. For greater distances one measures the time of flight of a short light pulse (perhaps a few tens of nanoseconds) emitted by a Nd:YAG or CO_2 laser and reflected by the object.

The interferometric measurements of distance usually rely on the use of a Michelson interferometer (Fig. 7.3). The laser beam is divided by a beam splitter into a measurement beam and a reference beam. The reference beam is reflected by a fixed mirror while the measurement beam is reflected by a mirror fixed to the object being measured. The two reflected beams are then recombined so as to interfere, and their combined amplitude is measured by a detector. When the position of the object is changed

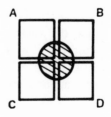

FIG. 9.1. Use of a quadrant detector for alignment. The beam (hatched area) is perfectly centered when the photocurrent signals from the four detectors *A*, *B*, *C*, and *D* are equal.

along the beam direction by $\lambda/2$, where λ is the laser wavelength, the interference signal will go from a maximum through a minimum and back to a maximum again. A suitable electronic fringe counting system can therefore give information on the displacement of the object. This measurement technique is usually employed in high-precision machine shops[26] and allows length measurements to be made with an accuracy of one part per million. It should be noted that this technique only measures distances relative to a given initial position. The advantage of the technique arises from the fact that it is fast, accurate, and is compatible with automatic control systems.

For greater distances, amplitude-modulation telemetry is used. Here the laser beam (He–Ne or GaAs) is amplitude modulated, and the length is determined from the phase difference between the emitted and return beam. The accuracy is again one part per million (i.e., 1 mm over a 1 km distance). This technique is used in geodesy and cartography. For distances greater than 1 km, the distance is determined from the time of flight of a short laser pulse (10–50 ns) emitted by Q-switched ruby or Nd:YAG lasers or by a TEA CO_2 laser. These applications are mostly of military interest (laser rangefinders) and will be discussed in a separate section. Among the nonmilitary applications, however, mention should be made of the measurement of the distance between the earth and the moon (with a precision of $\sim$20 cm) and the ranging of satellites.

The high degree of monochromaticity makes it possible to use lasers for velocity measurements, for both liquids and solids, by a technique known as Doppler velocimetry.[27] In the case of a flowing liquid, one illuminates the liquid using a laser beam, and the scattered light is then detected. Since the liquid is moving, the frequency of the scattered light is shifted, due to the Doppler effect, to a value slightly different from that of the incident light. This frequency shift is proportional to the velocity. Thus, by observing on a detector the beat signal between the scattered and incident light, one can measure the liquid velocity. The same technique can obviously be applied to moving solid objects. It has the advantage of being a noncontact measurement and, due to the high laser monochromaticity, of being very accurate over a wide velocity range (from a few centimeters per second to several hundred meters per second).

A particular velocity measurement that can be made by a laser is that of angular velocity. The instrument designed to do this is called the laser gyroscope,[28] and it consists of a laser whose resonant cavity has a ring configuration using three mirrors instead of the usual two (Fig. 9.2). This laser supports oscillation both for light propagating clockwise and anti-

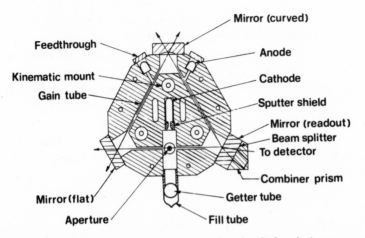

Mirror (curved)

Feedthrough

Anode

Kinematic mount

Cathode

Gain tube

Sputter shield

Mirror (readout)

Beam splitter

To detector

Combiner prism

Mirror (flat)

Getter tube

Aperture

Fill tube

FIG. 9.2. Ring laser using a quartz block construction (hatched region) to measure the component of angular velocity perpendicular to the plane of the figure. To measure the angular velocity components along three axes, one may use three ring lasers constructed within the same quartz block or three similar devices orthogonally mounted. (After Arono-witz.[28])

clockwise around the ring. The resonant frequencies for these two propagation directions are obtained from the condition that the length of the (ring) resonator be equal to an integral number of wavelengths. If the ring is rotating, then in the time needed for the light to complete a round trip, the resonator mirrors will move by a very small, but finite, angle. The beam that is rotating in the same direction as that of the resonator will see an effective resonator length slightly greater than that seen by the counterrotating beam. The frequencies of the two oppositely rotating beams are thus slightly different, and their difference frequency is proportional to the angular velocity of the resonator. By beating the two beams together, one can measure the angular velocity. A laser gyro allows this measurement to be made with a precision ($\sim 3 \times 10^{-3}$ deg/h) that is comparable with that available using the most sophisticated (and expensive) conventional gyroscopes.

Another field in which the laser's properties of directionality and monochromaticity are put to good use is that of ambient measurement of the concentration of various atmospheric pollutants. Conventional techniques require collection of the sample (which is not always easy) and subsequent chemical analysis. These techniques cannot therefore give real-time data and some of them do not lend themselves readily to automatic measurements in fixed locations. The laser technique consists of sending the light of an appropriate laser through the atmospheric region of interest and then detecting the backscattered light after collection by a telescope. The complete system is called an optical radar or LIDAR (from *li*ght *d*etection *a*nd *r*anging). The interaction of the laser light with the atmospheric pollutants involves several phenomena, namely, elastic scattering, Raman scattering, fluorescence, and absorption. Each of these phenomena has been used to detect and measure the concentration of a large variety of atmospheric pollutants (smoke, SO_2, NO_2, NO, etc.). The lasers used include ruby, frequency-doubled Nd:YAG, dye lasers, and various medium-infrared lasers (HF, DF, CO_2).

9.7 THERMONUCLEAR FUSION [29]

The peaceful harnessing of thermonuclear fusion would offer mankind an unlimited supply of energy. The central problem of energy production by thermonuclear fusion is the production and then confinement of a plasma containing heavy isotopes of hydrogen (deuterium, D, and tritium, T). The plasma must have a sufficiently high temperature ($T \simeq 10^8\,°K$) for the highly exothermic fusion reaction to take place ($D + T \rightarrow {}^4He + n$, where n indicates a neutron). There are two basic approaches to this problem of plasma production and confinement: (i) Magnetic confinement, this being the traditional approach, by which a plasma of relatively low density is confined for a relatively long time within a special "magnetic bottle." (ii) inertial confinement, which as far as peaceful uses of fusion are concerned, is the more recent approach. In this case the (D, T) plasma is rapidly heated while remaining confined by its own inertia. This means that, within the short time before the plasma becomes dispersed by the hydrodynamic modes, a sufficient amount of thermonuclear reaction takes place to release a large amount of energy (explosive reaction).

For the inertial confinement approach, lasers are considered one of the best means of providing the very fast heating up of the plasma. The liquid deuterium and tritium are contained in microspheres a fraction of a millimeter in diameter and are irradiated by a laser of sufficiently high energy and short duration as to induce the D + T reaction (microexplosion). From complex calculations, it has been shown that the efficient generation of thermonuclear energy requires not only the heating of the pellet to a very high temperature but also its compression to a density $\sim 10^4$ times greater than that of the liquid phase. This requires a spherically symmetric heating of the fuel microsphere, and a specially tailored time behavior of the laser pulse. Upon irradiation of the microsphere the following sequence of events takes place. First a low-density plasma is generated around the sphere by ablation of its outer material using a laser pre-pulse. This atmosphere is then irradiated in a more or less symmetric way by a second and more intense laser pulse. The consequent absorption of energy in the external atmosphere of the pellet leads to a strong material ablation due to surface evaporation which produces a violent implosion of the microsphere. At the time of maximum compression, the most intense part of the laser pulse should arrive so as to produce a significant fusion reaction.

A laser fusion reactor, in its simplest form, would consist of a combustion chamber at the center of which the microspheres would be injected at a rate of several per second and each one being induced by irradiation with a high-energy laser pulse to give an energetic microexplosion. During the fusion reaction the energy is released in the form of kinetic energy of ~ 14 MeV neutrons, charged particles, and e.m. radiation. To produce electrical energy, the fusion energy would be absorbed by a liquid lithium blanket surrounding the combustion chamber, and this would then be circulated to heat exchangers to produce steam which would be used in the conventional way. A laser suitable for producing these microexplosions should have a wavelength somewhere in the range from 250 to 2000 nm, energy per pulse of $1-3 \times 10^6$ J, pulse duration of $5-10 \times 10^{-9}$ s, and hence a peak power > 200 TW. It should have a repetition rate of a few hertz and hence an average power of ~ 10 MW. The laser should also have an efficiency $\sim 1\%$ and would therefore require an electrical power input of ~ 1 GW.

Despite the tremendous technical and engineering difficulties associated with such a problem, large laser fusion projects have been under development in several countries, but most notably, in the USA and USSR. All of these lasers use the *m*aster *o*scillator *p*ower *a*mplifier (MOPA) scheme. The laser pulse has its origin in a low-power oscillator, where its spatial and temporal characteristics can be accurately controlled. This light is then divided by beam splitters into several beams, each being then amplified. The amplified beams are then directed at the microsphere in a radially symmetric configuration. In the USA, lasers giving pulses with an energy of ~ 10 kJ and peak power of 20–30 TW have already been constructed. These involve Nd:glass and CO_2 laser systems with the number of beams being 6 for the CO_2 laser (Helios system) and 20 for the Nd glass (the Argus system). Systems giving an order of magnitude more energy and power are also under construction (the Nova laser, based on Nd:glass amplifiers, and the Antares system based on CO_2 laser amplifiers).

In the USSR a Nd:glass system with 256 beams (Dolphin system) has been constructed.

Some of these lasers have already been used to initiate thermonuclear reactions in several types of pellets. Large numbers of neutrons have been observed, this being indicative of the occurrence of the thermonuclear reaction, and compression of the pellet has also been observed. However, the results are still far from the so-called point of scientific breakeven, which is defined as the situation where the thermonuclear energy released is equal to the incoming laser energy. It is unanimously agreed that the construction of the nuclear reactor, if indeed it does prove feasible at all, will not be accomplished during this century.

9.8 INFORMATION PROCESSING AND RECORDING

In this section we describe some laser applications to the reading and writing of information (either in coded or analogue form). Some of these applications fall within the area of high technology, such as optical memories for computers, while others are more consumer oriented, such as point-of-sale scanners, and videodiscs.

The point-of-sale scanners read coded information shown on goods to be sold in supermarkets. The characters which identify the product are usually written in a ten bit code in the form of vertical bars of different width and spacing. The scanner consists of a He–Ne laser whose light is focused on the bar code. A photodetector measures the light reflected from dark and white regions of the bar code. Thus the scanner automatically reads the code and transmits the data to a computer to which the cash registers are linked. The computer recognizes the product, provides a printout of its name and price, and also updates the inventory. The use of these scanners has resulted in a considerable reduction of routine cash register operations.

Another consumer-oriented application, or rather, an actual consumer product, is the videodisc (and the audiodisc). A videodisc carries a recorded video program which can then be played back and displayed on a conventional TV set. The optical videodiscs are written by the manufacturer using special scribing equipment and then read by a laser (presently He–Ne in the domestic videodisc reader). A common way of recording involves cutting (along concentric tracks or a spiral track) pits of varying length and spacing. The pits have a depth of $\lambda/4$, where λ is the wavelength of the laser used for the reading process. During reading the laser beam is focused down so that it falls on one track only (the track width is somewhat smaller than the laser spot size, and the pitch approximately equals the spot size). When a pit falls within the laser spot the reflection is reduced due to the destructive interference between the light reflected from the sides and that reflected from the bottom of the slot. Conversely, the absence of a slot results in a strong reflection. In this way the TV information can be recorded digitally. The principle of the audiodisc is similar. The principal advantages of both laser video and audiodiscs are that a high density of recorded information can be achieved and that the reading process does not require contact with the substrate (hence the absence of wear). The main problem which has so far limited the sale of these systems is the high price, which in part is due to the price of the He–Ne laser. This problem may be overcome in the future using semiconductor rather than He–Ne lasers. A dramatic reduction in price would be expected for systems using semiconductor lasers.

Another application of lasers to the writing and reading of coded information is that of optical memories for computers.[30] The interest in such memories is again based on the high information density which can potentially be achieved. Laser techniques should in fact be able to provide an information density of $\sim 10^7$ bits/cm^2 since the beam can be focused to a spot size of the order of the wavelength. The writing technique consists of drilling small holes in an opaque material or in some way changing the transmission or reflection properties of a given substrate by using a laser of sufficient power (usually Ar^+ or He–Ne). This recorded information is then read by a low-power laser, or by the same laser at reduced power. The substrate is often made of photosensitive or heat-sensitive material and may even be a photographic film. However, none of these substrates can be erased. Erasable memories have been developed, based on the thermomagnetic, ferroelectric, or photochromic effects. Optical memories using the technique of holography (holographic memories) have also been developed. In conclusion, however, while the technical feasibility of optical memories has already been demonstrated, their economical viability appears to be, as yet, rather uncertain.

The last application to be discussed in this section is that of laser graphics. The technique involves focusing the beam, by a suitable scanning system, onto a page of photosensitive material while the laser intensity is amplitude modulated synchronously with the scanning, so as to produce drawings or lines of text. The signal to the modulator can be generated by a computer (as in the case of computerized nonimpact printing systems) or received as an electrical signal transmitted from a remote station (as in the case of facsimile transmission). In the latter case the signals can be generated by a suitable reading system, again using a laser. The reading equipment, in the remote station, consists of a (low-power) laser whose focused beam is scanned over the page being read. A photodetector monitors the varying amounts of light scattered by the dark and light regions on the page and transforms it into an electrical signal. Laser facsimile systems are now widely used in several newspaper companies for transmission of newspaper pages.

9.9 MILITARY APPLICATIONS

Military applications of lasers have always accounted for the largest share of the laser market as a whole. The most important applications at present are for: (i) laser rangefinders, (ii) laser target designators, and (iii) directed-energy weapons.

The laser rangefinder[31] is based on the same principle as that involved in a conventional radar. A short-duration laser pulse (usually of 10–20 ns duration) is aimed at the target and the backscattered pulse is monitored by a suitable optical receiver including a photodetector. The range is found by measuring the time of flight of this laser pulse. The main advantages of a laser rangefinder can be summarized as follows: (i) weight, cost, and complexity considerably less than for a conventional radar; (ii) ability to make range measurements even when the target is flying just above the land or sea surface. The main disadvantage is that the laser beam is strongly attenuated by the atmosphere under low-visibility conditions. Several types of laser rangefinders, with ranges up to ~15 km, are currently in use: (i) hand-held devices for infantry use (one of the latest models developed in the USA is of pocket size and weighs only 500 g, including the batteries); (ii) systems for use on board of tanks; (iii) anti-aircraft ranging systems. The first lasers used for these ranging applications were Q-switched ruby lasers. Laser rangefinders are now based mostly on Q-switched neodymium lasers (often Nd:YAG), although TEA CO_2 lasers provide an attractive alternative to neodymium lasers in some cases (e.g., tank rangefinders).

A second military application for lasers is as target designators. The principle is very simple: A laser, placed in a strategic position, illuminates the target. Due to the high laser brightness, the target, when viewed through a narrow band optical filter, will appear as a bright spot. The weapon (which may be a bomb, a rocket, or other explosive device) is equipped with an appropriate sensing system. In its simplest form this could be a lens to image the target onto a quadrant photodetector which controls the steering system of the weapon and thus guides it onto the target. In this way very high firing accuracy is obtainable (a laser pointing accuracy of ~ 1 m from a distance of 10 km appears feasible). The laser is usually Nd:YAG, while CO_2 lasers appear unsuitable owing to the complexity of the photodetectors required (which involve cryogenic temperatures). Target designation may be performed from an airplane, helicopter, or from the ground (e.g., using hand-held target designators).

Considerable effort is now being devoted, both in the USA and USSR, to develop lasers for use as directed-energy weapons. In this case large laser systems are envisaged, with a power perhaps in the megawatt range (for at least a few tens of seconds). An optical system directs the laser beam at the target (airplane, satellite, rocket) with the aim of causing irreversible damage to its sensing equipment or of causing such a damage to its surface that a failure eventually results from flight stresses. Ground-located laser stations appear less promising, owing to the so-called thermal blooming which would occur in the atmosphere. The atmosphere is heated by the beam (due to absorption), and this results in a negative lens which defocuses the beam. This problem can be avoided by having the laser on board an airplane flying at high altitude or in a satellite. Military secrecy about these developments has ensured that the available information on this subject is sparse and fragmentary.[32] It seems, however, that beams with cw power of 5–10 MW (for a few seconds) with pointing optics of 5–10 m diameter may be involved. The most promising lasers for this application are the chemical lasers (HF or DF). Chemical lasers are particularly interesting for airborne laser stations since the required energy can be stored compactly in the form of chemical energy of suitable reactants.

9.10 HOLOGRAPHY[33–35]

Holography is a revolutionary technique which allows three-dimensional (i.e., complete) pictures to be taken of a given object or scene. The word is derived from the Greek words "holos" (complete) and "graphos" (writing). Invented by Gabor in 1948 (it was proposed then as a means of improving the resolution of electron microscopes), the technique became a practical proposition and really demonstrated its potential only after the invention of the laser.

The principle of holography is shown in Fig. 9.3. A laser beam (the laser is not shown in the figure) is divided by the partially transmitting mirror S into two beams, A (reflected) and B (transmitted). Beam A falls

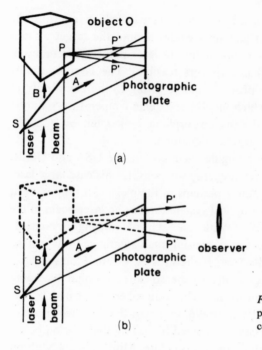

(a)

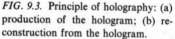

(b)

FIG. 9.3. Principle of holography: (a) production of the hologram; (b) reconstruction from the hologram.

directly onto a photographic plate, while beam *B* illuminates the object to be "holographed." Accordingly, part of the light scattered by the object will also fall onto the photographic plate, as indicated by the rays *P'* of Fig. 9.3a. As a result of the beam coherence, an interference pattern (which, in general, is very complex) will be formed on the plate due to the superposition of the two beams (beam *A*, often called the reference beam, and the beam scattered by the object). If the film is then developed and examined under sufficiently high magnification, these interference fringes can be observed (a typical distance between two successive dark fringes is about 1 μm). The interference pattern is very complicated, and when the plate is examined by eye it does not appear to carry any image resembling the original object. These interference fringes do, however, contain a complete record of the original object.

Now suppose the developed plate is returned to the position it occupied while being exposed, and the object under examination is removed (Fig. 9.3b). The reference beam *A* now interacts with the interference fringes on the plate and reproduces beyond the plate a diffracted beam which is exactly the same as that *P'* originally scattered by the object in Fig. 9.3a. An observer looking at the plate as shown in Fig. 9.3b will therefore "see" the object beyond the plate as if it were still there.

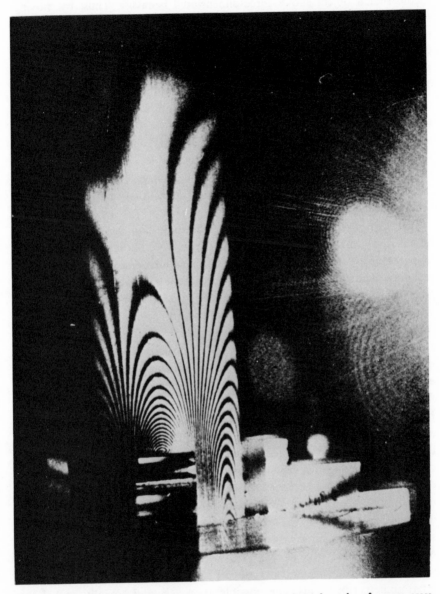

FIG. 9.4. Reconstruction of a double-exposure hologram taken of a tube of square cross section held by a vice. The tube has been slightly squeezed by the vice between the two exposures. (After Beesley.[36])

One of the most interesting characteristics of holography is that the reconstructed object shows three-dimensional behavior. Thus, by moving one's eye from the viewing position shown in Fig. 9.3b, it is possible to see the other sides of the object. Note that, to record a hologram, one needs to satisfy three main conditions: (i) The degree of coherence of the laser light must be sufficient for the interference fringes to be formed on the plate. (ii) The relative positions of object, plate, and laser beam must not change during exposure of the plate (often a few seconds in practice). In fact, the relative positions must change by less than half the laser wavelength if one is to avoid smearing out the interference. The laser, object, and plate should therefore be mounted on a vibration-isolated table. (iii) The resolution of the photographic plate must be high enough to record the interference fringes (films with a resolution of at least 2000 lines/mm are generally required).

Holography, as a technique of recording and reproducing three-dimensional images, has had its greatest success so far in the field of holographic art rather than in scientific applications. The holography-based technique known as holographic interferometry is, however, used in scientific applications as a means of recording and measuring strains and vibrations of three-dimensional objects. The principle of holographic interferometry is best explained by an example. With reference to Fig. 9.3b, suppose the object is replaced exactly in its original position. Then the observer will see two beams: (1) the beam P' diffracted by the hologram (as explained already); (2) the beam scattered by the object, as a result of being illuminated by the laser beam B which is partially transmitted by the plate. If the object is now subjected to a deformation from its original shape, the observer will see fringes displayed on the object due to the interference of these beams (1) and (2). The fringes indicate contours of equal object displacement along the viewing direction. The difference in the value of displacement for two adjacent fringes is half the wavelength of the laser used for the reconstruction process (if a visible He–Ne laser is used, this difference is ~ 0.3 μm). The technique is called holographic interferometry since the displacement measurement is made by interfering two beams, of which one (at least) is generated by the hologram. The technique has many variations, with the method described above (called real-time holographic interferometry) being just one possible approach, and actually one of the least used. The methods most frequently used are in fact the following: (i) static, double-exposure, holographic interferometry. Here two holograms of the same object are taken on the same plate, the first before, and the second after, the deformation. After developing the film the plate is

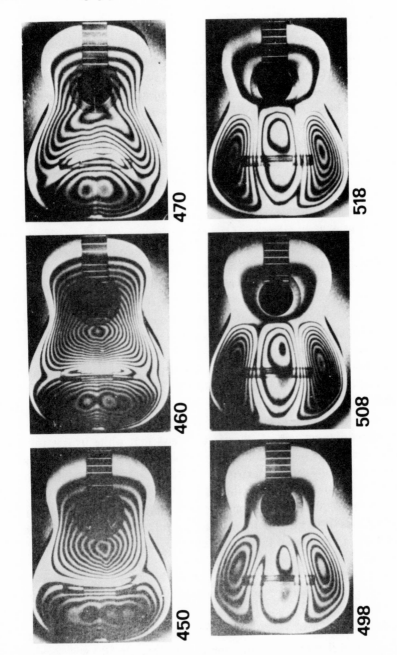

FIG. 9.5. Reconstructions of time-averaged holograms of a vibrating guitar. The frequency of vibration (Hz) is indicated at the side of each picture. (Photograph from the work of K. A. Stetson at the Institute of Optical Research, Stockholm.)

returned to its original position, but with the object removed (Fig. 9.3b). The object need no longer be present since the plate now contains both images from before and after the deformation, and hence also the corresponding interference pattern. As an example, Fig. 9.4 shows the reconstruction from such a hologram where the object is a tube of square section which has been squeezed in a vice between the two exposures. The interference fringes produced as a result of the deformation are clearly visible. (ii) Dynamic, time-averaged, holographic interferometry. This technique is particularly suited to vibrating objects. In this case a single hologram is taken, but with an exposure time larger than the vibration period. Thus a continuous ensemble of images corresponding to all object positions during the vibration is recorded in the same hologram. In this case, the reconstructed image of the object shows interference fringes on its surface which are indicative of the mode of vibration. As an example, Fig. 9.5 shows the fringe patterns observed on a vibrating guitar and their dependence on the vibration frequency (indicated at the side of each picture).

To work out the modes of vibration from these pictures we first note that one would expect to find a white fringe corresponding to those points which are not moving (i.e., for the nodal regions of the vibration). We further note that, for any vibrating point, the reconstructed image of that point will consist of a weighted average of the interference between images of that point during the vibration period. The weight is greatest for those images corresponding to the point at its maximum displacement (where the point spends more time). It therefore follows that white fringes (of lower intensity) will also be observed for those points for which the difference in displacement between the extremes of the vibration (along the viewing direction) is equal to an integral number of wavelengths.

The applications of holographic interferometry are very numerous and cover various fields ranging from measurements of strains and vibrations to the detection of material failure and contour mapping of objects.[35].

9.11 CONCLUDING REMARKS

We have seen that lasers, even 20 years after their discovery, are still full of youthful promise. Many applications are already well established, while others with great potential hold promise for the near future. Probably another ten or twenty years will be needed before the great revolution initiated by the laser can be fully appreciated.

So the laser, which in its early days used to be referred to as a bright solution in search of a problem, can already now be proclaimed as the bright solution to numerous problems in physics, chemistry, biology, medicine, electronics, and engineering.

REFERENCES

1. N. Bloembergen, *Nonlinear Optics* (Benjamin, New York, 1965).
2. S. A. Akhmanov and R. V. Khokhlov, *Problems of Nonlinear Optics* (Gordon and Breach, New York, 1972).
3. L. Fabelinskii, *Molecular Scattering of Light* (Plenum Press, New York, 1968).
4. *Picosecond Phenomena*, Vol. II, ed. by R. M. Hochstrasser, W. Kaiser, and C. V. Shank (Springer-Verlag, Berlin, 1980).
5. *High-Resolution Laser Spectroscopy*, ed. by K. Shimoda (Springer-Verlag, Berlin, 1976).
6. V. S. Letokhov and V. P. Chebotayev, *Nonlinear Laser Spectroscopy* (Springer-Verlag, Berlin, 1977).
7. D. L. Rousseau *et al.*, in *Raman Spectroscopy*, ed. by A. Weber (Springer-Verlag, Berlin, 1979), pp. 203–250.
8. J. W. Nibler and G. V. Knighten, in *Raman Spectroscopy*, ed. by A. Weber (Springer-Verlag, Berlin, 1979), pp. 253–295.
9. V. S. Letokhov and C. Bradley Moore, in *Chemical and Biochemical Applications of Lasers*, Vol. 3, ed. by C. Bradley Moore (Academic Press, New York, 1977), pp. 1–147.
10. A. McCray and P. D. Smith, in *Laser Applications*, Vol. 3, ed. by M. Ross (Academic Press, New York, 1977), pp. 1–39.
11. *Chemical and Biochemical Applications of Lasers*, Vol. 4, ed. by C. Bradley Moore (Academic Press, New York, 1978).
12. G. Spiro, in *Chemical and Biochemical Applications of Lasers*, Vol. 1, ed. by C. Bradley Moore (Academic Press, New York, 1974), pp. 29–67.
13. F. P. Mullaney *et al.*, in *Laser Applications in Medicine and Biology*, ed. by M. L. Wolbarsht (Plenum Press, New York, 1974), Vol. 2, pp. 151–200.
14. M. W. Berns, in *Laser Applications in Medicine and Biology*, ed. by M. L. Wolbarsht (Plenum Press, New York, 1974), Vol. 2, pp. 1–39.
15. R. M. Dwyer and M. Bass, in *Laser Applications*, ed. by M. Ross (Academic Press, New York, 1977), Vol. 3, pp. 107–134.
16. N. F. Gamaleya, in *Laser Applications in Medicine and Biology*, Vol. 3, ed. by M. L. Wolbarsht (Plenum Press, New York, 1977), pp. 1–170.
17. *Laser in Photomedicine and Photobiology*, ed. by R. Pratesi and C. A. Sacchi (Springer-Verlag, Berlin, 1980).
18. B. F. Hochheimer, in *Laser Applications in Medicine and Biology*, ed. by M. L. Wolbarsht (Plenum Press, New York, 1974), Vol. 2, pp. 41–74.
19. *Laser Surgery*, ed. by I. Kaplan (Jerusalem Academic Press, Jerusalem, 1976).
20. R. C. J. Verschueren, *The CO_2 Laser in Tumor Surgery* (Van Goreum and Co., Amsterdam, 1976).
21. J. F. Ready, *Industrial Applications of Lasers* (Academic Press, New York, 1978), Chapters 13–16.
22. A. Schachrai and M. Castellani Longo, *Ann. CIRP* **28**, 2 (1979).
23. *Opt. Eng.* **17**, No. 3 (1978).
24. Special Issue on Optical-Fiber Communications, *Proc. IEEE*, October 1980, **68**, (10).

25. J. F. Ready, *Industrial Applications of Lasers* (Academic Press, New York, 1978), Chapters 8, 11, 12.
26. G. O. Herrmann, F. Berry, and C. Dowdy, Interferometry in the machine shop, *Laser Focus*, May 1980, p. 78.
27. E. Drain, Doppler Velocimetry, *Laser Focus*, October 1980, pp. 68–79.
28. F. Aronowitz, in *Laser Applications*, ed. by M. Ross (Academic Press, New York, 1971), Vol. 1, pp. 134–199.
29. W. F. Krupke, E. V. George, and R. A. Haas, in *Laser Handbook*, Vol. 3, ed. by M. L. Stitch (North-Holland, Amsterdam, New York, Oxford, 1979), pp. 627–752.
30. J. F. Ready, *Industrial Applications of Lasers*, (Academic Press, New York, 1978), Chapter 21.
31. M. L. Stitch, in *Laser Handbook*, Vol. 2, ed. by F. T. Arecchi and E. O. Schulz-Dubois (North-Holland, Amsterdam, 1972), pp. 1745–1804.
32. *Aviat. Week Space Technol.*, July 28 (1980), pp. 32–66.
33. *Handbook of Optical Holography*, ed. by H. J. Caulfield (Academic Press, New York, 1979).
34. *Holographic Nondestructive Testing*, ed. by R. K. Erf (Academic Press, New York, 1974).
35. C. M. Vest, *Holographic Interferometry* (John Wiley and Sons, New York, 1979).
36. M. J. Beesley, *Lasers and Their Applications* (Taylor and Francis Ltd., London, 1972).

Appendixes

A) SPACE-DEPENDENT RATE EQUATIONS

The purpose of this appendix is to develop a rate-equation treatment in which the spatial variation of both the pump rate and the cavity field are taken into account. As a result of these spatial variations, it is to be expected that the population inversion will also be space dependent. For a four-level laser we can then write

$$\frac{\partial N_2}{\partial t} = W_p(N_t - N_2) - WN_2 - \frac{N_2}{\tau} \tag{A.1a}$$

$$\frac{dq}{dt} = \int_a WN_2 \, dV - \frac{q}{\tau_c} \tag{A.1b}$$

where the meaning of all the symbols used has already been given in Chapter 5. Equation (A.1a) expresses a local balance between pumping, stimulated emission, and spontaneous decay processes. Note that a partial derivative has been used on the left-hand side of the equation on account of the expected spatial dependence of N_2. The integral term on the right-hand side of equation (A.1b) extends over the volume of the active material and accounts for the contribution of stimulated processes to the total number of cavity photons q. This term has been written on the basis of a simple consideration of balance, using the fact that each individual stimulated process produces a photon. From (2.53c) and (2.61) we can express the stimulated rate W in terms of the transition cross section σ and the energy density ρ of the cavity field, viz.,

$$W = \frac{c\sigma}{\hbar\omega} \rho \tag{A.2}$$

where $c = c_0/n$ is the velocity of light in the active material. Note that ρ is taken to be dependent on both position **r** and time t, i.e., $\rho = \rho(\mathbf{r}, t)$, where its spatial variation is due to that of the cavity mode. If we now set $N_2 \simeq N$, where N is the population inversion, and assume $N_2 \ll N_t$, equations (A.1) with the help of (A.2) give

$$\frac{\partial N}{\partial t} = W_p N_t - (c\sigma)\frac{\rho}{\hbar\omega} N - \frac{N}{\tau} \tag{A.3a}$$

$$\frac{d\rho}{dt} = (c\sigma)\int_a \frac{\rho}{\hbar\omega} N\, dV - \frac{q}{\tau_c} \tag{A.3b}$$

Equations (A.3) are the rate equations for a four-level laser which apply when spatial variations are taken into account. Note that since both W_p and ρ depend on position, so also must N, and it cannot therefore be taken outside the integral in (A.3b). Note also that N would still show spatial dependence even if W_p were a constant. The physical reason for this spatial dependence of N has already been discussed in connection with Fig. 5.6: The standing wave field leads to spatial hole burning.

We will now solve equations (A.3) for the case where the laser is oscillating on a single mode. The spatial variation of field for this mode will be represented by the (dimensionless) quantity $U = U(\mathbf{r})$. We will consider a resonator of length L in which an active material of length l and refractive index n is inserted. The energy density ρ of the mode, outside and inside the active material, can be written respectively as

$$\rho_0 = B\hbar\omega U^2 \tag{A.4a}$$

$$\rho_n = nB\hbar\omega U^2 \tag{A.4b}$$

where $B = B(t)$ specifies the time variation of the energy density. Thus we have

$$q = \int_c \frac{\rho}{\hbar\omega}\, dV = B\left[n\int_a U^2\, dV + \int_{c_1} U^2\, dV \right] \tag{A.5}$$

where the first integral extends over the entire volume of the cavity, while the two integrals in the square brackets extend over the volume of the active material and over the remaining volume of the cavity, respectively. The form of (A.5) suggests that we define an effective volume in the cavity, V, as

$$V = \int_{c_1} U^2\, dV + n\int_a U^2\, dV \tag{A.6}$$

Note that this characteristic volume V depends on the way U is defined. With the help of (A.4b), (A.5), and (A.6), equations (A.3) can be recast in the more convenient form

$$\frac{\partial N}{\partial t} = W_p N_t - \left(\frac{c_0 \sigma}{V}\right) q U^2 N - \frac{N}{\tau} \tag{A.7a}$$

$$\frac{dq}{dt} = \left(\frac{c_0 \sigma}{V}\right) \int_a N U^2 \, dV - \frac{1}{\tau_c} q \tag{A.7b}$$

One way of solving (A.7) is to define a set, $\langle N_k \rangle$, of average values of N as follows:

$$\langle N_1 \rangle = \int_a N U^2 \, dV \bigg/ \int_a U^2 \, dV = \int_a N U^2 \, dV / V_a \tag{A.8a}$$

$$\langle N_2 \rangle = \int_a N U^4 \, dV \bigg/ \int_a U^2 \, dV \tag{A.8b}$$

$$\langle N_k \rangle = \int_a N U^{2k} \, dV \bigg/ \int_a U^2 \, dV \tag{A.8c}$$

where the integrals extend over the volume of the active material and where we have put

$$V_a = \int_a U^2 \, dV \tag{A.9}$$

V_a being the mode volume in the active material. From (A.7), using (A.8), we find

$$\dot{q} = \left[\left(\frac{c_0 \sigma}{V} V_a\right) \langle N_1 \rangle - \frac{1}{\tau_c}\right] q \tag{A.10a}$$

$$\langle \dot{N}_1 \rangle = \langle W_{p1} \rangle N_t - \left(\frac{c_0 \sigma}{V}\right) q \langle N_2 \rangle - \frac{\langle N_1 \rangle}{\tau} \tag{A.10b}$$

$$\langle \dot{N}_2 \rangle = \langle W_{p2} \rangle N_t - \left(\frac{c_0 \sigma}{V}\right) q \langle N_3 \rangle - \frac{\langle N_2 \rangle}{\tau} \tag{A.10c}$$

$$\cdot \quad \cdot \quad \cdot \quad \cdot \quad \cdot \quad \cdot \quad \cdot \quad \cdot$$

where we have written

$$\langle W_{p1} \rangle = \int_a W_p U^2 \, dV \bigg/ \int_a U^2 \, dV \tag{A.11a}$$

$$\langle W_{p2} \rangle = \int_a W_p U^4 \, dV \bigg/ \int_a U^2 \, dV \tag{A.11b}$$

If one looks at equations (A.10b) and (A.10c) one sees that the equation for $\langle N_1 \rangle$ contains $\langle N_2 \rangle$, the equation for $\langle N_2 \rangle$ contains $\langle N_3 \rangle$, etc. Consequently, we have an infinite hierarchy of equations for the variables $\langle N_k \rangle$. To solve this set of equations we have somehow to truncate this hierarchy, and this will be shown in the examples which follow.

As a first example we will consider a symmetric resonator consisting of two spherical mirrors whose radius of curvature is much greater than the cavity length L. The mode spot size w will then be approximately constant along the resonator length and can be taken to be equal to the value w_0 at the resonator center. For a TEM_{00} mode we can therefore write [compare with (4.33)]

$$U = \left[\exp - (r/w_0)^2\right]\sin(kz + \psi) \tag{A.12}$$

where $k = \omega/c$ and ψ is a constant-phase term whose value is such that U vanishes at the mirror position [e.g., $-(kL/2) + \psi = 0$]. Note that, as indicated by (A.12), the cavity field varies spatially both in the longitudinal direction (z coordinate) and in the transverse direction (r coordinate). In this case according to (A.6) and (A.9) the volumes V and V_a are given by

$$V = \tfrac{1}{4}\pi w_0^2 L' \tag{A.13a}$$

$$V_a = \tfrac{1}{4}\pi w_0^2 l \tag{A.13b}$$

where $L' = L + (n - 1)l$ is the effective length of the resonator [see (5.7a)]. With a mode such as that defined by (A.12), the hierarchy of equations (A.10) can be truncated at the mth equation by noting that

$$\langle N_{m+1}\rangle = \frac{\int (NU^{2m})U^2 \, dV}{\int U^2 \, dV} \simeq \frac{\int (NU^{2m}) \, dV}{\int U^2 \, dV} = \langle N_m\rangle \tag{A.14}$$

The approximation (A.14) follows from the fact that, for large m, the function $U^{2m} = \exp(-2mr^2/w^2)[\sin(kz + \psi)]^{2m}$ can be approximated by a set of unnormalized Dirac δ functions located at the points $(r = 0, z = z_i)$, z_i being the points where the maxima of $\sin^{2m}(kz + \psi)$ occur. In fact at these points $U^2(0, z_i) = 1$, and it can therefore be taken out from the integral. Thus, to first order, we can put $\langle N_2\rangle \simeq \langle N_1\rangle$ in (A.10b), and so we have

$$\langle \dot{N}_1\rangle = \langle W_{p1}\rangle N_t - \left(\frac{c_0\sigma}{V}\right)q\langle N_1\rangle - \frac{\langle N_1\rangle}{\tau} \tag{A.15a}$$

$$\dot{q} = \left[\left(\frac{c_0\sigma V_a}{V}\right)\langle N_1\rangle - \frac{1}{\tau_c}\right]q \tag{A.15b}$$

Equations (A.15) are now identical to (5.13) provided $\langle N_1\rangle$ is identified with N, the cavity mode volume V is given by (A.13a), the mode volume within the active material, V_a, is given by (A.13b), and $\langle W_{p1}\rangle$ is identified with W_p. So the approach used here allows a more precise definition of the quantities involved in equations (5.13).

As a second example we consider the case of a homojunction semiconductor laser (Fig. 6.32) in which the transverse dimension of the region occupied by the cavity field is much larger than the transverse dimension of the active region itself. If we take the x axis to be in the direction of current flow across the junction and the z axis (as usual) in the direction of propagation of the laser light, then for single longitudinal mode oscillation, we can write

$$U = \left[\exp(-x/w)^2 \right] \sin(kz + \psi) \tag{A.16}$$

where w is assumed to be much larger than the dimension l_x of the active region, Note that, for simplicity, U is assumed to be independent of the y coordinate. If we assume that the cavity mirrors are simply the two end faces of the semiconductor, we have

$$V = nV_a = n \int U^2 \, dx \, dy \, dz = \frac{n\sqrt{\pi}}{2} \left(w l_y l_z \right) \tag{A.17}$$

where l_y and l_z are the dimensions of the active region along the y and z axes respectively. Since $l_x \ll w$, we can write

$$\langle N_{m+1} \rangle = \frac{\int (N U^{2m}) U^2 \, dV}{\int U^2 \, dV} \simeq \langle N_m \rangle \tag{A.18}$$

where the argument advanced for putting $\langle N_{m+1} \rangle \simeq \langle N_m \rangle$ is similar to that given in connection with equation (A.14). Thus, to first order, equations (A.15) still apply and the average pump rate $\langle W_{p1} \rangle$, according to (A.11a) and (A.16), is

$$\langle W_{p1} \rangle = \left(\frac{2}{\sqrt{\pi}} \frac{l_x}{2w} \right) W_p \tag{A.19}$$

where W_p is the actual pump rate in the active region and has been taken to be constant over this region. Thus we see that the effective pump rate $\langle W_p \rangle$ is smaller than the actual pump rate W_p by a factor $2/\pi^{1/2}$ times the ratio of the thickness l_x of the active region to the width $2w$ of the mode.

As a final example we consider the case of a laser oscillating in many modes. In this case, as an approximation, we can still use equations (A.3) provided we take ρ to be constant and q to represent the total number of cavity photons. This is the same as saying that U^2 is assumed constant. In this case, if we set $U^2 = 1$, we find from (A.8b) that $\langle N_2 \rangle = \langle N_1 \rangle$, and equations (A.15) again apply provided that we set

$$V_c = AL' \tag{A.20}$$

$$V_a = Al \tag{A.21}$$

$$\langle W_p \rangle = \int W_p \, dV / V_a \tag{A.22}$$

where A is the cross-sectional area occupied by the laser beam in the active material.

B) PHYSICAL CONSTANTS

Planck constant h	6.6256×10^{-34} J s
$\hbar$	1.054×10^{-34} J s
Electronic charge e	1.60210×10^{-19} C
Electron rest mass m	9.1091×10^{-31} kg
Velocity of light in vacuum c_0	2.99792458×10^8 m/s
Boltzmann constant k	1.38054×10^{-23} J/°K
Bohr magneton β	9.2732×10^{-24} A m^2
Permittivity of vacuum ε_0	8.854×10^{-12} F/m
Permeability of vacuum μ_0	$4\pi \times 10^{-1}$ H/m
Energy corresponding to 1 eV	1.60210×10^{-19} J
Frequency corresponding to an energy spacing of $kT(T = 300°$K)	208.5 cm^{-1}
Energy of a photon with wavelength $\lambda = 0.5$ μm	3.973×10^{-19} J
Ratio of the mass of the proton to the mass of the electron	1836.13
Avogadro's number (molecules per gram-molecule)	6.0248×10^{23} (g-molecule)$^{-1}$
Radius of first Bohr orbit, $a = (4\pi\hbar^2\varepsilon_0 / me^2)$	0.529175×10^{-8} cm
Stefan–Boltzmann constant σ_{SB}	5.679×10^{-12} W cm^{-2}(°K)$^{-4}$

Answers to Selected Problems

CHAPTER 1

1.3. Take $\lambda = 0.55$ μm as the middle of the visible range. This corresponds to a frequency (in wavenumbers) $w = (1/\lambda) = 18,180$ cm^{-1}. Since $kT = 208$ cm^{-1} (see Appendix B), then $(E_2 - E_1)/kT = 65$ and $N_2^e = 7 \times 10^{-29} N_1^e$.

1.4. $E_2 - E_1 = kT = 208$ cm^{-1} [$\lambda \simeq 48$ μm, far infrared].

1.5. $(N_2 - N_1)_c = 5.2 \times 10^{16}$ cm^{-3}.

1.6. $D = 500$ m.

CHAPTER 2

2.1. $N_{\Delta \nu} = 8\pi V \Delta \lambda / \lambda^4 = 2 \times 10^{12}$ modes!

2.2. $\lambda = 0.48$ μm, green.

2.3. $\tau_{sp} = 4.78$ ms, $\phi = 0.63$.

2.4. Since the YAG molecular weight is 594, there are $\sim 1.38 \times 10^{20}$ Nd^{3+} ions/cm^3 in the ground $^4I_{9/2}$ level. Of these, only $\sim 46\%$ lie in the lowest level.

2.5. $\sigma = 8.8 \times 10^{-19}$ cm^2.

2.7. $\tau_{sp} = 5.75$ ns, $\tau_{nr} = \tau/(1 - \phi) = 38.5$ ns.

2.10. The single-pass gain for the onset of ASE is $G = 547$, which corresponds to $N_c \simeq 2.4 \times 10^{18}$ Nd^{3+} ions/cm^3 and a stored energy of $E \simeq 1$ J.

2.13. $B = 0.3$ cm$^{-1} = 9$ GHz.

2.14. $\Delta \nu = 4B = 36$ GHz.

2.18. $\rho = I/c$.

2.20. $\Delta\omega_{hole} = 4/T_2 = 2\Delta\omega_0$.

CHAPTER 4

4.2. $\Delta\nu = c_o/2L = 150$ MHz.

4.3. $N = \Delta\nu_0^*(4L/c_o) \simeq 23$.

4.4. $w_0 = [\lambda L/\pi]^{1/2} = 2.6$ mm, $w_s = \sqrt{2}\, w_0 = 3.67$ mm.

4.5. From Fig. 4.18 $N = 0.8$, hence $2a = 1.38$ mm.

4.7. $L_e = 2.65$ m, $w_0 = [L_e\lambda/2\pi]^{1/2} = 0.46$ mm, $w_s' = 0.5$ mm.

4.9. $z_1 = (R_2 - L)L/(R_1 + R_2 - 2L) = 0.857$ m, $L_e = 1.48$ m, $w_0 = 0.35$ mm, w_{s1} $= 0.533$ mm, $w_{s2} = 0.46$ mm.

4.10. $L = R_1 + R_2$.

4.11. From Fig. 4.29 we get $M = 1.35$, $2a_2 = 2[2L\lambda N_{eq}/(M-1)]^{1/2} = 4.26$ cm, $2a_1 > 5.75$ cm, $R_1 = 7.7$ m, $R_2 = 5.7$ m.

4.12. $\gamma = 45\%$

4.13. $w_I = w_0/\sqrt{2}$.

4.17. $\begin{vmatrix} 1 & L \\ 0 & 1 \end{vmatrix}$.

4.18. $\begin{vmatrix} 1 & 0 \\ -(2/R) & 1 \end{vmatrix}$.

CHAPTER 5

5.1. $V_a = \pi w_0^2 l/2$.

5.2. $\gamma = 1.61$.

5.4. From Fig. 4.21b ($g = 0.8$), we get $N \simeq 1.9$ and $a \simeq 1.1$ mm.

5.5. $L = 3$ m.

5.7. $P_{th} = 18$ kW, $P_1 = 17$ kW.

5.9. $\gamma = 5 \times 10^{-4}$; $x = 1.1$.

5.11. $\Delta\tau_p = 0.74$ ns.

5.12. A single pulse rather than a periodic sequence of pulses is obtained when the sum is replaced by the integral.

5.13. The overall amplitude reflectivity is $r = -r_1 + t_1r_2t_2\exp(i\phi) + t_1r_2r_1 \cdot$

$r_2 t_2 \exp(i2\phi) + \cdots$, where the minus sign of the first term accounts for the point made in the footnote of p. 161 and where $\phi = (4\pi n d'/\cos\theta'\lambda)$ with $n \sin\theta' = \sin\theta$. Summing the geometrical series we get $r = -r_1 + \{t_1 r_2 t_2 \exp(i\phi)/[1 - r_1 r_2 \exp(i\phi)]\}$. The overall power reflectivity is $R = |r|^2$. When $r_1 = r_2 = r$ and $t_1 = t_2 = t$ and if $\exp(i\phi) = 1$, we have $r = 0$ (note that $t^2 = 1 - r^2$).

5.14. $(2\pi l/\lambda_0)(n_{x'} - n_{y'}) = \pi/2$, where l is the crystal length (see theory of quarter-wave plate in an optics book).

5.15. $V = \lambda/4n_0^3 r_{63}$.

5.16. $V \simeq 2915$ V.

5.17. From (5.60) and (5.65) [$x = 4.545$]: $N_i/N_p = 1.89$ and $N_f/N_i = 0.237$. Then $E = (\gamma_1/2)(N_i/N_p)\eta_E(A/\sigma)\hbar\omega \simeq 19$ mJ, i.e., $\langle P \rangle = 189$ W (to be compared with the theoretical value of $P = 202$ W in cw operation). $\Delta\tau_p = 79.5$ ns.

5.18. $E = (\gamma_1/2\gamma)[(N_i - N_f)/2](V_a \hbar\omega)$. Compare with (5.59) and explain the factor 2 difference. $\Delta\tau_p$ is again given by (5.61). If W_p is approximated to $W_{cp} \simeq 1/\tau$, then the initial inversion is given by $N_i = N_t(x - 1)/(x + 1)$.

5.19. $E = 1.9$ J, $\Delta\tau_p = 10$ ns, $P_p = 190$ MW.

CHAPTER 6

6.5. It is of the order of the natural linewidth $\Delta\nu_{nat} \simeq 1/2\pi\tau = 160$ MHz, where $\tau \simeq \tau_{4s} \simeq 1$ ns with τ_{4s} being the lifetime of the $4s$ state.

6.6. It is of the order of the linewidth due to collision broadening. Assuming the partial pressures 1.5 Torr for CO_2, 1.5 Torr for N_2, and 12 Torr for He, we have $\Delta\nu_c = 7.58(\psi_{CO_2} + 0.73\psi_{N_2} + 0.6\psi_{He})p(300/T)^{1/2} = 74$ MHz at $T = 300°K$.

6.10. Since $\omega_1 = (k_1/M_O)^{1/2}$, where M_O is the mass of the oxygen atom, we get $k_1 = 16.8 \times 10^5$ dyn/cm. Neglecting the O–O interaction, we find (after a lengthy but straightforward calculation) that $\omega_3^2 = (k_1/M_O)[1 + (2M_O/M_c)]$, where M_c is the mass of the carbon atom. From this expression, using the value of k_1 found above, we get $\nu_3 = 2555$ cm^{-1} (compared with the experimental value of 2349 cm^{-1}).

6.12. For N_2 we have $\omega_{N_2} = (2k_1/M_N)^{1/2}$, where M_N is the mass of the nitrogen atom, while for CO we have $\omega_{CO} = (k_1/\mu)^{1/2}$, where the reduced mass μ only differs by about 2% from $M_N/2$.

6.13. As explained in Section 2.9.2 the J value corresponding to the most heavily populated level is such that $(2J + 1) = (2kT/B)^{1/2}$. From this we get $B = 0.3$ cm^{-1}.

6.14. Since only levels with odd J values can be populated in the upper state, the frequency spacing between two successive rotational lines is $4B \simeq 1.2$ cm^{-1} (actually the level separation is experimentally found to be about 2 cm^{-1}).

6.15. Assuming a level separation of 2 cm^{-1} we get $p \simeq 16$ atm. According to Fig. 6.13 the width of the gain curve is spanned by J' values ranging from $J' \simeq 11$ to $J' \simeq 41$. The width is thus $2B\Delta J' \simeq 60B \simeq 18$ cm^{-1}.

6.16. $\tau_p \simeq 1.8$ ps.

CHAPTER 7

7.1. From (7.5) we get $I = A^2(t)$, while from (7.4) $V^r = 2A \cos[\psi(t) - \langle\omega\rangle t]$; thus $\langle(V^r)^2\rangle = 2A^2 = 2I$.

7.3. For $\tau > 0$ we have $\gamma^{(1)} = \exp(i\omega\tau)$ (see problem 7.2) if there is no intervening collision during the time τ. Since the probability for this to occur is $P(\tau) = \exp(-\tau/\tau_c)$, we have $\gamma^{(1)} = \exp[(-\tau/\tau_c) + i\omega\tau]$. For $\tau < 0$ we must have $\gamma^{(1)}(-\tau) = \gamma^{(1)}(\tau)$; thus the expression for the correlation function valid for any time is $\gamma^{(1)} = \exp[(-|\tau|/\tau_c) + i\omega\tau]$.

7.5. $I_c = I[1 + \exp(-|\tau|/\tau_c)\cos\omega\tau]$, where I is the intensity of the e.m. wave entering the interferometer and $\tau = 2(L_3 - L_2)/c_0$. $V_P(\tau) = \exp(-|\tau|/\tau_c)$.

7.6. $\Delta L = \lambda = 10.6$ μm, $L_c = c\tau_{co} = c/4\pi\Delta\nu_{osc} = 238.7$ m.

7.8. $\theta = \theta_d \simeq \lambda/D = 0.116 \times 10^{-3}$ rad for a diffraction-limited beam. $A_c \simeq (\lambda/\theta)^2 = 9.8 \times 10^{-3}$ mm^2.

7.10. No, because the coherence area A_c (see Problem 7.8) is much smaller than $\pi w_1^2/4$.

7.12. We have in fact that $\Gamma^{(1)} = A_1 A_2^* \exp(i\omega\tau) = V_1(t + \tau)V_2^*(t)$.

CHAPTER 8

8.2. $f = 14.3$ cm, $L_1 \simeq 85$ cm, $L_2 \simeq 15$ cm.

8.3. $f = 68.1$ cm.

8.6. $\Gamma(l) = 2.28$ J/cm^2, $G = 7.12$, $E_{out} = 712$ mJ; $E_{stored} = \Gamma_s A \ln G_0$, where A is the cross-sectional area of the laser rod. Thus $E_{stored} = 727$ mJ, $\eta = 612/727 = 84.2\%$.

8.7. $E_{in} \simeq 170$ J, $E_{stored} = 538$ J.

8.8. $\Gamma_s = \hbar\omega/2\sigma z = 85$ mJ/cm^2, $E_{out} = 49.8$ J, $G = 2.93$, $E_{stored}/V = N\hbar\omega = N_j\hbar\omega/z = g_0\hbar\omega/\sigma z = 6.8$ J/liter.

8.11. *Hint*: Begin by showing that the refractive index of the extraordinary wave

$n^e(\theta)$ can be written as $n^e(\theta) = n^o n^e / [(n^o)^2 \sin^2 \theta + (n^e)^2 \cos^2 \theta]^{1/2}$. Then use (8.56).

8.12. $\theta_m = 44°$.

8.15. $I = 144 \text{ W/cm}^2$, $P = 11 \text{ mW}$.

8.16. From (8.91) we get $l_{SH} = \lambda_0 n / 2\pi d(2IZ) = 2.4$ cm, where $Z = 1/\epsilon_0 c_0 = 337 \ \Omega$ is the free-space impedance and I is the intensity of the incident wave. Hence $\eta = [\tanh(l/l_{SH})]^2 = 59.6\%$.

Index